216
Advances in Polymer Science

Advances in Polymer Science
Recently Published and Forthcoming Volumes

Self-Assembled Nanomaterials II
Nanotubes
Volume Editor: Shimizu, T.
Vol. 220, 2008

Self-Assembled Nanomaterials I
Nanofibers
Volume Editor: Shimizu, T.
Vol. 219, 2008

Interfacial Processes and Molecular Aggregation of Surfactants
Volume Editor: Narayanan, R.
Vol. 218, 2008

New Frontiers in Polymer Synthesis
Volume Editor: Kobayashi, S.
Vol. 217, 2008

Polymers for Fuel Cells II
Volume Editor: Scherer, G. G.
Vol. 216, 2008

Polymers for Fuel Cells I
Volume Editor: Scherer, G. G.
Vol. 215, 2008

Photoresponsive Polymers II
Volume Editors: Marder, S. R., Lee, K.-S.
Vol. 214, 2008

Photoresponsive Polymers I
Volume Editors: Marder, S. R., Lee, K.-S.
Vol. 213, 2008

Polyfluorenes
Volume Editors: Scherf, U., Neher, D.
Vol. 212, 2008

Chromatography for Sustainable Polymeric Materials
Renewable, Degradable and Recyclable
Volume Editors: Albertsson, A.-C., Hakkarainen, M.
Vol. 211, 2008

Wax Crystal Control · Nanocomposites Stimuli-Responsive Polymers
Vol. 210, 2008

Functional Materials and Biomaterials
Vol. 209, 2007

Phase-Separated Interpenetrating Polymer Networks
Authors: Lipatov, Y. S., Alekseeva, T.
Vol. 208, 2007

Hydrogen Bonded Polymers
Volume Editor: Binder, W.
Vol. 207, 2007

Oligomers · Polymer Composites Molecular Imprinting
Vol. 206, 2007

Polysaccharides II
Volume Editor: Klemm, D.
Vol. 205, 2006

Neodymium Based Ziegler Catalysts – Fundamental Chemistry
Volume Editor: Nuyken, O.
Vol. 204, 2006

Polymers for Regenerative Medicine
Volume Editor: Werner, C.
Vol. 203, 2006

Peptide Hybrid Polymers
Volume Editors: Klok, H.-A., Schlaad, H.
Vol. 202, 2006

Fuel Cells II

Volume Editor: Günther G. Scherer

With contributions by
M. J. M. Abadie · B. C. Benicewicz · O. Diat · G. Gebel
P. V. Kostoglodov · D. Y. Likhachev · J. Mader · C. Marestin
G. Maier · J. Meier-Haack · R. Mercier · P. N. Pintauro · A. L. Rusanov
T. J. Schmidt · V. Y. Voytekunas · R. Wycisk · L. Xiao

The series *Advances in Polymer Science* presents critical reviews of the present and future trends in polymer and biopolymer science including chemistry, physical chemistry, physics and material science. It is adressed to all scientists at universities and in industry who wish to keep abreast of advances in the topics covered.
As a rule, contributions are specially commissioned. The editors and publishers will, however, always be pleased to receive suggestions and supplementary information. Papers are accepted for *Advances in Polymer Science* in English.
In references *Advances in Polymer Science* is abbreviated *Adv Polym Sci* and is cited as a journal.

Springer WWW home page: springer.com
Visit the APS content at springerlink.com

ISBN 978-3-540-69763-3 e-ISBN 978-3-540-69765-7
DOI 10.1007/978-3-540-69765-7

Advances in Polymer Science ISSN 0065-3195

Library of Congress Control Number: 2008933501

Cover design: WMXDesign GmbH, Heidelberg
Typesetting and Production: le-tex publishing services oHG, Leipzig

Printed on acid-free paper

9 8 7 6 5 4 3 2 1 0

springer.com

Volume Editor

Dr. Günther G. Scherer
Paul Scherrer Institut
Labor für Elektrochemie
5232 Villigen, Switzerland
guenther.scherer@psi.ch

Editorial Board

Advances in Polymer Science
Also Available Electronically

For all customers who have a standing order to Advances in Polymer Science, we offer the electronic version via SpringerLink free of charge. Please contact your librarian who can receive a password or free access to the full articles by registering at:

springerlink.com

If you do not have a subscription, you can still view the tables of contents of the volumes and the abstract of each article by going to the SpringerLink Homepage, clicking on "Browse by Online Libraries", then "Chemical Sciences", and finally choose Advances in Polymer Science.

You will find information about the

- Editorial Board
- Aims and Scope
- Instructions for Authors
- Sample Contribution

at springer.com using the search function.

Color figures are published in full color within the electronic version on SpringerLink.

Preface

The concept to utilize an ion-conducting polymer membrane as a solid polymer electrolyte offers several advantages regarding the design and operation of an electrochemical cell, as outlined in Volume 215, Chapter 1 (L. Gubler, G. G. Scherer). Essentially, the solvent and/or transport medium, e.g., H_2O, for the mobile ionic species, e.g., H^+ for a cation exchange membrane, is taken up by and confined into the nano-dimensional morphology of the ion-containing domains of the polymer. As a consequence, a phase separation into a hydrophilic ion-containing solvent phase and a hydrophobic polymer backbone phase establishes. Because of the narrow solid electrolyte gap in these cells, low ohmic losses reducing the overall cell voltage can be achieved, even at high current densities.

This concept was applied to fuel cell technology at a very early stage; however, performance and reliability of the cells were low due to the dissatisfying membrane properties at that time. The development of perfluoro sulfonate and carboxylate-type membranes, in particular for the chlor-alkali process, directly fostered the further development of proton-conducting membranes and, as a consequence, also the progress in this type of fuel cell technology (polymer electrolyte fuel cell, PEFC).

Within the past 20 years, tremendous progress has been achieved in PEFC technology, in particular since the automotive industry has joined forces to further develop this energy conversion technology with its advantages in efficiency and environmental friendliness. This development has brought about a much deeper understanding of the various functions of the polymer electrolyte in the cell, particularly under duty cycle conditions of automotive applications. As a further and utmost prerequisite, the cost issue came to every one's attention.

Many national and international research programs have recently initiated work on proton-conducting polymer membranes for fuel cell applications. The contributions in these two volumes aim to summarize some major efforts, without claiming to be exhaustive.

Hence, M. Eikerling, A. A. Kornyshev, and E. Spohr start out in Volume 215, Chapter 2 with a general description of proton-conduction in polymer membranes, elucidating the influence of water and charge-bearing species in the polymer environment. Y. Yang, A. Siu, T. J. Peckham, and S. Holdcroft give an overview on implications of design approaches for synthesis of fuel cell membranes in Volume 215, Chapter 3. Some recent progress in the most prominent class of these membranes, the perfluoro sulfonic acid-type membranes, is described in Volume 215, Chapter 4 from an industrial perspective by M. Yoshitake and A. Watakabe. The development of the radiation grafting process to yield fuel cell membranes is covered in Volume 215, Chapter 5 by S. Alkan Gürsel, L. Gubler, B. Gupta, and G. G. Scherer, based on their long-term experience working in this area. The requirement for operating cell temperatures above 100 °C has led to the approach of composite membranes, combining the advantageous properties of inorganic and polymeric proton conductors (D. J. Jones, J. Rozière, in Volume 215, Chapter 6) to control the water content at these temperatures. On the basis of the promising properties of polymeric aromatic engineering materials and their modification to proton-conducting membranes, G. Maier and J. Meier-Haack review the state-of-art in sulfonated aromatic polymers in Volume 216, Chapter 1. High-temperature applications are also in the focus of the next two contributions. Polymer blends with phosphoric acid allow operating temperatures well above 100 °C, with advantages in water management and electrocatalysis (CO-tolerance), as pointed out in the contribution by J. Mader, L. Xiao, T. J. Schmidt, and B. C. Benicewicz in Volume 216, Chapter 2. A similar approach was followed, introducing the phosphonic acid group directly onto the polymer chain, by A. L. Rusanov, P. V. Kostoglodov, M. J. M. Abadie, V. Y. Voytekunas, and D. Y. Likhachev in Volume 216, Chapter 3. Two new classes of polymers and their properties are addressed in the last two Chapters 4 and 5 in Volume 216. R. Wycisk and P. N. Pintauro describe their view on polyphosphazene-based membranes for fuel cell applications, while C. Marestin, G. Gebel, O. Diat, and R. Mercier report on their and others' work on polyimide-based membranes.

As documented in and expressed by these various contributions, the topic "Polymers for Fuel Cells" is a vast one and concerns numerous synthetic and physico-chemical aspects, derived from the particular application as a solid polymer electrolyte. In this collection of contributions, we have emphasized work which has already led to tests of these polymers in the real fuel cell environment. There exist other synthetic routes for proton-conducting membrane preparation, which are not discussed in this edition. Furthermore, certain polymers are utilized as fuel-cell structure materials, e.g., as gaskets or additives (binder, surface coating) to bipolar plate materials. These aspects are not covered here.

In summary, we endeavored to bring together contributions from several expert groups who have worked in this area for many years to summarize the current state-of-the-art. There still exist many challenges down the road to bring at least some of these developments to commercial fuel cell technology. For an ultimate success, a comprehensive polymer *materials* approach has to be adopted to rationalize all the various aspects of this highly interdisciplinary task.

The editor wishes to thank all the authors for their contribution and the Paul Scherrer Institut for its support of membrane work over many years.

Villigen, May 2008 Günther G. Scherer

Contents

Contents of Volume 215

Fuel Cells I

Volume Editor: Scherer, G. G.
ISBN: 978-3-540-69755-8

Adv Polym Sci (2008) 216: 1–62
DOI 10.1007/12_2008_135

Published online: 29 April 2008

Sulfonated Aromatic Polymers for Fuel Cell Membranes

Gerhard Maier[1] (✉) · Jochen Meier-Haack[2] (✉)

[1]polyMaterials AG, Innovapark 20, 87600 Kaufbeuren, Germany
gerhard.maier@polymaterials.de

[2]Leibniz Institute of Polymer Research Dresden, Hohe Strasse 6, 01069 Dresden, Germany
mhaack@ipfdd.de

Abstract Aromatic and heteroaromatic polymers are well known for their often excellent thermal and chemical stability as well as their good mechanical properties and high continuous service temperatures. Therefore, they have long been considered promising candidates for the development of proton-conducting membranes for fuel cells, especially for applications above 80 °C. Typically, sulfonic or phosphonic acid groups are introduced to provide acidic sites. While it is possible to introduce these groups by post-modification of the preformed polymers, the preferred method in many cases is modification of the monomers and subsequent polymer synthesis, because this allows better control of the number, distribution, and position of the acidic sites. Compared to perfluorosulfonic acid polymers, such as Nafion, proton-conducting membranes based on aromatic hydrocarbon polymers tend to exhibit excellent conductivities in the fully hydrated state and significantly reduced crossover, especially of methanol in DMFC applications. However, this often comes at the expense of a higher degree of swelling and a greater loss of conductivity with decreasing water content, which is considered a severe drawback e.g., for automotive applications. Recent approaches to improving the property profile of hydrocarbon membranes include block copolymers, rigid rod polymers, and the attachment of acidic groups via side chains.

Keywords Block copolymers · Hydrocarbon membranes · Proton exchange membranes · Sidechain functionalized polymers · Sulfonated poly(arylene ether)s · Sulfonated rigid rod polymers

1 Introduction

Aromatic polymers are generally considered to be well suited as a basis for polymer electrolyte membranes in Polymer Electrolyte Membrane Fuel Cell (PEMFC) or Direct Methanol Fuel Cell (DMFC). Many aromatic polymers are used for technical applications under very demanding conditions and, depending on the exact chemical structures, their property profiles can be adjusted to a wide range of requirements. Polyimides (PI), polyaramides (PA), polyamidimides (PAI), polyarylates (often liquid crystalline fully aromatic polyesters) (LCP), poly(arylene ether ketone)s (PAEK), poly(arylene ether sulfone)s (PAES), poly(ether imide)s (PEI), polycarbonates (PC), poly(phenylene oxide) (PPO), poly(phenylene sulphide) (PPS), and poly(benzimidazole) (PBI) are examples of engineering and high performance polymers which are being used commercially for technical applications. In addition to the commercially available materials, a huge variety of aromatic polymers with high thermal stability, including poly(*para*-phenylene) derivatives (PPP) and heteroaromatic polymers, such as poly(quinoline)s (PQ), poly(phenylquinoxaline)s (PPQ), poly(oxadiazole)s, poly(benzoxazole)s (PBO), and many others, has been prepared in many labs. Some of the most prominent properties, which many classes of aromatic polymers have in common, are high thermal stability, including high heat distortion temperature, high continuous service temperature, and/or high decomposition temperature, as well as excellent mechanical properties, including high elastic modulus and high tensile and impact strengths.

High decomposition temperature is often identified with high oxidative stability, and, in combination with good mechanical properties, this is considered a desirable starting point for the development of a proton conducting membrane. However, the conditions the membrane must endure in a fuel cell are quite different from the conditions for which these polymers were originally intended. In addition to oxidizing conditions at the cathode, the membrane is also exposed to reducing conditions at the anode, both in the presence of active catalysts, and aqueous acidic hydrolyzing conditions, in a situation where the membrane is swollen in water or methanol/water mixtures. As a proton conductor, the polymer is substituted with acidic groups, and – with very few exceptions – proton transport requires considerable amounts of water to be present within the membrane. The acidic groups ensure swelling of the polymer membrane in water, and at the same time they provide a high number of charge carriers (protons) within the membrane by dissociation. In contrast, engineering or high performance polymers swell very little in water, often significantly less than 1%. Thermal stabilities are typically studied in dry air, and under such conditions where decomposition temperatures as high as 600 °C can be found, e.g., for polyimides or certain heteroaromatic polymers such as polyquinolines. However, these stabilities do not appear to translate into similar stabilities under wet conditions. Figure 1 shows decomposition temperatures of selected polymers in dry air compared with those determined in the presence of water vapor [1].

Two observations are obvious from the graphs in Fig. 1: first, the decomposition temperatures in dry atmosphere are much higher than those under wet conditions; second, they also vary much more under dry conditions than under wet conditions. Consequently, the potential stability of a polymer under fuel cell conditions can probably not be judged with sufficient confidence from studies of the dry polymer.

Besides thermal stability, the introduction of acidic groups into the parent polymer structure dramatically changes the chemical and physical properties. Absorption of water in PEMFC applications or water methanol mixtures in DMFC applications was mentioned already: it is undesired and very small in most engineering plastics and high performance polymers, but it is essential and can amount to several hundred percent in proton conduct-

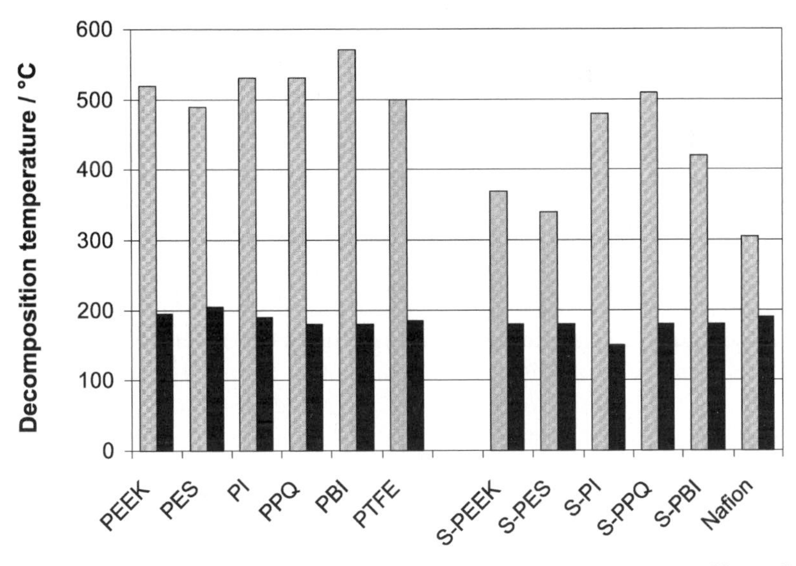

Fig. 1 Decomposition temperature (defined as 5% weight loss) of various sulfonated and non-sulfonated polymers in Helium (*cross hatched columns*) and saturated water vapor (*closed columns*); data from [1]

ing membranes. Also, crystallinity can be reduced partially or completely by sulfonation, resulting in dramatic changes of mechanical properties and loss of stability against solvents. Depending on the number of acidic groups per polymer chain, the polymers can even become soluble in water, water methanol mixtures, or other polar solvents. While this is partly desirable, because some polymers such as poly(arylether ketone)s (e.g., PEEK) would otherwise be difficult to process into membranes, it needs to be controlled in order not to loose the mechanical integrity of the membranes. It is also immediately clear that any polymer which is swollen to a significant extent in a solvent (e.g., water) is more prone to chemical attack than if it were present as a bulk injection molded article.

Besides hydrolysis, degradation by free radicals is a concern in fuel cell applications. These issues must be considered when selecting suitable polymer structures for membrane materials. Ideally, during operation of a PEMFC, no free radicals should be present. In reality, however, there are several possibilities for the formation of radicals. The membrane is never completely impermeable for oxidant (oxygen) and fuel (hydrogen, methanol). Thus, there is always the possibility of the presence of a small amount of oxygen at the anode, and hydrogen or methanol at the cathode. Direct reaction could proceed through one-electron processes, involving free radical intermediates, such as hydroxyl or hydroperoxy radicals, which could then attack the membrane polymer. Crossover of oxidant and fuel through the membrane also results in non-zero concentration of both within the membrane, resulting in the possibility of direct reaction and again formation of free radicals throughout the membrane. An interesting paper on this subject was published recently by a group of Chinese researchers [2]. In order to identify the origin of the radicals which decompose the membrane, the following set of experiments was performed: the polarization curve of a fuel cell with a sulfonated polystyrene membrane was followed over time, while it was run continuously with hydrogen/oxygen at 80 °C, and fully humidified at a current density of 1 A/cm^2. Degradation was very strong after 228 h, as shown by a decrease of the cell voltage at 300 mA/cm^2 from over 700 mV to approximately 150 mV. After disassembly of the cell, the membrane thickness was found to have decreased from 160 μm to 137 μm. The analysis of the sulphur content along a cross-section of the membrane by energy dispersive X-Ray analysis (EDAX) showed a homogeneous distribution in a virgin membrane, while, after the fuel cell test, the sulphur content was strongly reduced at the cathode side of the membrane. In addition, the infrared (IR) spectroscopy showed a loss of aromatic groups. The authors concluded that the degradation begins at the cathode side and then progresses inwards, consuming the polymer. In order to prove this, another membrane was prepared from sulfonated polystyrene coated with recast Nafion® on the cathode side. A fuel cell experiment with this membrane performed under the same conditions as before did not show any indications of degradation for 240 hours. This allows several conclusions:

Firstly (and obviously), the degradation does indeed start at the cathode, probably by the formation of free radicals by imperfect reaction at the catalyst. Secondly, formation of radicals at the anode or within the membrane is very low or absent. This is quite interesting, since in the steady state of operation, the concentration of fuel which permeates from the anode through the membrane to the cathode should exhibit a linear profile, with the lowest concentration at the cathode. If the degradation does start at the cathode side of the membrane despite the low concentration of fuel there, it is likely that the catalyst is in some way involved in the formation of the undesired radicals, as assumed by the authors of the paper discussed above [2]. Consequently, improving the performance of the catalyst for the cathode reaction may also reduce the formation of radicals, possibly resulting in significantly enhanced membrane lifetimes. Thirdly, radicals which are formed at the cathode do not diffuse quickly throughout the membrane, at least not on the time scale of a few hundred hours. Otherwise, degradation would not be located at one side of the membrane, and it would not be possible to prevent it by simply coating this side with a non-degrading polymer. It is clear that these considerations deserve more investigation, since, if the observations can be confirmed and generalized, they could point to ways to significantly improve the lifetime of non-fluorinated membranes.

Long-term stability is a major concern, but there is also a long list of other properties which are required for successful use of a membrane electrolyte in a fuel cell. Depending on the intended use of the fuel cell, the importance of the various properties changes. For instance, the use in automotive applications requires very high performance at low catalyst loading (for cost), no loss of performance at reduced or even absent humidification, and a lifetime above 5000 h under quickly and constantly changing power levels, including many start-stop cycles, even under freezing conditions. Operation at elevated temperatures, i.e., in the range of 110–130 °C, has been cited as important as well, but may not be the immediate focus of the automakers anymore [3] because of other problems associated with high temperature operation, which do not originate from the membrane. Also, while the current price of commercially available membranes, such as Nafion®, is prohibitive for use in automobiles, projections by General Motors (GM), based partly on data from DuPont, indicate that, at amounts required once a significant number of automobiles is built with fuel cells as power source, it is likely that the price target can be met [3]. Thus, the present high cost of perfluorinated proton conducting materials is not necessarily a fundamental problem that needs to be solved by science, but rather an economical issue. Potential low cost or operation at high temperatures alone should not be a sufficient justification for membrane work. For successful alternative membrane materials, proton conductivity (fully humidified as well as under dry conditions) must not be less than that of Nafion®, and water absorption must not be higher than that of Nafion®.

The situation is different when the use in DMFC is concerned. Here, one of the main issues is low methanol crossover, and, related to this, low swelling of the membrane in methanol/water mixtures. In this field, membranes based on aromatic polymers have an advantage over Nafion® due to their generally low methanol crossover rate. In addition, the chemical structure of hydrocarbon polymers can be adjusted relatively easily (at least in the lab), allowing for optimization of swelling in methanol/water mixtures. Humidification, operation under dry conditions, and freezing are not as much of a problem, since methanol/water mixtures can be used as fuel. If the intended use is for portable electronic devices, cost per kW power is less critical (compare your laptop battery: it delivers probably approximately 20 W for 3 h at a price of 150 $, amounting to 7500 $/kW). The system complexity and, hence, the size are much more of a problem.

The third application for fuel cells, which is often suggested in the literature, are small power plants for decentralized generation of electrical power and heat. Such systems would likely be operating continuously, under constant conditions. Size and weight are not the most important concerns, but cost is important. For the membrane, the positive side is probably operation under constant conditions. However, the demands for lifetime are probably extreme, amounting to much more than 10 000 h and up to 50 000 h.

2
Aromatic Polymers for Proton Conducting Membranes

Considering all this, almost all classes of high performance polymers have been used as basis for the development of proton conducting polymers for fuel cell applications. Figure 2 shows an overview of the structures.

The most common approach for the synthesis is sulfonation of a preformed, often commercially available, polymer. Sulfonic acid groups are selected as the source of protons for three simple reasons: they are easy to introduce into aromatic rings; they dissociate more readily than typical carboxylic acids, resulting in a larger number of charge carriers; and, unlike phosphonic acids, they do not easily form anhydrides on dehydration, resulting in easy and quick rehydration of sulfonated polymers in contrast to phosphonated ones.

Synthesis of the polymers by post-sulfonation is straightforward whenever there are electron-rich phenyl rings present in the polymer backbone or in side chains. This is, for example, the case in many poly(arylene ether ketone)s and poly(arylene ether sulfone)s, poly(phenylene oxide), poly(phenylene sulphide), certain poly(*para*phenylene)s, poly(phenylquinoxaline)s, and others. However, the exact chemical structure, such as the exact level of activation or deactivation of the aromatic groups, or the presence of acid labile groups, as well as the solubility of the starting material and the end product, deter-

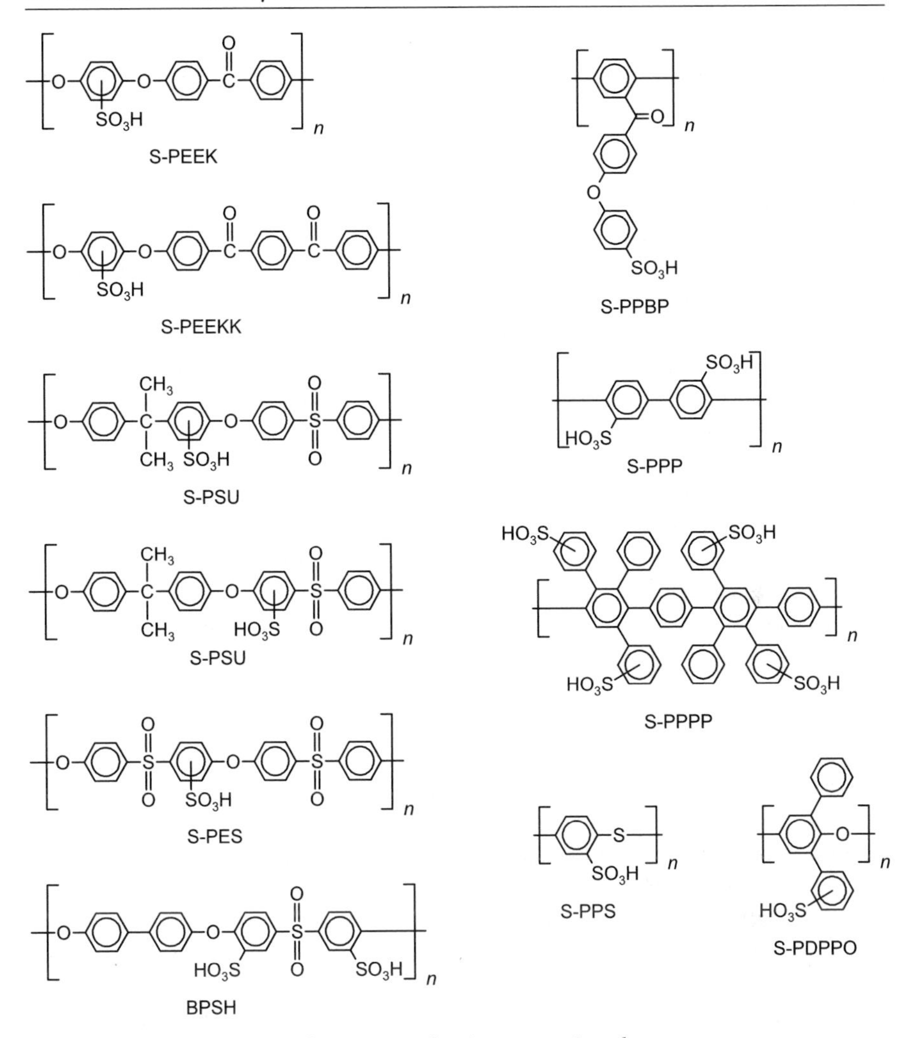

Fig. 2 Chemical structures of proton-conducting aromatic polymers

mine the type of sulfonating agent which is required, and to some degree also the extent of sulfonation. Concentrated sulphuric acid, oleum (SO_3 dissolved in sulphuric acid), and chlorosulfonic acid are the most commonly used reagents. In ideal cases, reaction time and temperature allow to control the degree of sulfonation. Aromatic polyethers, such as PEEK, are examples for this kind of behavior, a general representation of which is shown in Fig. 3. The plot in Fig. 3 is based on data from various authors summarized in [1].

Sometimes, when the aromatic groups of the polymer are too activated, simple post sulfonation can lead to crosslinking. The arylsulfonic acid groups formed by sulfonation are then able to act as sulfonating agent themselves,

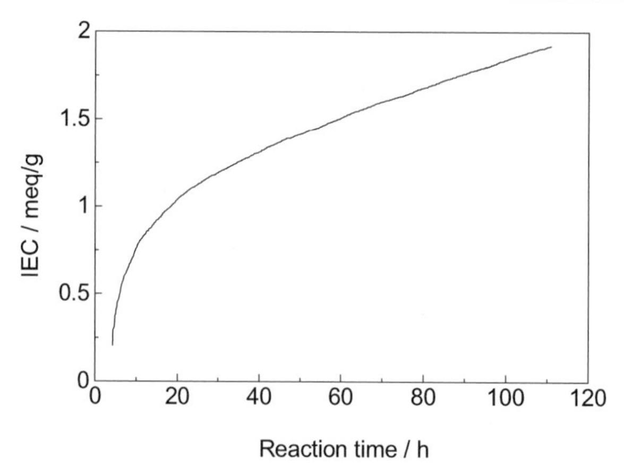

Fig. 3 Degree of sulfonation of PEEK in dependence of time. Sulfonation by concentrated sulphuric acid at room temperature (curve taken from [1])

and some of them can form sulfone bridges between different polymer chains. Thus, very activated polymers can be sulfonated only to relatively low degrees of sulfonation before crosslinking begins. Higher sulfonation of these polymers requires alternate routes which proceed via less activated systems. One example is the sulfonation of poly(phenylene sulphide) via the sulfonium salt, which sufficiently deactivates some of the phenyl rings to allow sulfonation of the others without crosslinking (see section 3.7).

Another example is the sulfonation of poly(ether sulfone)s via lithium organic intermediates, which creates special sites for sulfonation [4, 5]. Although this latter technique was developed for a different reason, it offers the possibility to create sites that are active for sulfonation but not for crosslinking (Fig. 4).

Sulfonation of preformed polymers will always lead to sulfonation of the most activated sites, typically in electron-rich phenyl rings which are substituted by ether or thioether groups. Carbonyl or sulfone groups, fluorinated groups, and some heteroaromatic groups (six-membered rings with one or more heteroatoms, or five-membered rings with two or more heteroatoms) typically deactivate. Unfortunately, sulfonation is reversible. Under acidic conditions, especially at higher temperatures, sulfonic acid groups are lost from activated sites. Therefore, long-term stability of several thousand hours of operation requires sulfonation in less activated, preferably deactivated sites. In principle, this is possible with the common sulfonating agents. However, it requires more drastic conditions (e.g., oleum at temperatures above 150 °C), which can often not be tolerated by the polymer chains. If electron-rich and electron-poor phenyl rings are present in the same polymer, as in poly(arylene ether sulfone)s and poly(arylene ether ketone)s, sulfonation of

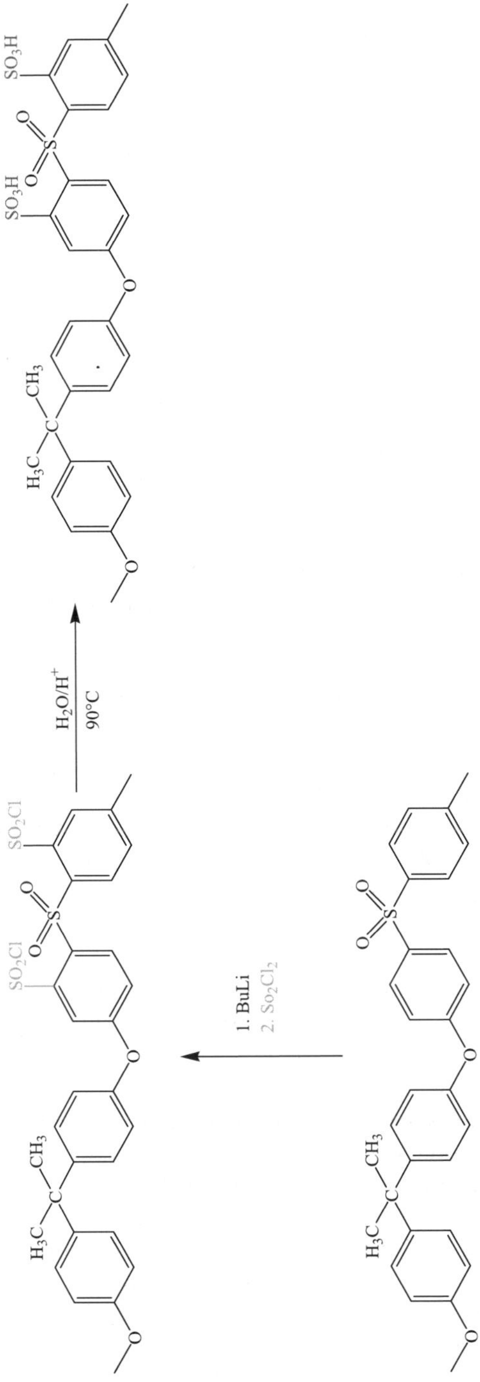

Fig. 4 Sulfonation via lithiation [4, 5]

the deactivated ketone or sulfone sites requires conditions which lead to partial degradation of the polymer chain.

There are two approaches to introduce sulfonic acid groups in deactivated positions. One was developed by Kerres [4, 5] and is the one shown in Fig. 4. It proceeds via organometal intermediates, because these groups are more easily introduced into the electron-poor sites than into electron-rich ones. An exchange of the metal for sulfonic acid groups then introduces sulfonation in those electron-poor sites that were originally occupied by metal atoms.

The other approach is to move the sulfonation step from the polymer stage to the monomer stage. Its most prominent disadvantage is obvious: if the sulfonation is to be achieved at the monomer stage, no preformed commercially available polymers can be used. Rather, the polymer must be synthesized using sulfonated monomer(s). While synthesis of the aromatic polymers shown in Fig. 3 is well established, sulfonation of the monomers often changes the properties. Solubilities, tolerance of the sulfonic acid (or salt) groups towards polymerization conditions, and/or tolerance of any catalysts towards the sulfonic acid or salt groups must be considered. Despite the difficulties this can cause sometimes, this route is very attractive, since it allows more control over position and degree of sulfonation than post-sulfonation of preformed polymers. For example, it has been used very successfully for the synthesis of poly(arylene ether)s and poly(*para*-phenylene)s.

The following chapters give an overview of the most recent developments in the field of aromatic proton conducting membranes. Additional information, especially on previous literature, can be found in a number of preceding reviews [1, 6–12].

3 Poly(Arylene Ether Ketone)s

3.1 General

Poly(arylene ether ketone)s are polymers which consist of aromatic units, joined together by ether and ketone groups. All aromatic groups, e.g., 1,4-phenylene, 1,3-phenylene, 1,2-phenylene, biphenylene, naphthylene, heteroaromatic rings, polycyclic ring systems, can be included, and actually have been included at least in laboratory samples. In addition to ether and ketone groups, other linking groups are also possible, e.g., alkylene, perfluoroalkylene, thioether, cyclic groups, etc. Until about ten years ago, there was quite a selection of different commercially available poly(aryl ether ketone)s. All of them consisted exclusively of 1,4-phenylene rings, joined by ether and ketone groups. Figure 5 shows examples of the structures.

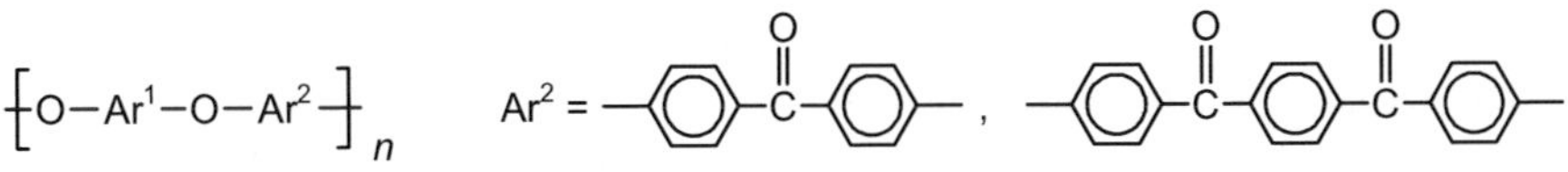

$\left[O-Ar^1-O-Ar^2 \right]_n$

General structure

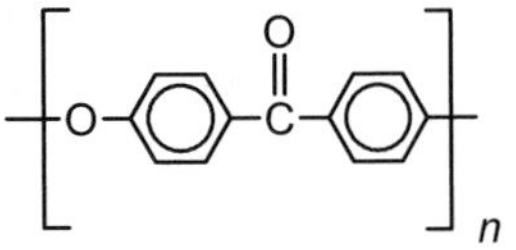

$Ar^2 =$

PEK; e.g. Victrex PEEK P22 (Victrex); Stilan (Raychem)

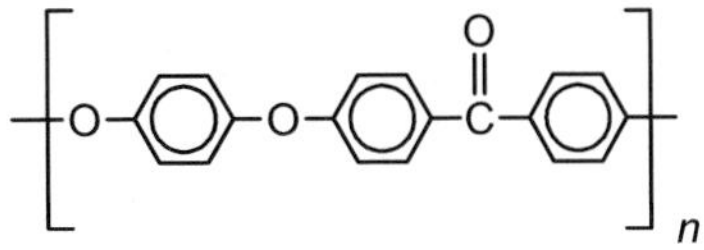

PEEK; e.g. Victrex PEEK (Victrex), Gatone PEEK (Gharda), Kadel (Solvay), CoorsTek PEEK (CoorsTek), Ketron (Quadrant)

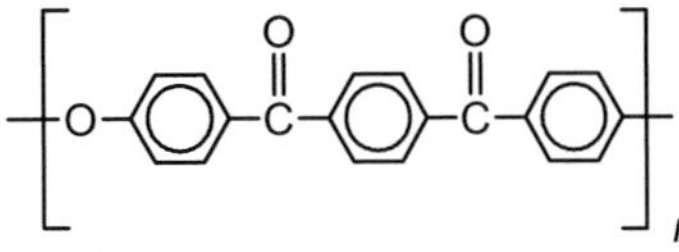

PEKK; e.g. CoorsTek PEKK (CoorsTek)

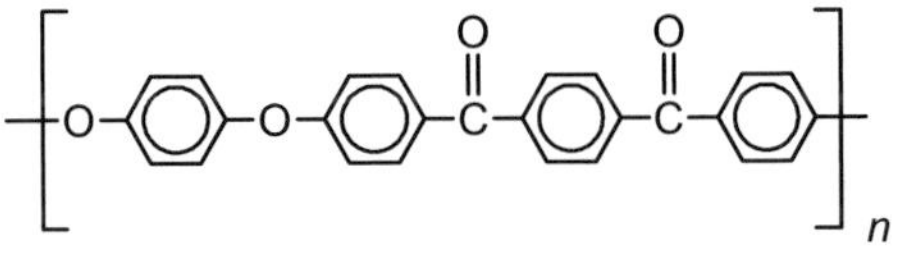

PEEKK; e.g. Hostatec (Hoechst)

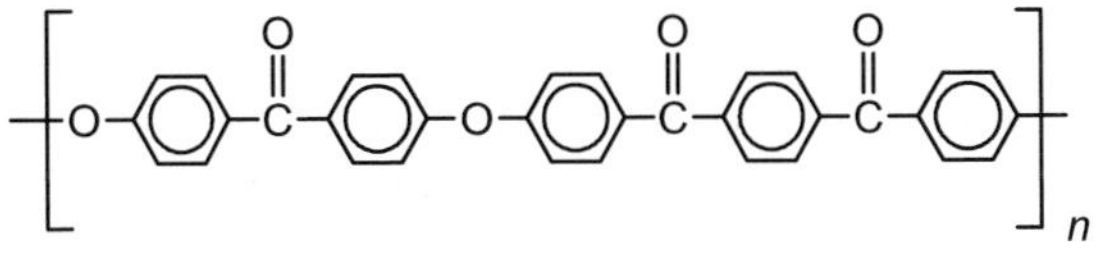

PEKEKK; e.g. Ultrapek (BASF)

Fig. 5 Examples of structures of poly(aryl ether ketone)s which are or were commercially available. Ar^1 is typically a *para*-substituted bisphenol, but can in principle be any aromatic system. Many more structures have been prepared in the lab, e.g., PEEEK, PEEKEK, etc.

All commercially available poly(arylene ether ketone)s follow the same scheme of composition as those shown above. They are semicrystalline, and since the only difference between these structures is the sequence of ether and ketone groups, they are often not only miscible, but some of them also cocrystallize. The crystallinity of the poly(arylene ether ketone)s is their most important feature, since it determines their physical properties, especially the high melting temperature (around 330–400 °C) and resulting high heat distortion temperature and high continuous service temperature, the excellent chemical stability and ESC-resistance, and the superior mechanical properties. Within the subgroup of poly(aryl ether ketone)s which follow the structural principles shown in Fig. 5, an unofficial but widely used nomenclature has been established. It assumes that all aromatic units in the polymer chain are 1,4-phenylene groups, and that there are no substituents. Then, the ether and ketone groups are simply listed in their order of appearance along the polymer chain, as indicated in Fig. 5, for example, "poly(ether ether ketone) (PEEK)" or "poly(ether ketone) (PEK)". This does not work with biphenyl, 1,3-phenylene, and heteroaromatic groups present.

Presently, however, only three types of poly(arylether ketone)s are still produced commercially: PEEK, PEK, and PEKK. Most work for proton conducting membranes has been done based on PEEK, but the others have been used as well [1, 6, 7, 13–32].

Sulfonated PEEK (S-PEEK) becomes water soluble above an ion exchange capacity of 1.8 mmol/g [1], corresponding to a degree of sulfonation of approximately 60% (60% of all repeating units have one sulfonic acid group). For application in fuel cell systems, especially those which are intended for intermittent use with start-stop cycles and varying load, the water level in the membrane will vary. Strong changes in swelling under changing humidity levels would result in strongly varying stress within the cell (or rather stack of cells) and may damage the membrane electrode assembly, e.g., by compressing the gas diffusion layer and/or delamination of membrane and catalyst layer. Therefore, swelling of the membrane in water must be relatively low (preferably not more than that of Nafion®: 16% at 80 °C and 100% r.h. by weight, 32% by volume). Consequently, a membrane based on S-PEEK would have to have an IEC considerably lower than 1.8 mmol/g in order to avoid excessive swelling. However, this results in relatively low proton conductivity even in the fully hydrated state. Figure 6 shows a comparison of the conductivites of Nafion® 117 and S-PEEK with an IEC of 1.6 mmol/g at 100 °C in dependence of the relative humidity. Even at 100% relative humidity, the conductivity of the S-PEEK membrane is lower than that of Nafion® 117 by a factor of 5. Pre-treatment of the membrane at higher temperatures in water or sulphuric acid can decrease this difference to a factor of about 2. As the humidity is reduced, the gap increases, since the conductivity of the S-PEEK drops much faster than that of Nafion® (Fig. 6).

Water uptake was studied by Kreuer [7, 31] and compared to Nafion®. For an interpretation of these data, one needs to consider that the sulfonic acids involved are strong acids, which will dissociate when possible. This requires a certain amount of water, depending on the degree to which the ions after

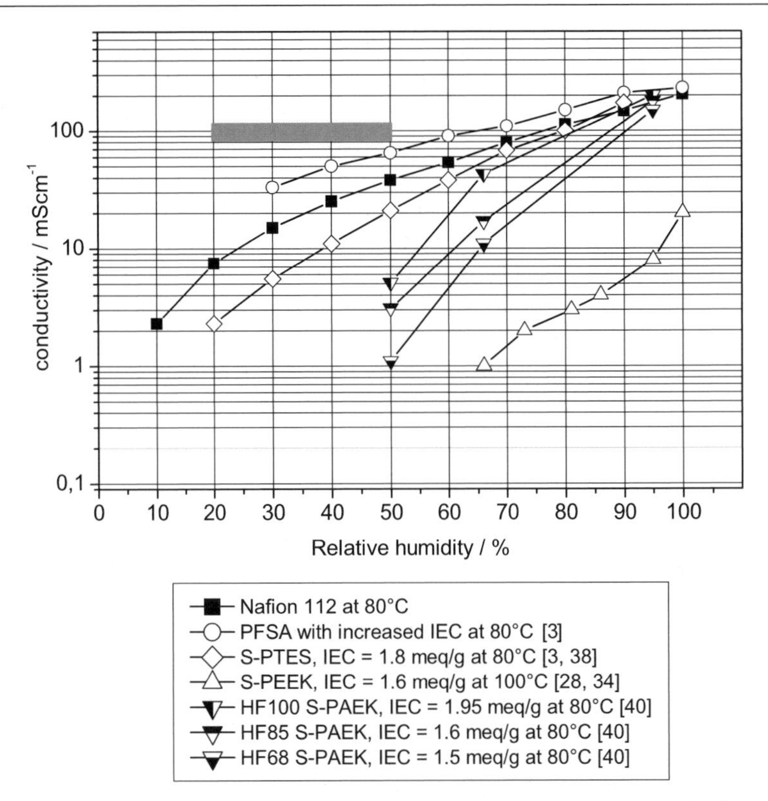

Fig. 6 Humidity dependence of the conductivities of Nafion® and aromatic hydrocarbon membranes [3, 28, 34, 38, 40]

dissociation need to be hydrated for stabilization, or, in other words, the acid strength. Thus, water in the membrane has different "functions", resulting in different "states" of water: water required for the primary hydration of the dissociated acid, loosely bound water, and "bulk" water as a second phase [7, 24]. In order to correct for different degrees of sulfonation, the ratio λ of moles of water per mole of sulfonic acid groups is introduced. A plot of λ versus temperature for S-PEEKK of different IEC and Nafion® 117 is shown in Fig. 7.

Nafion® 117 in liquid water takes up more water per sulfonic acid group than S-PEEKK of IEC values between 0.78 mmol/g and 1.78 mmol/g up to a certain temperature, which depends on the IEC value of the S-PEEKK. At this temperature, which is 65 °C for IEC = 1.78 mmol/g, 80 °C for IEC = 1.4 mmol/g, 100 °C for IEC = 0.78 mmol/g, the water content of the S-PEEKK membranes increases tremendously. Nafion® shows similar behavior only at a temperature of 140 °C. Until this temperature is reached, its molar water content is almost constant at $\lambda = 20$. The excess swelling of S-PEEKK at temperatures of 100 °C or less causes severe problems in using these materials as membranes in fuel cells.

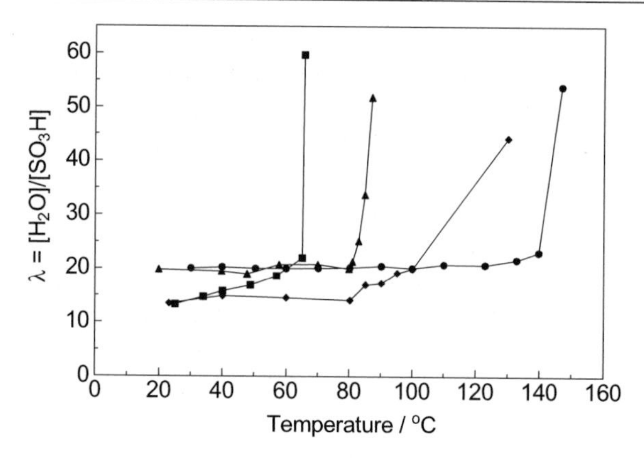

Fig. 7 Molar water uptake of different polymers vs. temperature [7,24], ● Nafion, ■ S-PEEKK (IEC = 0.78 mmol/g), ▲ S-PEEKK (IEC = 1.4 mmol/g), ◆ S-PEEKK (IEC = 1.78 mmol/g)

Since the densities of Nafion® and S-PEEKK are different, data should be compared based on volume (mmol/mL) rather than weight (mmol/g). Table 1 compares values of λ with weight-based and volume-based water uptakes for Nafion® and S-PEEKK.

A value of $\lambda = 10$ for Nafion® 117 shows that Nafion® takes up only 16% (w/w) of water at room temperature, corresponding to an estimated volume increase of 32% (w/v). This remains constant up to more than 120 °C (Fig. 7). In comparison, S-PEEKK with an IEC = 1.78 mmol/g takes up 32% (w/w) of water based on weight, corresponding to an estimated volume increase of 43% (w/v) at $\lambda = 10$. This increases to $\lambda = 20$ (64% water uptake based on weight, 86% volume increase) at 65 °C, and shoots up to $\lambda = 60$ (192% water uptake based on weight, 260% volume increase) at 70 °C (Fig. 7). This difference in swelling behavior has been attributed to the lower hy-

Table 1 Water uptake for Nafion® and S-PEEKK at room temperature [a]

		Nafion® 117	S-PEEKK		
IEC mmol/g		0.89	0.78	1.4	1.78
IEC mmol/mL		1.75	1.05	1.88	2.39
Density g/mL		1.97	1.34	1.34	1.34
Water uptake %	$\lambda = 10$	16	14	25	32
per weight	$\lambda = 20$	32	28	50	64
Water uptake %	$\lambda = 10$	32	19	34	43
per volume	$\lambda = 20$	64	38	68	86

[a] Data based on DuPont Product Information (Nafion®) and [7,31] (S-PEEKK)

drophobicity of the PEEKK polymer chains as compared to the perfluorinated chains of Nafion®, and the resulting less strict microphase separation into hydrophilic aqueous domains dispersed in a hydrophobic continuous phase of sulfonated poly(aryl ether ketone)s in contrast to sulfonated perfluoropolymers [7].

S-PEEKK with a degree of sulfonation of 70% (IEC = 1.69 mmol/g) [31] and S-PEEK with a degree of sulfonation of 65% (IEC = 2.13 mmol/g) [33] can supposedly match or even surpass the conductivity of Nafion® at 80 °C in the fully hydrated state. In another source [26], however, the conductivity of S-PEEK with 65% sulfonation is reported to be about a factor of 10 lower than that of Nafion® [6, 26]. A S-PEEK with a degree of sulfonation of 88% (IEC = 2.48 mmol/g) is reported to show a conductivity which is lower than that of Nafion® by a factor of approximately 3 at 100 °C and 85% r.h., but does meet or beat that of Nafion® at 160 °C and 75% r.h. [28, 34]. However, the water uptake of this polymer is so high that no conductivities could be measured above 90% relative humidity. Yet another source claims a conductivity of 400 mS/cm at 80 °C and 90% r.h. for a S-PEEK with 85% sulfonation [6]. At a degree of sulfonation of 11% (IEC = 1.6 mmol/g), S-PEEK was reported to have a conductivity up to 45 mS/cm at 20 °C in the fully hydrated state, depending on pre-treatment conditions. In another study, a sulfonated poly(ether ketone) (S-PEK) with an IEC = 1.71 mmol/g was found to have a conductivity of 100 mS/cm at 90 °C and full hydration, while Nafion® 112 was measured at 88 mS/cm in the same study. Variations in post-treatment of the same membrane can also lead to large differences in conductivity and water uptake [35]. Apparently, conductivity data vary strongly, even under presumably similar conditions. Thus, data can only be compared safely within a study, while a comparison between studies may lead to misinterpretations. As shown above, the values of conductivity reported for S-PEEK under similar conditions can vary by up to three orders of magnitude. Rozière and Jones [1] attribute this to the solvent used for casting, with large differences found between those data reported by authors casting from NMP [28, 31, 34] and those casting from DMF or DMAc [26, 29]. The effect was studied by Kaliaguine et al. [36]. It is related to the stability of the solvents. DMF and DMAc may decompose, producing dimethylamide, which then may be protonated, resulting in dimethylammonium counter ions for the sulfonic acid groups. This is likely to change the solubility characteristics significantly, which can change the transport properties of the membranes, even if the ammonium ions are afterwards exchanged for protons. A strong influence of the casting conditions on transport characteristics has also been frequently found in gas separation membranes, where the distribution of free volume and the relaxation of the polymer chains plays a role for transport properties. A look at the thermodynamics [37] of the interaction between polymer and solvent and the resulting implications for film formation by solvent casting with subsequent drying shows that solvent, starting concentration, and temperature profile cannot be chosen at random. Under unfavourable conditions, phase separation may occur, resulting in films with so much internal stress that they do not even have good mechanical properties. Swelling properties and even transport properties may be affected as well if polymer chains are not relaxed within the film. The same polymer, cast under different conditions or from a different solvent, may exhibit excellent properties. However, such influences should be transient (aging is often observed for gas separation membranes). Therefore, it should be possible to find pre-treatment conditions for proton conducting membranes which eliminate all influences of history.

Nevertheless, simple sulfonated poly(aryl ether ketone)s do not appear to be a challenge for the commercially available perfluorosulfonic acid polymers when swelling behavior at desired operating temperatures is considered besides conductivity. Swelling is important for technical applications, e.g., in fuel cell powered cars. Based on considerations concerning stack design, swelling of the membrane in the presence of liquid water at the operating temperature (80–100 °C) must probably be less than 100% by volume [3]. Considerations such as internal pressure and the resulting requirements of strength of the stack components, potential damage to the gas diffusion media by swelling membranes, and fatigue issues due to swelling-deswelling cycles during the typical operation cycles of a car seem to indicate an expected limit in this range. At the same time, the need to reduce system complexity leads to the desire to operate the fuel cell without external humidification. Therefore, the conductivity of the membrane is ideally no less than 100 mS/cm at humidities between 20% r.h. and liquid water [3]. So far, no membrane can satisfy these requirements. Unfortunately, for many materials the humidity dependence of the conductivity is not reported in the literature. Some exceptions are shown in Fig. 6.

It can be seen that with decreasing humidity the proton conductivity drops much more dramatically for S-PEEK than for Nafion®. Figure 7 also shows data for a sulfonated poly(thioether sulfone) S-PTES [3, 38] (see Fig. 8 for the chemical structure), which exhibits much higher conductivities than the S-PEEK. Its conductivity even surpasses that of Nafion® 112 at humidities above 85%, but then drops off more steeply with decreasing humidity. However, it should be noted that the S-PTES shown here with an IEC of 1.8 mmol/g dissolves in boiling water within two hours [3] and is therefore not suitable for operation in a fuel cell under conditions with strongly varying load, where liquid water at high temperatures can be present. Still, this comparison shows

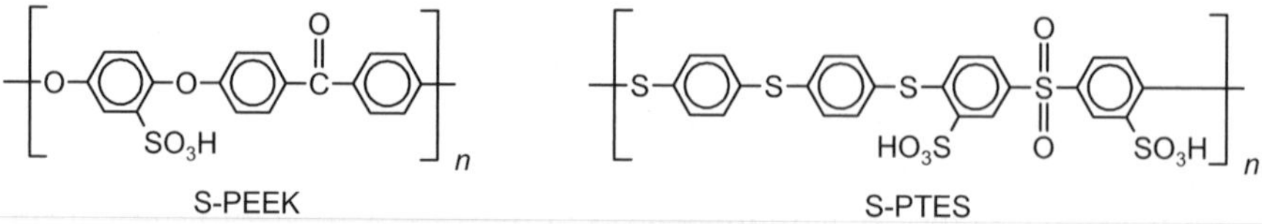

Fig. 8 Chemical structures of S-PEEK and S-PTES

that the chemical structure of polymer chains has a strong effect on proton conductivity. One obvious difference between the S-PEEK and the S-PTES is the position of the sulfonic acid groups (Fig. 8).

The S-PEEK of this example was prepared by sulfonation of PEEK (post-sulfonation) [28, 34]. Its sulfonic acid groups are therefore attached to the most electron-rich phenyl rings of the polymer chain: those between the ether groups. Apart from the fact that the sulfonic acid groups are somewhat labile in this position, their position on electron-rich phenyl rings also reduces their acidity because sulfonation is reversible under certain conditions. In contrast, the sulfonic acid groups of the S-PTES were introduced into the monomer dichlorodiphenylsulfone before polymerization [38]. They are in the less electron-rich phenyl rings attached to sulfone groups, and hence they are more acidic than those of the post-sulfonated S-PEEK. Molecular modelling [39] indicates that the acidity governs the number of water molecules required for an acid group to dissociate and release a proton (as a hydrated $H_3O^{\oplus}$ ion). This is one of the reasons cited for the high conductivity of Nafion® and other perfluorosulfonic acid polymers in comparison with hydrocarbon polymers: perfluoroalkyl sulfonic acids are more acidic than aromatic sulfonic acids. This can be observed in a plot of conductivity versus water content (Fig. 9a).

While at a water uptake of 30% (corresponding to a relative humidity of 85%) the conductivity of S-PTES meets that of Nafion® 112, it is only half of that of Nafion® at smaller water contents (lower relative humidity). Unfortunately, similar data are often not available for other polymers described in the literature to compare and confirm any interpretation for different chemical structures. Some data are available, although these were not obtained in the same way. (The data shown in Fig. 9a) were obtained by determination of the water uptake of a film when exposed to an atmosphere with controlled moisture content at 80 °C and measurement of the conductivity under the same conditions. The data shown in Fig. 9b), taken from the literature for various sulfonated hydrocarbon polymers, were obtained by swelling the polymer films in water (varying times and temperatures, unfortunately) and measuring the conductivity in water (varying temperatures, unfortunately). Nevertheless, it would be really interesting to have data which are really comparable for many different polymer systems, because the information from the plots in Fig. 9 could be interpreted as an "efficiency" with which the polymer can "use" the water present in the film to transport protons, i.e., how much water is required for a desired conductivity in different systems? For technical applications, there is a maximum degree of swelling of the membrane which can be tolerated. From plots as in Fig. 9 one could conclude which polymer system will offer the highest conductivity under this limitation.

Other modifications of the chemical structure of the repeating units of poly(aryl ether ketone)s also show potential for significant improvement. For example, the introduction of fluorinated aromatic rings and bulky groups re-

a)

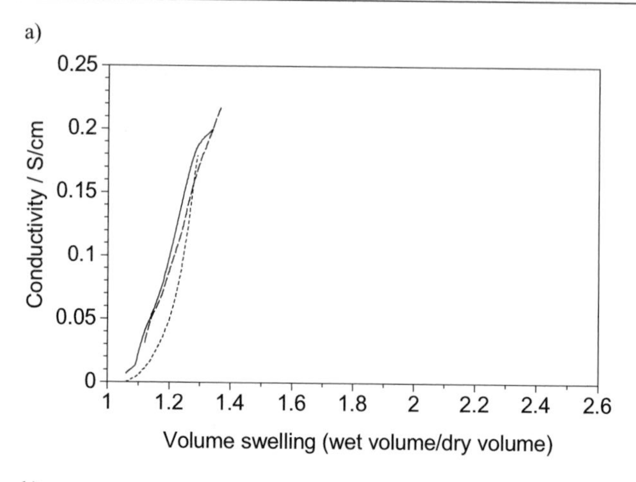

b)

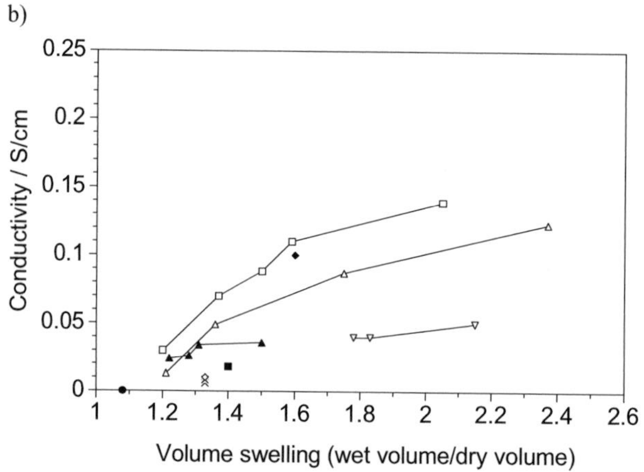

Fig. 9 Conductivity in dependence of water content. **a** – – – Nafion® 112, — a modified poly(perfluoro sulfonic acid), and - - - S-PTES [3]; **b** various hydrocarbon polymers: -♦- sulfonated poly(phosphazene) [1, 192], -◇- benzylsulfonated PBI (80% sulfonation) [1], -□- BPSH [166], -■- sulfonated polyimide blockcopolymer [61, 64], -●- S-PEEK (65% sulfonation) [6]; sulfonated poly(*para*-phenylene)s: -x- sulfonated poly(2-(4′-phenoxy)benzoyl-1,4-phenylene (65% sulfonation) [6], -▽- sulfonated poly(2-benzoyl-1,4-phenylene) (IEC/mmol/g = 2.6; 2.8; 4.0) [79], -▲- sulfonated poly(2-(4′-phenyl)benzoyl-1,4-phenylene) blockcopolymers (IEC/mmol/g = 0.7; 0.75; 0.94; 1.2) [78], -Δ- sulfonated phenylated poly(*para*-phenylene) (IEC/mmol/g = 0.98; 1.4; 1.8; 2.2) [77]

sults in membranes with much higher conductivity at reduced humidity [40] than conventional sulfonated PEEK or PEEKK (Figs. 6, 10).

The degree of sulfonation was varied by incorporation of varying amounts of 6F-bisphenol instead of 9,9-bis(4′-hydroxyphenyl)fluorene, and sulfonation under conditions where only the fluorene units were sulfonated, not the 6F-bisphenol units. HF100 S-PAEK contains no 6F-bisphenol. HF85

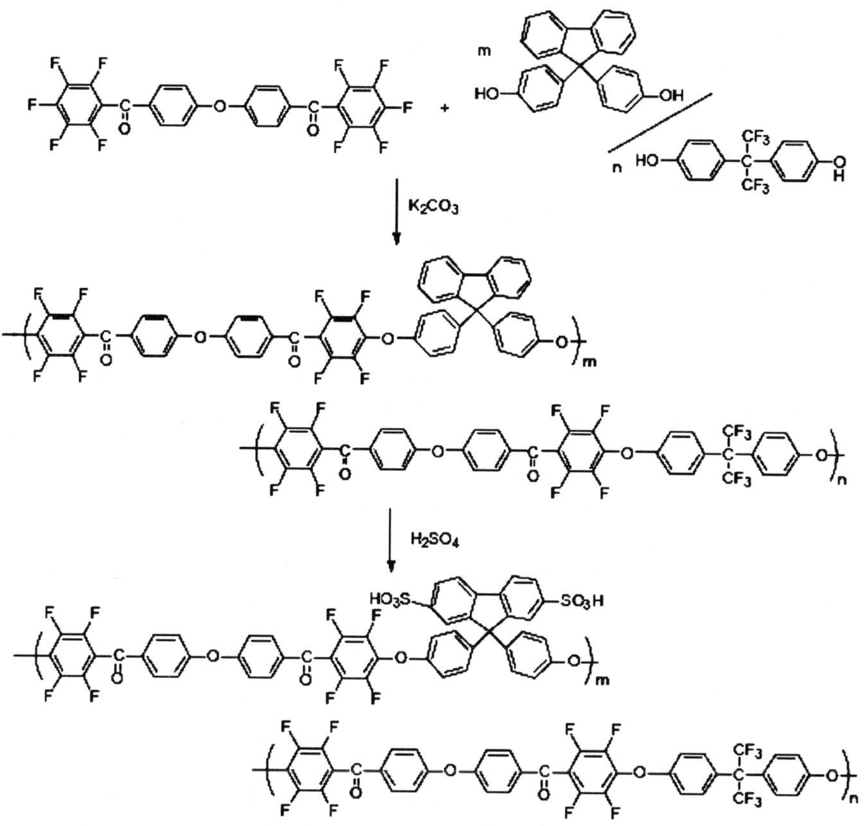

Fig. 10 Poly(aryl ether ketone)s with modified chemical structure [40]

S-PAEK and HF68 S-PAEK contain 15% and 32% 6F-bisphenol, respectively (Fig. 10). Especially, the polymer with the highest degree of sulfonation (HF100 S-PAEK) exhibits good conductivity at 50% relative humidity. However, it swells excessively [40]. Within the series described here, the polymer with 15% 6F-bisphenol appears to be the best compromise between swelling and low r.h. conductivity.

Instead of sulfonated fluorenone units as bulky groups, sulfonated naphthalene can also be used [41]. To this end, 6,7-dihydroxy-2-naphthalene sulfonic acid is used as building block to introduce the sulfonic acid groups (Fig. 11).

Although these materials do not exhibit the outstanding conductivities reported for the fluorine-based polymers described above, they do match or exceed the conductivity of Nafion® 117 at temperatures above 110 °C [41]. Interestingly, the water uptake of the polymer with IEC = 1.6 mmol/g and a conductivity of 60 mS/cm at 110 °C (comparable to Nafion® 117) is no more than 63% on immersion in liquid water at 80 °C for 24 h.

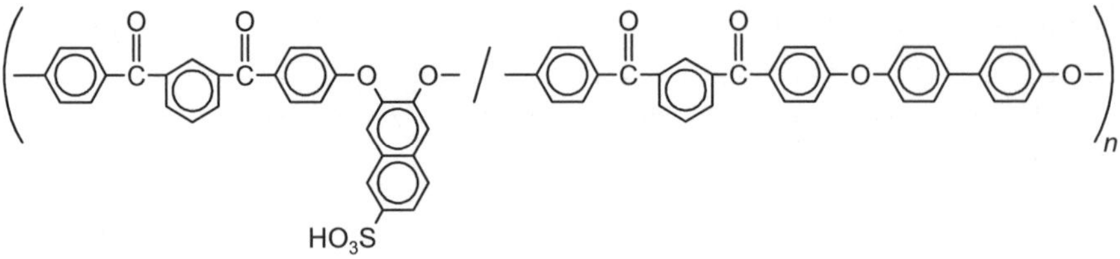

Fig. 11 Poly(aryl ether ketone)s with bulky groups [41]

3.2 Sulfonated Poly(Arylene Ether Ketone)s in DMFC

Sulfonated poly(aryl ether ketone)s appear to have their merits especially in DMFC, where one of the problems of current commercially available perfluorinated membranes is the relatively high methanol permeability, which translates to a reduced power efficiency of the fuel cell.

Sulfonated poly(arylene ether ketone)s have been shown to possess lower methanol permeability than Nafion® by a factor of 3–4 (corrected for membrane thickness) [42]. Incorporation of inorganic proton conductors, such as heteropolyacids, can reduce methanol permeability further to a factor of more than 20 (corrected for thickness) [42] compared to plain Nafion® 117, while maintaining almost the same proton conductivity as Nafion® 117 at room temperature (fully hydrated). Some results are shown in Table 2.

S-PEEK with tungstophosphoric acid (TPA) alone exhibits a higher methanol flow than unfilled S-PEEK, and the TPA is also extracted to a significant extent by liquid water. Interestingly, both methanol flow and bleeding of TPA are strongly decreased by the addition of ZrO_2. The zirconium oxide was not added as a preformed powder, but was prepared in situ by hydrolysis of zirconium tetra propoxide in the casting solution [42]. In the case

Table 2 Reduction of methanol crossover in S-PEK by the addition of inorganic additives (data from [42])

Polymer	D (μm)	MeOH flow (g/hm^2)	Reduction of MeOH flow	Conductivity[a] (mS/cm)
Nafion® 117	175	658	–	94
S-PEK, 1.71 mmol/g	70	411	4×	54
S-PEK:TPA[b] 90:10	75	978	1.6×	82
S-PEK:ZrO_2:TPA 64:8:28	108	47	23×	86

[a] fully hydrated, 25 °C, in 0.333 M H_2SO_4
[b] TPA: Tungstophosphoric acid, $H_3PW_{12}O_{40}$

of highly sulfonated S-PEEK (85% sulfonation, IEC = 2.4 mmol/g), which is soluble at room temperature in water/methanol (80/20), the incorporation of 14% of ZrO_2 stabilizes the membrane, so that even at 55 °C it does not dissolve [43].

A strong reduction in water uptake of S-PEEK can also be achieved by the addition of layered silicates (nano-clays, such as laponite and montmorillonite) [44]. The addition of 10% (w/w) of laponite or montmorillonite to S-PEEK (IEC = 1.7 mmol/g) reduces water uptake at 80 °C from more than 400% (w/w) to less than 50% (w/w). Methanol permeability is reduced by a factor of 2 to 3 as compared to the neat S-PEEK. The proton conductivity of the untreated composite membrane is reduced by a factor of more than 10, but a treatment of the composite membrane with 1 M sulphuric acid for 10 h brings the conductivity back up to the original value of $\approx$50 mS/cm at room temperature (fully hydrated). This is attributed to an exchange of the sodium ions in the layered silicate for protons [44].

3.3 Block Copolymers

A relatively recent approach to improve the property profile of sulfonated poly(aryl ether ketone)s is the use of block copolymers. It is clear that high conductivity of S-PAEK, especially at low relative humidity (and, hence, low water content), can only be achieved at very high sulfonation levels. However, such polymers take up excessive amounts of water, or are even water soluble (see sections above). A potential solution to this dilemma is crosslinking. Unfortunately, at crosslinking densities which suppress swelling sufficiently, the materials typically become brittle. In general, this is not surprising, since both effects are closely related to the length of the chain segments between two crosslinks. In order to significantly reduce the swelling, the segments must be so short that they cannot coil up significantly in the relaxed state. Otherwise, in the presence of water, uncoiling would allow for a large change in macroscopic volume and, hence, swelling. If the segments are so short that they can no longer coil up significantly, they will be mechanically fixed by the crosslinks, resulting in reduced mobility and, hence, loss of the ability to dissipate mechanical energy, causing brittleness.

Separate optimization of mechanical strength, high proton conductivity (i.e., high IEC), and low water uptake may be possible if the functions are separated in a block copolymer structure. The hydrophobic blocks can serve as matrix for mechanical strength and limited swelling, and the hydrophilic blocks with acidic groups can serve as proton transport pathways, comparable to the microphase separated morphology of Nafion®. While a large part of the literature deals with sulfonated block copolymers based on styrene (proton transport related: [45, 46]), there has also some work been done on block copolymers based on poly(arylene ether ketone)s [47–49], poly(arylene ether sulfone)s [50–58], polyimides [59–65], and even poly(*para*-phenylene)s [66].

The block architecture of the polymer chains significantly changes the physical properties. In the case of poly(arylene ether ketone)s, for instance, it was shown that, while randomly sulfonated copolymers with IEC >1.6 mmol/g

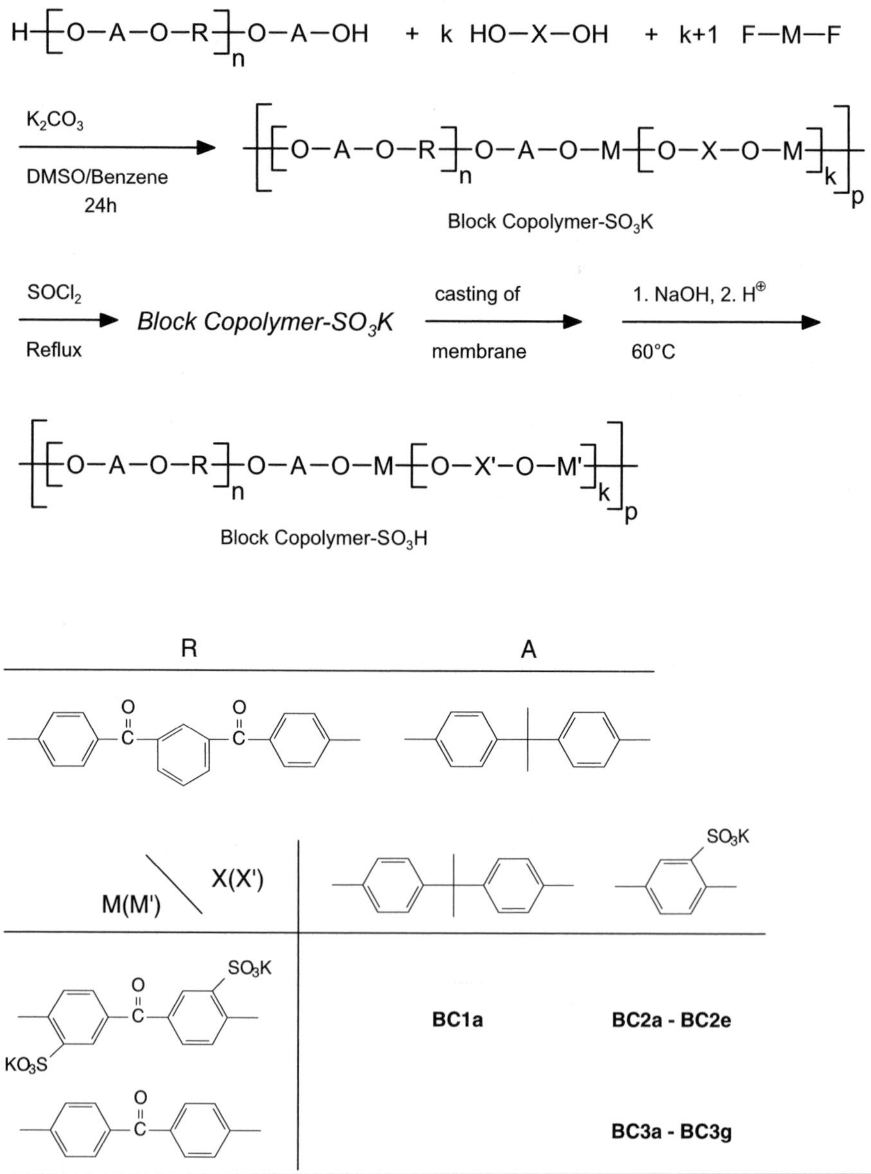

Fig. 12 Sulfonated poly(arylene ether ketone) block copolymers [48, 49]

Table 3 Sulfonated poly(arylene ether ketone) block copolymers [48, 49]

Polymer	IEC (titration) (mmol/g)	Water uptake (%) at 60 °C	at 100 °C	Conductivity (mS/cm) r.t.	at 60 °C
1	1.86	53	250	95	166
2	1.48	18	230	81	129
3	1.36	15	210	66	116
4	0.68	4	11	10	24

were completely or at least partly soluble in water even at room temperature [67], sulfonated block copolymers consisting of the exact same components were insoluble even in boiling water, even at IEC values up to 1.86 mmol/g [48, 49]. Figure 12 shows the general structure of these block copolymers.

The block copolymers exhibit conductivities up to 166 mS/cm at 60 °C, fully hydrated, at a water uptake up to 250% in boiling water. Table 3 shows some properties of these materials.

Even higher conductivities were found for block copolymers consisting of sulfonated poly(aryl ether sulfone) blocks alternating with fluorinated aromatic polyether blocks (Fig. 13), although they had relatively high water uptake even at room temperature. Table 4 shows conductivites and water uptake data.

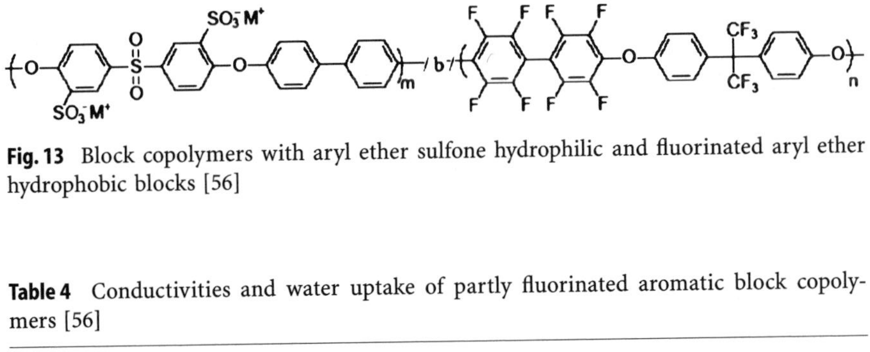

Fig. 13 Block copolymers with aryl ether sulfone hydrophilic and fluorinated aryl ether hydrophobic blocks [56]

Table 4 Conductivities and water uptake of partly fluorinated aromatic block copolymers [56]

Polymer	Molar mass hydrophilic block (g/mol)	Molar mass hydrophobic block (g/mol)	IEC (mmol/g)	Conductivity at room temperature (mS/cm)	Water uptake at room temperature (%)
1	5000	2800	2.3	320	470
2	5000	5000	1.5	120	130
3	15000	15000	1.46	160	260

Clearly, separation of the requirements of "high IEC" and "good mechanical properties" in different blocks by combining segments with high degree of sulfonation with totally unsulfonated, hydrophobic segments is an attractive concept, despite the increased synthesis effort.

3.4 Blends of S-PAEK with Inert Polymers

Blending sulfonated proton conducting polymers with non-sulfonated inert polymers is another approach to separate the requirements for high proton conductivity on one hand and good mechanical properties on the other. Recently, sulfonated poly(arylene ether ketone)s have been blended with various elastomers [68], PVdF [69] [70], polyacrylonitrile [71], polyethylene [72, 73], and phenolic resins [74]. In most cases, it is possible to approach the conductivity of Nafion® type materials under fully hydrated conditions, and at the same time reduce the water uptake and swelling in comparison to the sulfonated parent polymers. Also, mechanical properties can be improved. However, in order to achieve high conductivities, a high degree of sulfonation of the parent polymer is necessary, raising the issue of long-term durability and potential extraction of soluble fractions.

3.5 Poly(*para*-phenylene)s

Another class of chemically exceptionally stable polymers are the poly(*para*-phenylene)s. In addition, they are stiff, rod-like polymers. This feature usually tends to cause insolubility of polymers, which is an attractive approach: rod-like polymers can be expected to be insoluble in water at much higher IEC values than polymers consisting of flexible chains, such as poly(arylene ether)s. Since proton conductivity depends on IEC, such polymers have the potential to form membranes with high conductivity at relatively low degrees of swelling and good dimensional and mechanical stability.

Derivatives which are soluble in organic solvents can be prepared from substituted monomers. Considering the very demanding conditions in fuel cells, fully aromatic structures are preferred. An attractive system is based on the coupling of substituted dichlorophenyl monomers using a Nickel catalyst, with the most popular monomer being 2,5-dichlorobenzophenone. The resulting polymers are typically soluble in dipolar aprotic solvents and can be sulfonated similar to poly(arylene ether ketone)s and poly(arylene ether sulfone)s [6, 26, 75–82]. Figure 14 shows the chemical structures involved.

The poly(*para*-phenylenes)s can be considered rigid rod polymers, which distinguishes them from all other types of polymers typically used for proton conducting membranes, possibly with the exception of certain polyimides. A striking consequence of the rigid rod structure is the fact that the poly(*para*-

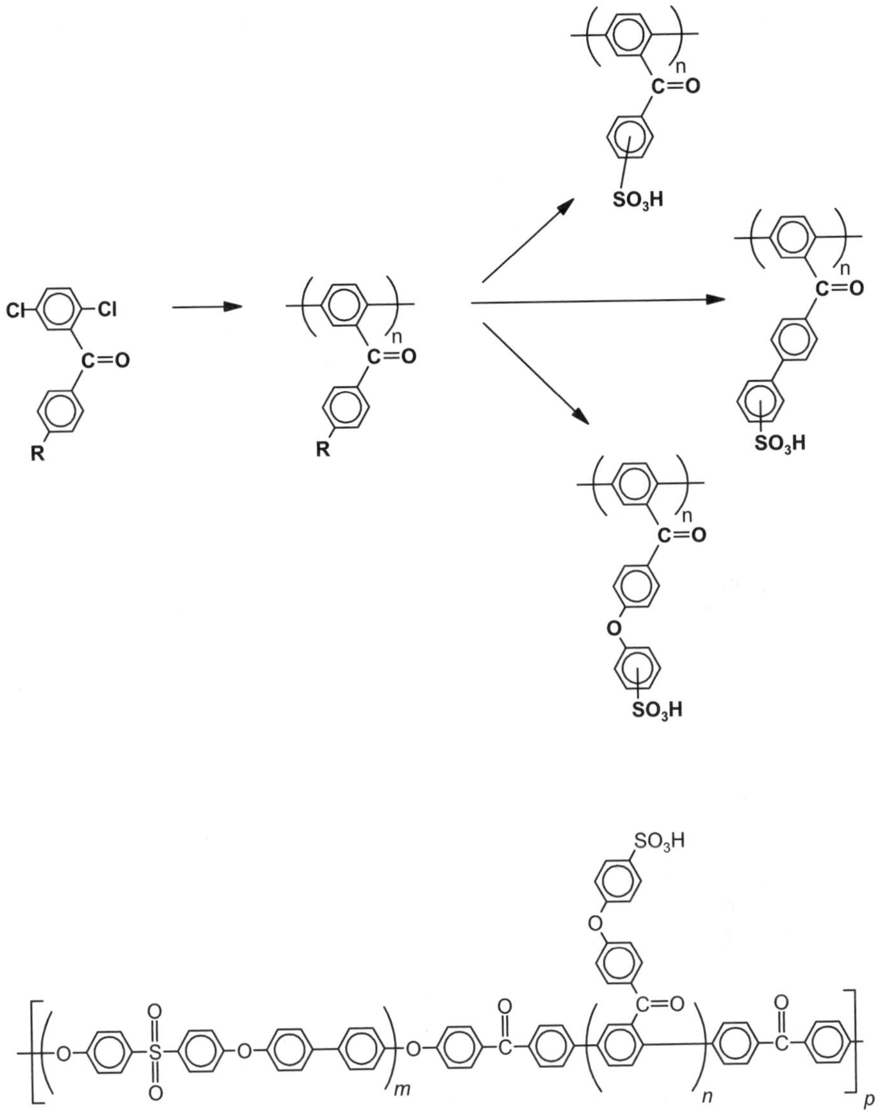

S-P1

Fig. 14 Structures and synthesis of poly(*para*-phenylene)s for proton transport membranes

phenylenes)s can be sulfonated to very high IECs without rendering them soluble in water. For example, poly(2-benzoyl-1,4-phenylene) is still insoluble in water at an IEC of 4.0 mmol/g. Its water uptake at room temperature in liquid water is 115%. Sulfonation up to IEC of 6.5 mmol/g is possible, although at this extreme degree of sulfonation the polymer does become water soluble.

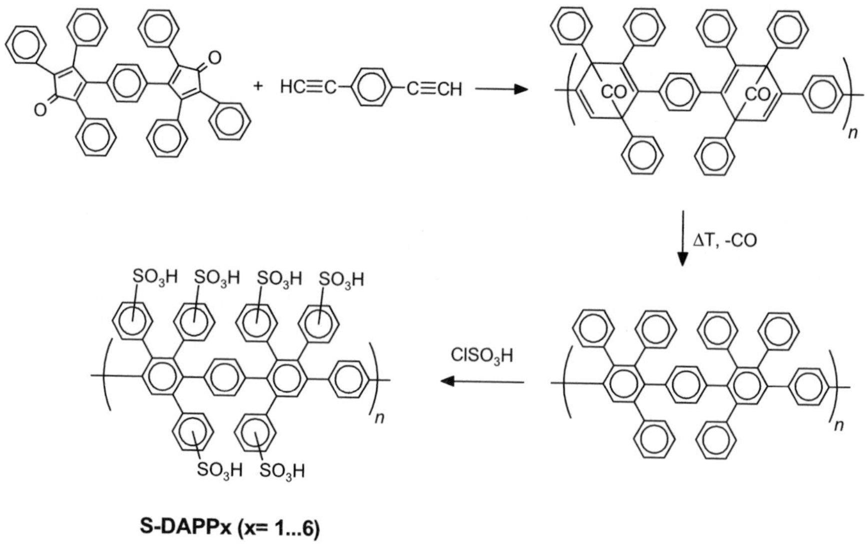

ΔT, -CO

$ClSO_3H$

S-DAPPx (x= 1...6)

Fig. 15 Poly(*para*-phenylene)s by Diels–Alder reaction [77]

Probably because of limited molar mass, the brittleness of these poly(*para*-phenylene)s can be compensated by incorporating them as sulfonated segments into block copolymers with poly(aryl ether sulfone)s [78]. As for the poly(aryl ether ketone) block copolymers, water uptake is reduced while relatively high conductivity (considering the IEC) is maintained [78].

Another method [77] for the synthesis of poly(*para*-phenylene)s utilizes a reaction which was already successfully used by Stille [83–85] in the 1960s: Diels-Alder reaction (Fig. 15).

Diels-Alder additions are generally regioselective, resulting in predominantly *para* linkages of the polymer chain. However, in sterically crowded systems regioselectivity may be lost to some degree, which introduces some *meta* linkages. Thus, these polymers may not be as strictly rod-like as the polymers prepared by catalytic aryl coupling. Their phenyl substituents provide a large number of potential sites for sulfonation, which can be achieved with chlorosulfonic acid. Above an IEC of 2.2 mmol/g, corresponding to approximately two sulfonic acid groups per repeating unit, they were reported to undergo excessive swelling and form a hydrogel in water [77].

A third pathway to soluble poly(*para*-phenylene)s is Ullmann coupling of sulfonated dibromo biphenyl monomers [80] (Fig. 16).

For the coupling reaction, it is useful to exchange the protons of the sulfonic acids for organic cations, such as pyridinium or trimethyl benzylammonium. In order to modify the polymer properties, some of the sulfonic acid groups can be used to introduce aromatic substituents by sulfone formation.

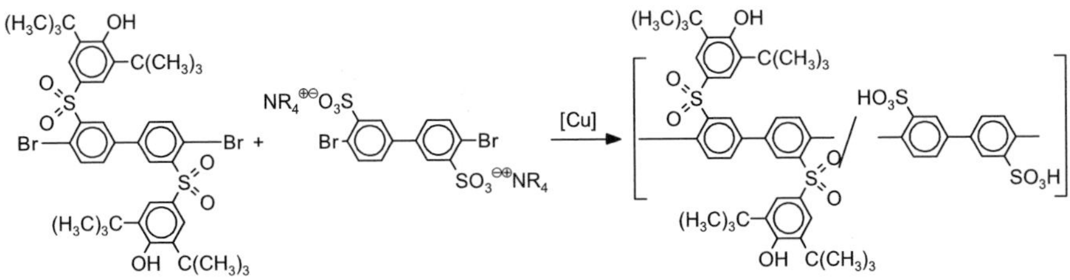

Fig. 16 Poly(*para*-phenylene)s by Ullmann coupling [80]

Table 5 IEC, water uptake and conductivity of selected poly(*para*-phenylene)s [77–80]

Polymer	IEC (mmol/g)	Water uptake at 30 °C in liquid water % (w/w)	Conductivity at room temperature in liquid water (mS/cm)
Nafion® 112	0.91	25	100
S-DAPP4 (Fig. 15)	2.2	137	123
S-DAPP3 (Fig. 15)	1.8	75	87
S-DAPP2 (Fig. 15)	1.4	36	49
S-P1 (Fig. 14)	4.0	115	50
S-P1 (Fig. 14)	2.8	83	40
Copolymer 5% di-*tert.*butylphenol (Fig. 16)	5.7	(50% v/v from 22% r.h. to 100% r.h.)	250
Copolymer 5% di-*tert.*-butylphenyl (Fig. 16)	6.1	(87% v/v from 22% r.h. to 100% r.h.)	700

Table 5 summarizes characterization data of some poly(*para*-phenylene)s. Values for Nafion® are added for comparison.

An interesting comparison was drawn between sulfonated PEEK and sulfonated poly(2-(4-phenoxy)benzoyl-1,4-phenylene) (Fig. 17). They are almost structural isomers, except for one more ether bridge in PEEK. Yet, at comparable degrees of sulfonation, the poly(*para*-phenylen) derivative showed up to 2 orders of magnitude higher proton conductivity when measured under the same conditions [6], and the proton conductivity did not drop off at temperatures up to 110 °C (Fig. 17).

3.6 Non-Rigid Rod Polyphenylenes

In principle, all dichloro aromatics can be coupled with the same Nickel based system which has been used for the synthesis of the poly(2-benzoyl-

a)

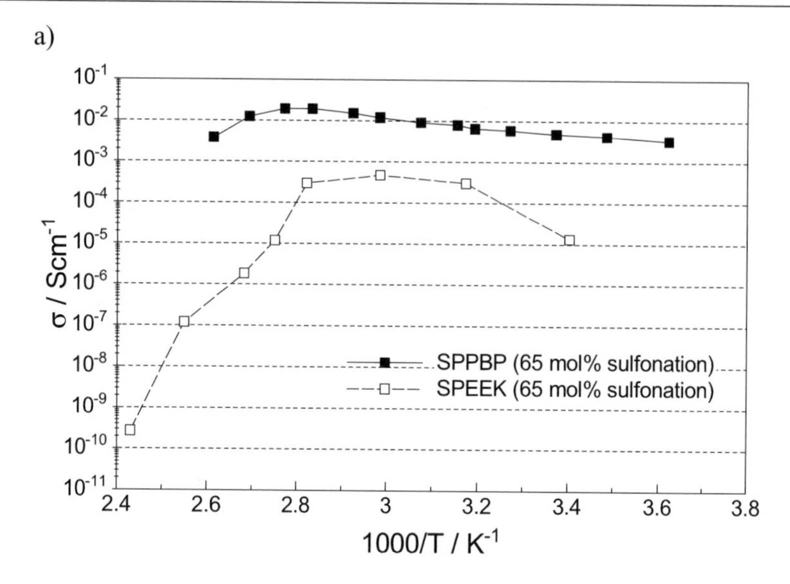

b)

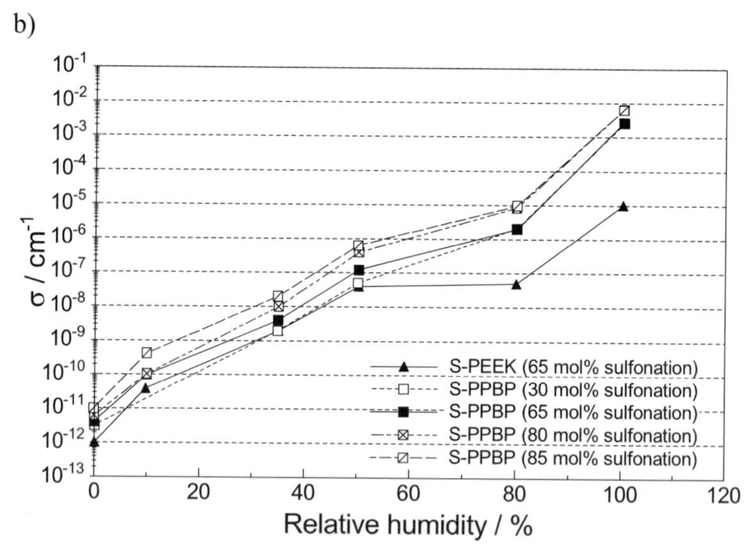

Fig. 17 Comparison of the proton conductivity of S-PEEK and S-PPBP **a** vs. temperature and **b** vs. relative humidity [6]

1,4-phenylene) described above. However, whenever the coupling is not exclusively *para*, the resulting polymers are no rigid rods.

For example, Frey and Mülhaupt et al. [86] studied copolymers from 1,3-dichlorobenzene and 4,4′-dichlorodiphenyl sulfone, which had either carboxylic or sulfonic acid side groups. Figure 18 shows the structures.

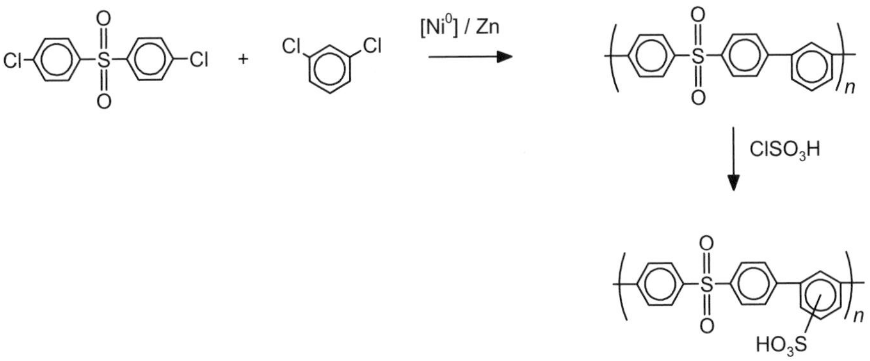

Fig. 18 Non-rigid rod sulfonated polyphenylenes [86]

With sufficiently high IEC, these polymers had higher proton conductivity than Nafion® 117 over a temperature range from 20–110 °C (fully hydrated, in a gas-tight sample holder). Water uptake was very high: more than 500% at 90 °C. Interestingly, a blend of a carboxylated and a sulfonated polymer of this type with an overall IEC value of 4.8 mmol/g still showed a conductivity comparable to that of Nafion® 117, but had a water uptake of only 132% at 90 °C.

3.7 Poly(Phenylene Sulphide)

Poly(phenylene sulphide) itself is a semicrystalline material with excellent mechanical properties and good chemical stability. Due to its electron-rich structure, it can be sulfonated relatively easily up to a degree of sulfonation corresponding to 0.85 sulfonic acid group per phenyl ring (IEC = 4.5 mmol/g) [87]. Even higher sulfonation, up to two sulfonic acid groups per phenyl ring (IEC = 7.4 mmol/g), cannot be achieved directly due to crosslinking, but is accessible though a cationic intermediate, which prevents crosslinking [88, 89]. The proton conductivity of this material was found to be 20 mS/cm at 20 °C and 95% relative humidity [89].

4 Polysulfones

Over the past three decades, several aromatic poly(arylene ether sulfone)s have been commercialized. These polymers show unique combinations of chemical and physical properties, including high stability against hydrolysis, high thermal stabilty, high stability against oxidation and UV-light, high glass transition temperature, and good transparency, when amorphous. First

A)

x (diphenyl ether) + 2 x ClO_2S–(diphenyl ether)–SO_2Cl

$FeCl_3$
120°C - 190°C
- HCl

x (phenoxyphenyl)–SO_2Cl

$FeCl_3$
120°C - 190°C
- HCl

B)

x Cl–(4,4'-dichlorodiphenyl sulfone)–Cl

x NaO–(bisphenol A)–ONa
DMAc/Toluene
160°C 24h
- 2x NaCl

x HO–(bisphenol A)–OH
K_2CO_3
DMAc/Toluene
160°C 24h
- 2x KCl

C)

TMSO–(bisphenol A)–OTMS

2 CsF
- 2 Me_3SiF

CsO–(bisphenol A)–OCs

F–(4,4'-difluorodiphenyl sulfone)–F
- 2 CsF

◀ **Fig. 19** Synthesis of poly(ethersulfone)s by **A** Friedel–Crafts sulfonylation and **B** by nucleophilic polycondensation in solution using phenolates and **C** melt polycondensation using trimethylsilyl derivatives of bisphenols

attempts to synthesize polysulfones were succesfully carried out by a Friedel-Crafts sulfonylation reaction of arylene disulfonyl chlorides, e.g., diphenyl ether-4,4′-disulfonyl chloride with diaryl ethers, e.g., diphenyl ether, or by self-condensation of 4-phenoxy benzene sulfonyl chloride in the presence of $FeCl_3$ [90, 91] (Fig. 19A). While the former reaction bears the risk of side reactions, namely, sulfonylation not only in *para* but also in *ortho* position, the latter gives only the desired linear all-*para* product.

Meanwhile, most commercial polysulfones (PSU) and poly(ether sulfone)s (PES) are obtained from conversion of suitable aromatic dihalides with bisphenols by nucleophilic displacement polycondensation (Fig. 19B). Generally, 4,4′-dichlorodiphenyl sulfone (DCDPS) is reacted with alkali salts of bisphenols [92, 93]. The reaction is carried out in solution using *N*-methyl-2-pyrrolidone (NMP), *N,N*-dimethyl acetamide (DMAc), or dimethyl sulfoxide (DMSO) as the solvent. Occasionally, the more reactive, but also more expensive, 4,4′-difluorodiphenyl sulfone might be used for experimental purposes. Usually, the electronegativity of the sulfone linkage is sufficient to increase the reactivity of the aromatic chloride in DCDPS (Fig. 19).

Alternatively, the bistrimethylsilyl ethers of the bisphenols can be used instead of the alkali salts. This approach has the advantage, since the formation of water and, thus, the risk of a hydrolytic cleavage of C–F bonds is avoided. Furthermore, the purification of the silylated bisphenols can be achieved by simple vacuum distillation. The use of silylated bisphenols also allows for the preparation of poly(arylene ether)s in the melt (T 130–300 °C) in the presence of catalytic amounts of CsF or KF, thus, avoiding the removal of large amounts of inorganic salts and solvents [94–96] (Fig. 19C).

Polysulfones (PSU) and poly(ether sulfone)s (PES) have been widely used as membrane materials for ultrafiltration, pervaporation [97–99], or electrodialysis [100], due to their chemical und thermal stability, high glass transition temperature (T_g), which is in the range of 180 °C to values well above 200 °C, as well as their good film-forming properties and solubility in dipolar aprotic solvents, such as NMP, DMAc, or DMSO.

Besides the classical polyethersulfone (Fig. 19A) derived from the reaction of 4,4′-dihalodiphenyl sulfone and 4,4′-hydroxydiphenyl sulfone or self-condensation of 4-halo-4′-hydroxydiphenyl sulfone and polysulfone (Fig. 19B) derived from the reaction of bisphenol A (2,2-*bis*-(4-hydroxyphenyl) propane) and 4,4′-dihalodiphenyl sulfone, a large number of polysulfones have been either commercialized or prepared for research purpose by variation of the bisphenol moieties.

5
Functionalization of Poly(Ether Sulfone)s

Functionalized (e.g., sulfonated or phosphonated) poly(arylene ether sulfone)s can be attained by two different routes. The most suitable and most often applied way to obtain functionalized and, in particular, sulfonated polymers is the post-treatment (sulfonation) of a given polymer [101]. While sulfonated poly(arylene ether sulfone)s are easily accessible by electrophilic or nucleophilic substitution at the aromatic ring, the phosphonation is much more complicated and is described in the literature less often than the sulfonation [102–104]. Phosphonic acids and, in particular, arylphosphonic acids are not of such strong interest as proton conducting membrane materials in fuel cells, because of their lower acidic activity compared to the corresponding sulfonic acids. Therefore, a much higher concentration of phosphonic acid groups in the polymer is required in order to get a high proton conductivity. Furthermore, synthetic routes for the preparation of phosphonated polymers are rather limited as compared to sulfonic acid derivatives. On the other hand, arylphosphonic acids show a higher thermostability and are not susceptible to "dephosphonation". In addition, phosphonated polymers are considered to be promising candidates for membranes with sufficient proton conductivities at low humidification levels or even in the absence of humidity.

The second route to obtain functionalized polymers is given by the use of monomers already bearing functional groups, e.g., sulfonic acid groups. This method has the advantage that (a) the site of functionalization, (b) the number of functional groups, and (c) the distribution of functional groups, either randomly or blockwise, can be easily controlled in the polymer chain.

Both methods, the post-sulfonation of preformed poly(arylene ether sulfone)s and the preparation of functionalized polymers by the use of sulfonated monomers, have been widely described in the literature and will be discussed in the following sections.

5.1
Post-Sulfonation and Post-Phosphonation of Polysulfones

Depending on the chemical composition of the polymer backbone and the desired degree of sulfonation, various sulfonating agents with different reactivities can be selected and are commercially available (Table 6).

Although easy to carry out, post-sulfonation of poly(arylene ether sulfone)s bears some risks and disadvantages over the synthesis of the same type of polymer using sulfonated monomers. These are mainly degradation of the polymer backbone and the homogeneity. Iojoiu et al. studied the influence of various processing paramaters of sulfonation of different poly(arylene ether sulfone)s on the degree of sulfonation as well as on the material properties

Table 6 Sulfonating Agents

Sulfonating Agent	Reactivity	Reaction site	Comments
Chlorosulfonic acid	High	Electron-rich ring	Inexpensive, side-reactions (degradation, crosslinking)
Fuming sulfuric acid (Oleum)	High	Electron-rich ring	Inexpensive, crosslinking
Sulfuric acid	High	Electron-rich ring	Inexpensive, lowering of reactivity by reaction product water
Sulfur trioxide/ triethylphosphate (TEP)	Medium to high	Electron-rich ring	Inexpensive, reactivity might be controlled by TEP content
Trimethylsilylsulfonyl chloride	Medium	Electron-rich ring	Relatively expensive
Acetylsulfate	Low	Aliphatic double bonds	Inexpensive
BuLi + SO_2 or SO_3	High	Electron-poor ring	Expensive
BuLi + sultones, halogenoalkylsulfonic acids...	High	Electron-poor ring	Expensive

in a recent paper [105] (the reader is also referred to the literature cited in this paper for more detailed information). Most attractive seems to be the sulfonation in concentrated sulfuric acid (i.e. 98%) [78, 106–108] or chlorosulfonic acid [97, 99, 109–117], acting both as solvent and as sulfonating agent since both reagents are inexpensive and readily available. As reported by Blanco et al. [116], a rapid degradation occurs when sulfonating PSU, which makes this method at least questionable for a number of poly(ether sulfone)s. They proposed a degradation mechanism which involves the protonation of the ether oxygen [116]. A much higher stability was observed for PES-C (for chemical structure see Fig. 20). Furthermore, when using sulfuric acid for the sulfonation the byproduct water dilutes the reaction medium, thus decreasing its reactivity. The effect of acid concentration on the sulfonation kinetics has been studied for a poly(ether ether ketone), for example, by Huang et al. [118]. On the other hand, PES, which is not soluble in concentrated sulfuric acid, or PPSU are both soluble in halogenated solvents, which are well adapted to electrophilic substitution. Although the starting materials are perfectly soluble in these solvents, the sulfonated products obtained by the reaction with chlorosulfonic acid [18, 74, 97, 117, 119] or SO_3 (PPSU, [120]) are not, and they precipitate during the reaction, which leads to inhomogeneously sulfonated products and the degree of sulfonation is uncontrolled. To overcome this problem, Genova-Dimitrova et al. [30] suggested to add small amounts of DMF to the reaction mixture in order to keep the polymer in solution.

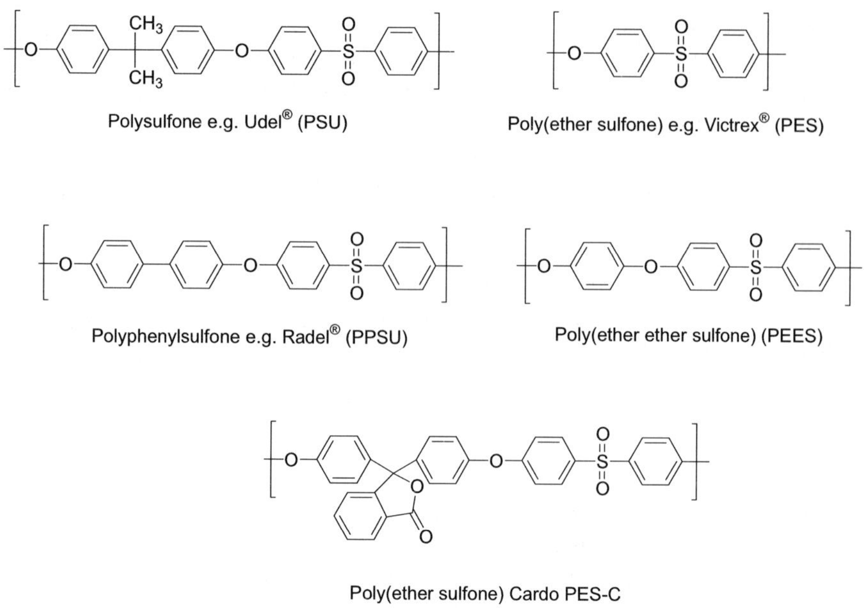

Polysulfone e.g. Udel® (PSU)

Poly(ether sulfone) e.g. Victrex® (PES)

Polyphenylsulfone e.g. Radel® (PPSU)

Poly(ether ether sulfone) (PEES)

Poly(ether sulfone) Cardo PES-C

Fig. 20 Commercially available poly(ether sulfone)s

In other studies, the sulfonation with SO_3-triethylphosphate complex in dichloromethane has been proposed and was described to be more reliable with a minimum risk of side reactions [101, 121–123]. The disadvantage of using SO_3-triethylphosphate complex is its toxicity and high reactivity of SO_3, as well as the exothermic reaction with triethylphosphate which makes it difficult to use.

An alternative approach to carry out the sulfonation reaction under homogeneous conditions is based on the use of trimethylsilyl chlorosulfonate as sulfonating agent and dichloromethane or dichloroethane as solvents [27, 30, 50, 120, 124, 127]. The reaction mixture remains homogeneous when kept anhydrous, due to the trimethyl silylester formed during the reaction. The ester further reduces the risk of side reactions, namely, crosslinking or degradation, as reported from sulfonation reactions with chlorosulfonic acid. Dyck et al. [120] reported a much more homogeneous reaction product and a much better control of the degree of sulfonation when using trimethylsilyl chlorosulfonate instead of SO_3-TEP complex for the sulfonation of PPSU. The sulfonation with the SO_3-TEP complex always led to the formation of two fractions: a water-soluble one with a high degree of sulfonation (>2.88 mmol/g) and a water-insoluble fraction with a degree of sulfonation <1.5 mmol/g). The degree of sulfonation was controlled by either reaction time or by the amount of trimethylsilyl chlorosulfonate added to the reaction mixture. The inhomogeneity is also reflected in the membrane properties as, for example, the methanol permeability. The SO_3-TEP sulfonated membranes showed a methanol permeability comparable to Nafion®, which are ca. two-fold higher than that of the more homogeneously sulfonated products yielded from the reaction with trimethylsilyl chlorosulfonate. While generally dense symmetric membranes are employed in fuel cell applications, Dyck et al. [120] described the preparation of an asymmetric membrane with a thin dense top layer and a support layer exhibiting a closed-cell structure. An asymmetric membrane with an IEC of 2.08 mmol/g showed a four-fold higher proton conductivity (ca. 55 mS/cm at 80 °C) compared to a symmetric membrane from the same material and of the same thickness. Park et al. [124] investigated the effect of thermal treatment of sulfonated membranes (sPSU) on membrane properties, such as water uptake, proton conductivity, and methanol permeability. The sulfonation was achieved by treatment of PSU with a 1:1 molar mixture of chlorosulfonic acid/chlorotrimethylsilane in 1,2-dichloroethane. The degree of sulfonation was controlled by the amount of sulfonating agent added to the reaction mixture. A maximum degree of sulfonation of 75%, corresponding to an IEC of 1.45 mmol/g was achieved. Thermal treatment of the membranes for 2 h at 150 °C, which is well below the glass transition temperatures (T_g) of the membrane materials (193–225 °C, depending on the degree of sulfonation) resulted in a pronounced lowering of the methanol permeability but it only slightly affected the water uptake and the proton conductivities, especially for membranes with low ion-exchange capacities (0.55–0.85 mmol/g).

Yang et al. reported on the preparation of polysulfone-*block*-PVDF copolymers [50, 51]. Again, the sulfonation of the polysulfone was conducted with trimethylsilyl chlorosulfonate and the degree of sulfonation was controlled by the amount of sulfonating agent added to the reaction mixture. As a result, the block copolymers showed higher proton conductivities at IECs <1.4 mmol/g than sulfonated homopolymers with comparable IECs. For lowest IEC polymers (ca. 0.8 mmol/g) the conductivity was enhanced by a factor of 4 over a temperature range from 30 °C up to 80 °C. With higher IECs no differences between the two different types of membranes were found. Since the λ-values (mol H_2O/mol SO_3H) were nearly identical for the two membrane types, the conductivity differences could not be associated with differences in λ. TEM micrographs revealed that in both polymers with low IEC ionic aggregates exist, being larger in the block copolymer (50–200 nm) than in the homopolymer. It was proposed that the presence of the hydrophobic block promotes the phase separation between sulfonated and non-sulfonated domains and thus the formation of ionic aggregates and an ionic network, leading to the observed conductivity enhancement. In high IEC membranes with fully developed ionic network, the relatively small hydrophobic blocks have only little or no effect on the conductivity.

An interesting approach to obtaining sulfonated poly(arylene ether sulfone)s was reported by Zhang et al. [52, 125]. They prepared poly(arylene ether sulfone)-*block*-polybutadiene copolymers and achieved a selective sulfonation of the remaining double bonds in the flexible polybutadiene block-segments by using acetylsulfate. Although IECs were only in the range from 0.107 mmol/g to 0.624 mmol/g, relatively high proton conductivities up to 30 mS/cm (IEC 0.624 mmol/g) at 25 °C were recorded. This was attributed to the fixation of the proton conducting groups to the flexible polybutadiene segments (T_gs ranging from – 37.7 °C to – 4.5 °C), which provides greater mobility and allows for easier formation of ionic pathways. In a second attempt to enhance the ion-exchange capacity of the block copolymers, both the poly(arylene ether sulfone) (PAES) segments as well as the polybutadiene segments were sulfonated. While the former were obtained by polymerization of sulfonated and non-sulfonated DCDPS with bisphenol A, the latter were obtained by selective sulfonation with acetylsulfate. The degree of sulfonation of the PAES segments were controlled by the ratio of DCDPS to SDCDPS. Double bonds, which still remained in the flexible block after the sulfonation process, were epoxidized to further reduce the gas permeability. The introduction of the sulfonated PAES segments resulted in a remarkable increase in intramolecular and intermolecular chain interactions, leading to higher T_g and T_m of the membranes. Furthermore, the microstructure of the membranes were found to correspond to DS of the PAES segments, as the size of polybutadiene domains increased with increasing DS up to 40 mol % SDCDPS (IEC = 0.622 mmol/g) of sulfonated PAES segments. The latter sample showed the highest proton conductivity (0.108 S/cm at 90 °C) of all samples under investigation in this study due to the formation of an interpenetrating ionic network. Further increase of the DS of sPAES resulted in a demixing of the different blocks. In order to get sulfonated polysulfones with the functional groups in the side-chain and a predefined degree of sulfonation, Meier-Haack et al. prepared polysulfones with phenylhydroquinone moieties in the backbone [126–128]. The sulfonation was conducted with various sulfonating agents, such as concentrated sulfuric acid or chlorosulfonic acid trimethylsilylester. Contrary to the sulfonation of similar poly(ether ether ketone)s [129, 130], the sulfonation of the poly(ether sulfone)s occurred always at the side-chain as well as at the main-chain. Despite the mixed and partly "oversulfonation" ($IEC_{titr.} > IEC_{theo.}$) samples with an IEC up to 1.5 mmol/g, (water uptake of 25 mol water/mol SO_3H) showed promising properties (proton diffusion, methanol diffusion) concerning future applications, e.g., in fuel cells. Furthermore, these materials showed no hydrolytic cleavage of the sulfonic acid group and no backbone degradation upon heating in water for 168 h at 130 °C.

5.2
Functionalization by Chemical Grafting Reactions

A promising alternative to the direct functionalization (sulfonation; phosphonation) of high performance polymers is offered by the chemical grafting reaction. This synthetic route involves first an activitation step, e.g., by irradiation (UV, ion beam, electron beam, γ-ray) or chemical treatment followed by conversion of the activated sites.

Belfort and co-workers, for example, described the photochemical modification of poly(ether sulfone) and sulfonated poly(sulfone) nanofiltration membranes for the control of fouling [131, 132]. Irradiation of sPES or PSU with UV light of a wavelength of 254 nm led to scission of the S–C bond in the polymer backbone, thus, forming a radical at that position. In the presence of *N*-vinyl pyrrolidone,for example, grafting occurs. With no vinyl monomer present, the radical terminates with a hydroxyl group. In both cases more hydrophilic membranes with lower fouling tendency were obtained. However, since the activation is realized by bond breaking within the polymer main-chain, the molecular weight and, therefore, the mechanical properties of the membrane material might be negatively influenced. This might also be the reason why this method has not so far been applied in the production of ion-exchange membranes for fuel cell applications.

Another method to functionalize poly(arylene ether sulfone)s is given with the metalation (lithiation) route, which has been intensively employed by several research groups [4, 5, 32, 102, 133–145], and has been the subject of a recently published review by Jannasch [146]. The activation is usually achieved by the treatment of the respective polymer with a strong base, such as butyllithium or alkali hydrides (NaH, LiH), at low temperatures (– 78 °C to – 30 °C) in an inert solvent, such as THF. Further reaction of the activated polymers with, for example, carboxylic acid, sulfuric acid, or phosphonic acid derivatives, leads to carboxylated, sulfonated, or phosphonated polymers, respectively. This method is not limited to the introduction of acidic groups. The activated polymers can be treated with numerous electrophilic substances, many of which are commercially available (Fig. 21). Amination and hydroxylation, for example, have been reported in the literature by Guiver and co-workers [135] and Kerres and co-workers [32]. Phosphonation of poly(arylene ether sulfone)s via the metalation route has been recently described by Jannasch and co-workers [102, 147]. The phosphonic acid group was linked directly to the polysulfone main-chain by the lithiation method. In the second work, Lafitte and Jannasch described the phosphonation of benzoyl-modified polysulfones in the side-chain. In the first step, the polysulfone was reacted with methyl iodobenzoat by the lithiation method to give iodobenzoyl-PSU. This product was further converted into benzoyl-difluoromethylphosphonate via a CuBr mediated cross-coupling reaction with diethyl(bromodifluoromethyl)phosphonate and subsequent hydrolysis with bromotrimethylsilane (Fig. 21). Membranes with an IEC of 1.79 mmol/g showed a very low water uptake of less than 10 wt. % in the range from 20 °C to 100 °C and a reasonable proton conductivity of 5 mS/cm at 100 °C.

SO_2

Oxidation

SO_2Cl_2 2. Hydrolysis

(Ar = Phenyl, Naphthyl)

A = F, SO_3Na

CO_2

$BrCF_2PO_3Et_2$

1.) $BrSiMe_3$ 2.) HCl

2. NaOH 3. HCl

$NaBH_4$

or

◀ **Fig. 21** Possible reactions leading to sulfonated or phosphonated polymers via lithiation route

Since the polymers to be activated need to have at least one acidic proton in the main chain and must be soluble in suitable solvents, the number of polymers which can be modified by this particular method is limited. So far only the modification of poly(arylene ether sulfone)s and polybenzimidazoles as suitable materials for fuel cell membranes has been reported in the literature [4, 5, 32, 102, 136–146, 148–151]. The attempt to lithiate poly(ether sulfone) under homogeneous conditions failed so far, due to poor solubility and the alternating electron-donating and electron-withdrawing linkages present in poly(ether sulfone), which result in an overall unfavorable balance in the polymer chain.

When dealing with sulfonated aromatic systems, one has to keep in mind that sulfonated aromatic rings are susceptible to desulfonation, depending on the electron density at the sulfonated site and on the conditions in the medium, such as elevated temperatures and the presence of acid. Since the direct sulfonation is an electrophilic reaction, the sulfonation occurs at electron-rich sites of the polymer backbone, which is the *ortho*-position to the ether linkage in polysulfones. The advantage of the metalation method over the direct sulfonation is that the activition takes place at electron-poor sites of the polymer backbone, in general, in *ortho*-position to the electron-withdrawing sulfone-linkage. Thus, these sulfonated polysulfones are expected to be less sensitive to desulfonation than those obtained by the direct sulfonation method. However, the site of conversion can be shifted from the electron-poor (*ortho*-position to the sulfone linkage) site to the electron-rich site (*ortho*-position to the ether linkage) by introduction of bromine atoms into the polymer main chain either by bromination or by using brominated monomers. At low temperatures, lithium-bromine rather than lithium-hydrogen exchange occurs exclusively [133]. While Guiver and co-workers used the lithiation method to modify NF, RO, and UF membranes, Kerres and co-workers employed this method for the preparation of ion-exchange membranes for electrochemical applications, such as fuel cells, by the introduction of sulfonic acid groups. This was accomplished by conversion of the lithiated species, for example, with SO_2, followed by oxidation of the obtained sulfinic acid giving the sulfonic acid. The sulfinic acid derivatives were also subject to further reaction with dihalogenoalkyls, e.g., diiodobutane, to achieve a covalent crosslinking (Fig. 22A) in order to reduce extensive swelling of highly sulfonated membrane materials [32].

Besides the covalent crosslinking, Kerres et al. investigated the properties of fuel cell membranes of ionically crosslinked polysulfonic acids. This type of crosslinking was achieved by blending sulfonated poly(arylene ether sulfone)s and poly(arylene ether ketone)s with basic polymers, such as polybenzimidazole, poly(ethylene imine), poly(vinyl pyridine), or amino functional-

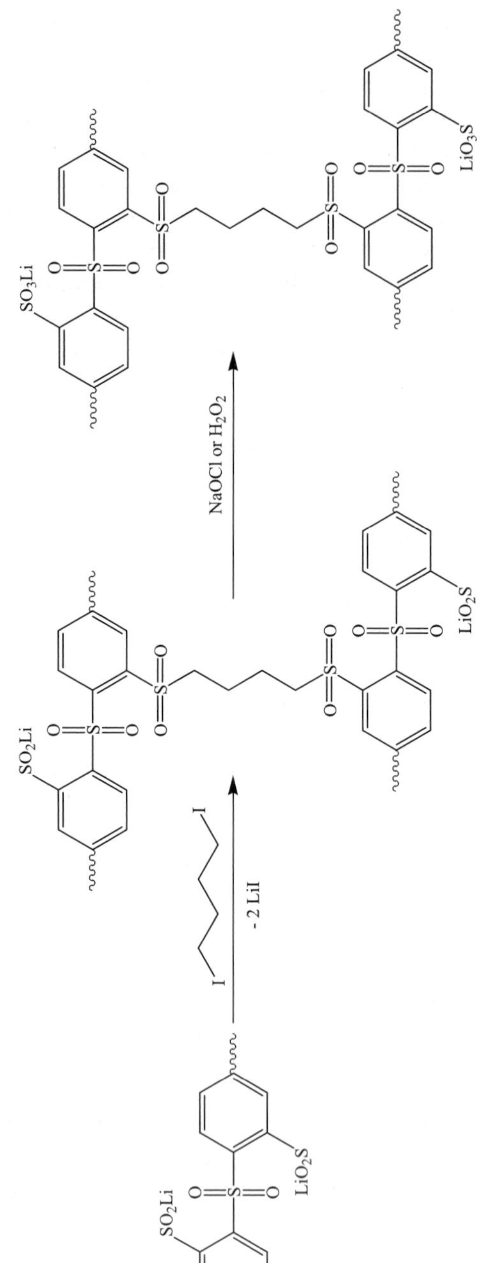

Fig. 22 Crosslinking of poly(ether sulfone)s by conversion of lithium sulfinate groups with diiodobutane

ized polysulfone [138, 140, 141]. The latter was obtained by conversion of the lithiated PSU, for example, with 4-acetylpyridine or *bis*(diethylamino)benzophenone. Ionically, covalently, and both ionically and covalently crosslinked membranes with IECs in the range from 1.1 to 1.25 mmol/g showed low swelling (40–45%) in water up to temperatures of 60 °C [141]. At higher temperatures the covalently crosslinked membrane showed highest dimensional stability, followed by the ionically crosslinked membrane, which was explained with the breakdown of the ionic crosslinks. Surprisingly, at least dimensionally stable was the "double crosslinked" sample. Due to incompatibilities between the sulfonated PEK and the sulfinated PSU, a phase separated morphology occurs in the resulting membrane with the ionically crosslinked polymers as continuous phase. Therefore, the swelling of this membrane is mainly determined by the weaker ionically crosslinked phase rather than by the dispersed covalently crosslinked particles. Both the covalently and the ionically crosslinked membranes showed in DMFC tests at temperatures ranging from 25 °C up to 110 °C quite similar behavior to Nafion® 105, but with much lower methanol crossover [141].

Jannasch and co-workers used the lithiation route to place the sulfonic acid groups into the side-chain rather than the main-chain. The activated polysulfone was reacted with either 2-sulfobenzoic acid anhydride [142] or 4-fluorobenzoyl chloride. The latter was further reacted with hydroxyaryl sulfonic acids, such as 4-sulfophenol sodium salt or 2-naphthol-7-sulfonic acid sodium salt [144] (Fig. 23). In another approach, lithiated PSU or PPSU were first reacted with SO_2, then further converted with a bromoalkyl sulfonic acid sodium salt, or with 1,3-propane sultone, or 1,4-butane sultone [143] (Fig. 21). Thus, the alkyl sulfonic acids are bonded to the polymer backbone via a sulfone linkage, which, in contrast to a phenyl-alkyl ether linkage, should be resistant towards acidic hydrolysis. The latter can be obtained by reacting sultones with hydroxyl groups in the polymer backbone as described, for example, by Kricheldorf and co-workers [152] and Orugi and co-workers [153, 154]. The introduction of the sulfonic acid group into the side-chains might have some advantages over the "traditional" main-chain sulfonation. Separating the ionic sites from the main-chain opens up the opportunity to manipulate and influence the phase separation into hydrophilic and hydrophobic domains, which occurs in the membranes while in contact with water. This may give rise to the formation of more stable morphologies, despite the presence of a highly swollen phase. Thus, the introduction of the ionic group into the side-chain may allow for higher IEC without compromising the resistance towards swelling. Sulfo-alkylated and sulfo-phenylated poly(arylene ether sulfone)s showed thermostabilities up to 300 °C, as revealed by TGA measurements. The proton conductivities of the sulfoalkylated poly(arylene ether sulfone)s were reported to be 77 mS/cm at 70 °C and 100% relative humidity for PSU carrying 0.9 sulfopropyl groups per repeating unit. The major drawback of these polymers is their high water uptake at temperatures above 75 °C, which was assigned to the plasticization effect of the alkyl chains on the PSU main chain. More promising with respect to swelling seems to be the grafting of arylsulfonic acids onto the PSU main-chain. Jannasch and co-workers reported a moderate swelling (15–18 H_2O/SO_3H) of sulfophenylated PPSU membranes (0.9 side chains/repeat unit; IEC = 1.63 mmol/g)

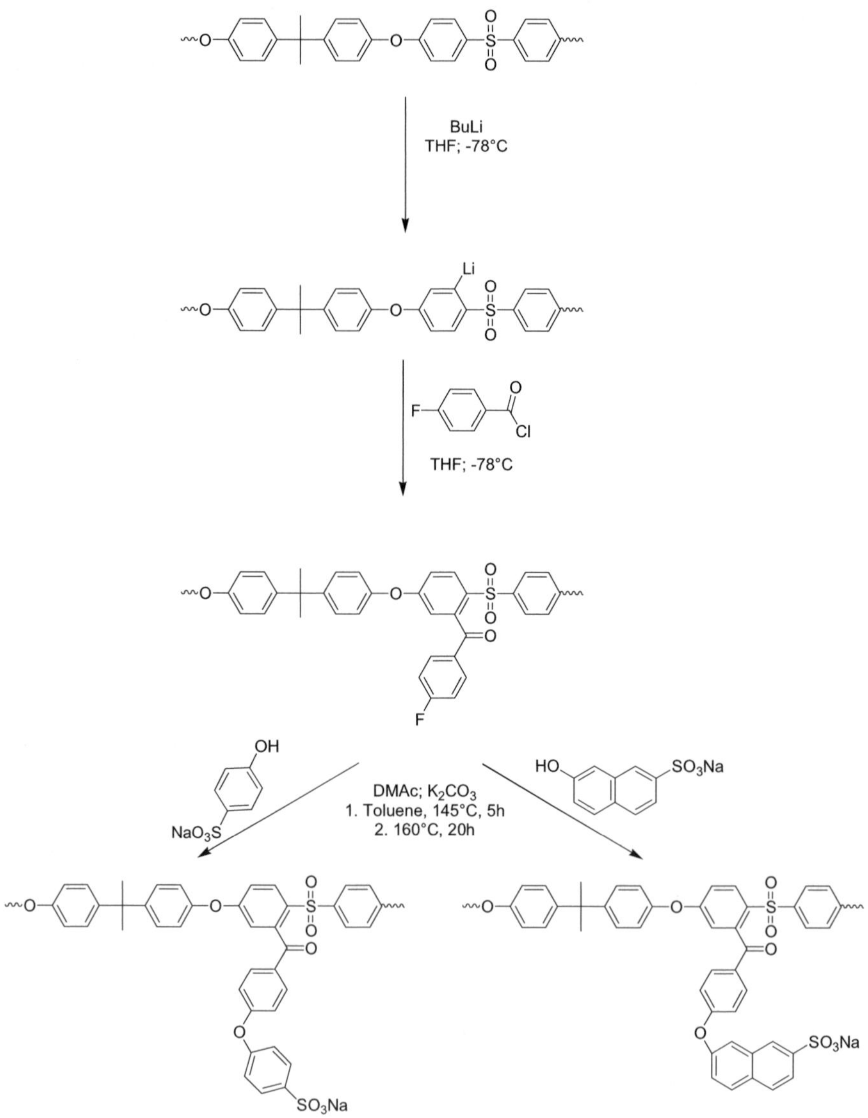

BuLi
THF; -78°C

Li

F Cl O

THF; -78°C

OH

NaO_3S

DMAc; K_2CO_3
1. Toluene, 145°C, 5h
2. 160°C, 20h

HO SO_3Na

SO_3Na

SO_3Na

Fig. 23 Introduction of functionalized side groups into poly(ether sulfone) by lithiation and subsequent nucleophilic addition

at temperatures up to 140 °C. A PSU membrane, having the same degree of grafting (IEC = 1.51 mmol/g) was dimensionally stable up to 120 °C. The proton conductivities of PPSU membranes (degree of grafting of 0.9) were in the range from 37 mS/cm at 30 °C to 42 mS/cm at 140 °C. Increasing the length and stiffness of the sulfonated side-chain by grafting of 4-sulfophenoxy benzoyl leads to lower proton conductivities at 120 °C, 11 mS/cm, and 32 mS/cm for a degree of grafting of 0.5 and 0.9, respectively. The introduction of 7-sulfo-2-naphthoxy benzoyl groups resulted in very brittle membranes, which were, therefore, not further characterized in terms of electrochemical properties [144] (Fig. 23). The same authors introduced side-chains consisting of higly sulfonated naphthoxybenzoyl (double sulfonated) or pyrenoxybenzoyl groups (3-fold sulfonated) to a polysulfone backbone [155]. This chemical structure leads to the formation of larger and more uniform ionic clusters than conventionally sulfonated polysulfones with sulfonic acid groups distributed along the main-chain. This particular architecture allows the formation of networks of water-filled nanopores, which facilitated high levels of proton conductivity up to 300 mS/cm at 100 °C (exceeding commercial Nafion®) at moderate levels of water uptake. Recently, Lafitte and Jannasch reported on the fuel cell performance of a 2-sulfobenzoyl-PSU membrane (IEC 1.2 mmol/g) in the temperature range from 60 to 110 °C [156]. A lifetime of at least 300 h at the operation temperature of 60 °C was stated for this membrane. The comparatively lower performance to Nafion® 117 was explained with non-optimized Nafion ®-containing electrodes.

5.3
Functionalized Poly(Arylene Ether Sulfone)s from Sulfonated Monomers

An alternative approach to the post-functionalization route is the use of functionalized monomers. This method has the advantage of overcoming the risks of side reactions, such as degradation or crosslinking connected with the post-functionalization route. Furthermore, it provides an easy route to control the degree and site of functionalization as well as the distribution of functional groups along the polymer chain. Several research groups have focused their work on this area. Most often a sulfonated 4,4′-dichlorodiphenyl sulfone, or 4,4′-difluorodiphenyl sulfone, or, in the case of poly(arylene ether ketone)s, sulfonated 4,4′-dihalobenzophenone, is used together with the non-sulfonated monomer and various types of bisphenols, such as bisphenol A, 4,4′-dihydroxy biphenyl, phenolphthalein, and others [67, 100, 157–171]. The sulfonated monomer can be easily obtained by sulfonation of the dihalo sulfone with fuming sulfuric acid (30% SO_3) at 90 °C for 6 h [157] (Fig. 24A). Here, the sulfonated monomer was used for the preparation of more hydrophilic polysulfones, but for no specific application. Other groups used the commercially available sodium salt of hydroquinone sulfonic acid as sulfonated monomer in the synthesis of sulfonated poly(arylene ether sul-

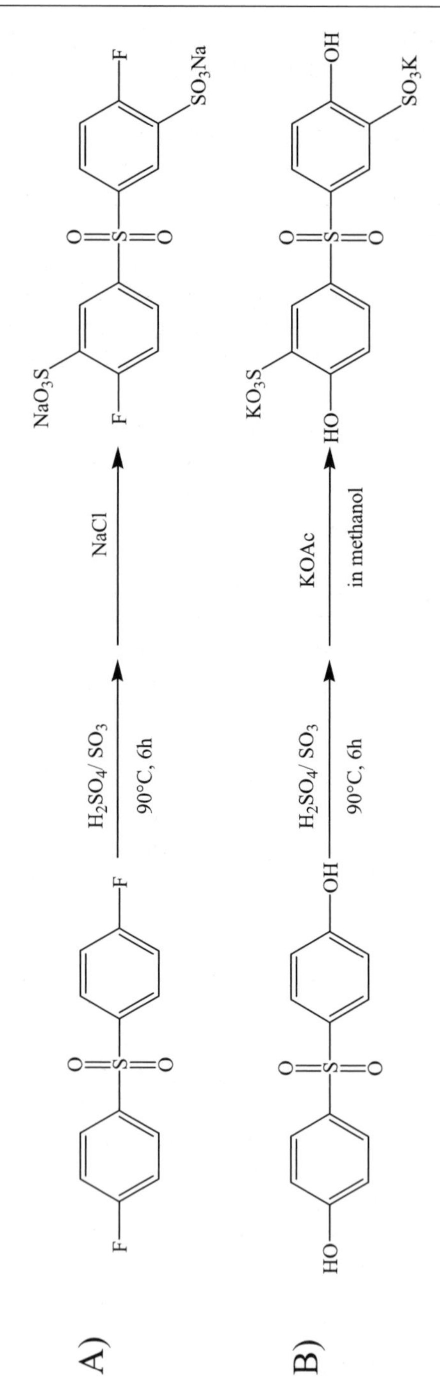

Fig. 24 Synthesis of **A** sulfonated 4,4′-difluorodiphenyl sulfone and **B** 3,3′-sulfonyl-*bis*-(6-hydroxybenzene sulfonic acid) (sulfonated bisphenol S)

fone)s [58, 67, 172, 173]. Kozlowski mentioned in a patent the incorporation of 3,3′-sulfonyl-*bis*-(6-hydroxybenzene sulfonic acid) (sulfonated bisphenol S; Fig. 24B) into polymers [174].

The sulfonation of the dihalide monomer has the advantage that the deactivated ring in *ortho*-position to the halogene is sulfonated, giving rise to higher stability with respect to desulfonation, and second two sulfonic acid groups are incorporated into the repeating unit. Furthermore, the reactivity of these monomers to nucleophilic displacement reactions is enhanced due to the electron-withdrawing effect of the introduced sulfonic acid groups. Although in *ortho*-position to the ether linkage in the polymer, the sulfonic acid groups should be more stable against desulfonation than in the hydroquinone units due to the electron-withdrawing and therefore deactivating effect of the sulfone or ketone linkage. In the hydroquinone sulfonic acid based poly(arylene ether)s, the sulfonic acid groups are positioned at an activated benzene ring with two ether linkages. However, as reported by Vogel et al., poly(arylene ether sulfone)s sulfonated at the hydroquinone moiety showed a high resistance against desulfonation at 130 °C in water [175].

Takeuchi [176] reported on the self-condensation of 2,6-*bis*(*p*-sodium sulfophenoxy)benzonitrile at 140 °C in phosphorous pentoxide/methane sulfonic acid, giving sulfonated hyperbranched polymers (Fig. 25). This product with an equivalent weight of 436 g/mol was soluble in DMSO and water and showed no film-forming properties. Membranes with an IEC ranging from 0.31–1.20 mmol/g were obtained by blending various amounts of the hyperbranched polymer with poly(vinyl alcohol). Although the matrix was crosslinked with glutaraldehyde, the membranes had a high water uptake between 59% and 101%. Interestingly, the membrane with the highest content of the hyperbranched polymer showed the second least water uptake (62%) of all samples under investigation.

Xiao et al. [177] reported on sulfonated poly(arylene ether ketone)s based on bisphenol S and 4,4′-difluorobenzophenone and sulfonated 4,4′-difluorobenzophenone (sodium salt). The ion-exchange capacity was varied between 1.23 mmol/g (30 mol % sulfonated monomer) and 1.73 mmol/g (45 mol % sulfonated monomer). These polymers were amorphous in nature and showed an increasing T_g ranging from 268 °C to 317 °C, with increasing amount of the sulfonated moiety in the polymer backbone. The water uptake of the membranes was reported to be much lower than that of Nafion® 115.

The electrochemical properties of cation-exchange membranes based on sulfonated poly(arylene ether sulfone)s (s-PES) were described in a paper by Kang et al. [100]. The intended applications for these membranes were general electro-membrane rather than fuel cell applications. The properties of the membranes prepared in this work were compared to commercially available ion-exchange membranes (Neosepta® CM-1, CMX, and CMB), as well as to sulfonated polysulfones obtained by post-sulfonation of Udel® 1700 with chlorosulfonic acid (s-PSU) (Fig. 26). The s-PES membrane materials were

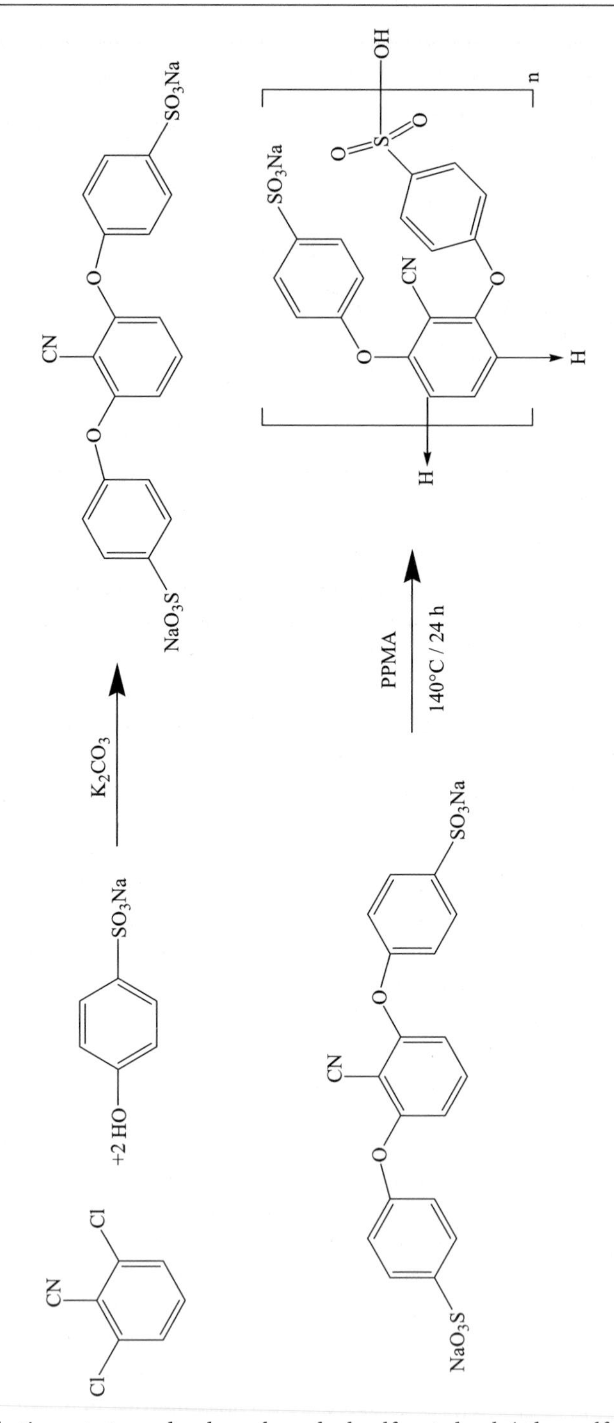

Fig. 25 Synthetic route towards a hyperbranched sulfonated poly(ether sulfone)

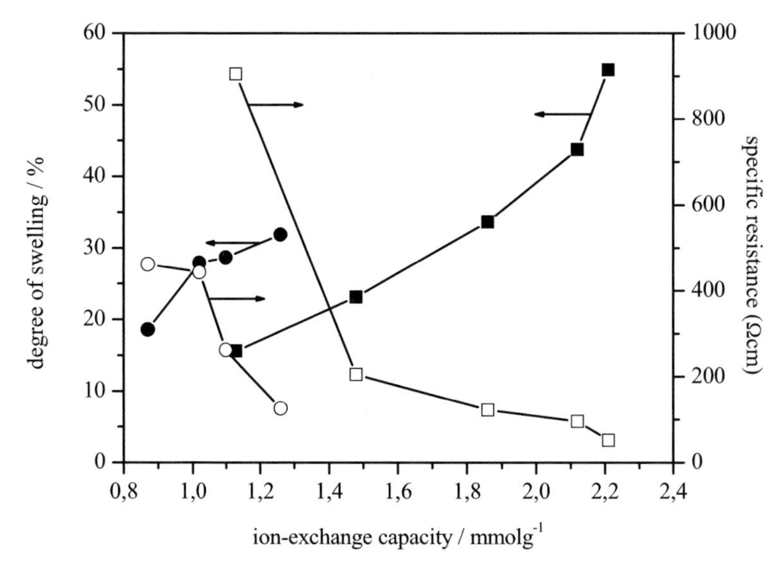

Fig. 26 Comparison of sulfonated poly(arylene ether sulfone) membranes prepared by post-sulfonation and direct polymerization (●, ○: s-PSU; ■, □: s-PES; *open symbols*: specific resistance, *filled symbols*: swelling). Data taken from [100]

composed of 4,4′-dichlorodiphenylsulfone, 4,4′-dihydoxybiphenyl, and varying amounts of sulfonated 4,4′-dichlorodiphenylsulfone, in order to adjust the ion-exchange capacity between 1.13 and 2.21 mmol/g. By variation of the sulfonation time, ion-exchange capacities of s-PSU between 0.87 mmol/g and 1.26 mmol/g were achieved. These values correspond to degrees of sulfonation ranging from 50% to 72%, taking into account that only one sulfonic acid group is introduced to the bisphenol A unit. Furthermore, it should be mentioned that the s-PSU material was sulfonated at the electron-rich bisphenol A moiety in the polymer backbone.

Despite having higher IEC, s-PES samples swelled much less than the s-PSU samples. Contrary to these findings, the specific resistances of s-PSU samples were significantly higher. The s-PES membranes with an IEC ranging from 1.86 mmol/g up to 2.21 mmol/g exhibited electrical properties comparable to the commercial membranes. The electrical resistances of these membranes were below 1.0 Ωcm^2. However, the mechanical properties were drastically weakened when the IEC exceeded 2 mmol/g. These findings were rationalized with the lower molecular weight of these polymers due to the low reactivity of the sulfonated monomer, which prevents the formation of high molecular weight samples. In these membranes the plasticizing effect of absorbed water is more pronounced than in a high molecular weight sample of lower IEC. However, no molecular weights for the prepared polymers were given in the paper. Chronopotentiometric measurements indicated excellent electrochemical properties, making the s-PES membranes suitable for electro-membrane application at high current ranges, especially electro-dialysis.

Shin et al. reported on sulfonated poly(arylene ether sulfone)s prepared from 4,4′-difluorodiphenylsulfone, hydroquinone, and hydroquinone sulfonic acid [67]. The degree of sulfonation was controlled by variation of the hydroquinone/hydroquinone sulfonic acid ratio. The IECs calculated from the monomer composition were in the range from 0.59 mmol/g to 2.47 mmol/g (determined 2.13 mmol/g). Despite the high IEC the water uptake was very low, corresponding to 8.8 to 10 water molecules per sulfonic acid group. Due to their good mechanical and thermal properties, these polymers are considered to be promising candidates for fuel cell applications. However, neither proton conductivities nor fuel cell test data were reported.

Lakshmi et al. [173], Meier-Haack et al. [172], and Taeger et al. [58] used the silyl-method for the preparation of sulfonated poly(arylene ether sulfone)s. Like Shin et al. [67], these authors used hydroquinone sulfonic acid as source for the proton exchange group. Polymers were prepared from the more reactive 4,4′-difluorodiphenyl sulfone rather than 4,4′-dichlorodiphenyl sulfone and various silylated bisphenol comonomers, such as hydroquinone, phenolphthalein, 2,6-dihydroxynaphthalene, bisphenol A [173], or 4,4′-dihydroxybiphenyl [58, 172, 173]. The use of silylethers and difluoro aromates allows for lower reaction temperatures (150 °C), as in the case of dichloro aromates and free bisphenols (180–190 °C). In the latter case, the water formed has to be removed by azeotropic distillation and the lower reactivity of the dichloro compound has to be taken into account.

Polymers with 75 or 60 mol% of hydroquinone sulfonic acid were prepared, resulting in membranes with theoretical IECs ranging from 1.32 mmol/g to 1.95 mmol/g [173]. The water uptake of the samples varied with the IEC as well as with the chemical structure of the polymer backbone. Samples with the highest IEC showed the highest water uptake. It was found that the water uptake could be reduced when bulky bisphenols, such as 2,6-dihydroxynaphthalene, phenophthalein, or 4,4′-dihydroxybiphenyl, were incorporated into the polymer backbone. On the other hand, the highest proton conductivities were detected for membranes derived from polymers with hydroquinone (3.3 mS/cm at 25 °C (IEC 1.95 mmol/g)) or bisphenol A (3.05 mS/cm at 25 °C (IEC 1.81 mmol/g)).

It is well accepted that the superior properties of Nafion® membranes result from a phase separation between the hydrophobic perfluorinated polymer backbone and the sulfonic acid bearing side-chains. This phase separation leads to the formation of ion-conducting channels in the nm-range. Also ion-exchange membranes derived from sulfonated styrene-ethylene/butylene-styrene triblock copolymers (s-SEBS) exhibit a phase-separated morphology, which led to proton conductivities similar to Nafion® but with a much lower methanol crossover [178].

Table 7 Ion exchange capacity and water uptake of poly(ether sulfone) membrane samples (taken from [58])

Sample [a]	Water/ uptake (%)	H_2O/SO_3H mol/mol	Ion exchange capacity (mmol/g) calculated from monomer comp.	titration	NMR
HPA	1.4	–	0	–	–
HPB	n.d. [a]	–	2.48	–	–
PAE-MBC 5/5	34.4	19	1.24	0.83	0.99
PAE-MBC 10/5	21.7	14	0.83	0.67	0.86
PAE-MBC 15/5	12.6	12	0.62	0.46	0.59
PAE-MBC 20/5	9.5	9	0.51	0.35	0.59
PAE-MBC 10/10	28.3	14	1.24	0.84	1.14
PAE-MBC 20/20	34.8	16	1.24	0.90	1.19
PAE-RC 1/1	33.2	16	1.24	1.15	1.14
PAE-RC 2/1	16.9	12	0.83	0.85	0.81
PAE-RC 3/1	11.8	11	0.62	0.45	0.61
PAE-RC 4/1	8.4	10	0.50	0.08	0.48
Nafion® 117	28.9	21	0.91 [b]	0.77	–

[a] not determined due to solubility in water
[b] [27]

Taeger et al. [58] and Meier-Haack et al. [172] investigated the properties s-PES block copolymers and the corresponding random copolymers. The aim of this work was to provide fully aromatic ion-exchange membranes with a phase-separated morphology. Both, block and random copolymers were prepared from 4,4′-difluorodiphenylsulfone, hydroquinone sulfonic acid, and 4,4′-dihydroxybiphenyl. The theoretical IECs were adjusted by monomer composition between 0 mmol/g (homopolymer from 4,4′-dihydroxybiphenyl and 4,4′-difluorodiphenylsulfone) and 2.48 mmol/g (homopolymer from hydroquinone sulfonic acid and 4,4′-difluorodiphenylsulfone).

The water uptake and ion-exchange capacities for poly(ether sulfone) membranes are given in Table 7. The ion-exchange capacities, determined by ^{1}H NMR spectroscopy are in good agreement with the values calculated from the initial monomer composition. However, using the titration method, the values are somewhat lower, which is rationalized by the fact that only a part of the sulfonic acids groups is accessible to the sodium ions. On the other hand, with NMR spectroscopy, all sulfonic acid groups are detected, regardless of whether they play an active role in the ion-exchange process or not. Therefore, the IEC determined by the titration method gives a more realistic value than that obtained from NMR spectroscopy concerning the behavior of the membrane in the fuel cells.

The water uptake shows a nearly linear dependence from the ion-exchange capacity of the membranes. It is in the same range as observed for the Nafion®

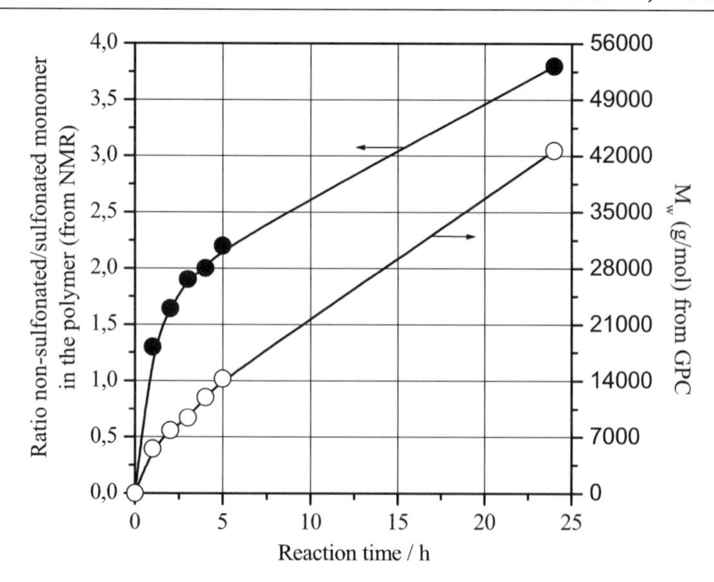

Fig. 27 Influence of reaction time on the monomer ratio in the polymer followed by NMR spectroscopy as illustrated by the preparation of PAE-RC 4/1 (taken from [58])

membranes. Unexpectedly, no difference between the water uptake of random and multiblock copolymers was observed in contrast to sulfonated polyaramide membranes [179]. From studies following the polymerization process of a random sulfonated copoly(ether sulfone), it was deduced, that during the preparation of random copolymers, a block-like structure is obtained due to differences in the reactivity of the monomers (Fig. 27). Therefore, it is likely that the morphology of the membranes prepared from random or block copolymers is very similar, resulting in a similar behavior of the membranes.

As expected, the diffusion coefficients increased with increasing IEC, due to higher hydrophilicity resulting in an enhanced swelling of the membranes. At comparable IEC, the methanol diffusion coefficients for poly(ether sulfone) membranes are one to two orders of magnitude lower than those of the Nafion® membrane, indicating improved barrier properties for methanol (lower methanol crossover) of poly(ether sulfone) membranes. Still, for most membranes the proton diffusion coefficients were also lower than those of the Nafion® membrane. The proton conductivities (Fig. 28A) are reflecting the findings of the diffusion coefficient measurements. Although by a factor of two lower than that of Nafion® 117, the fuel cell performances of the s-PES membranes were comparable to that of Nafion® (Fig. 28B).

The McGrath group [159–171] extensively studied the use of sulfonated 4,4′-dichlorodiphenylsulfone in combination with various bisphenols, such as bisphenol A, bisphenol 6F, hydroquinone, biphenol, and 4,4′-dichlorodiphenylsulfone, for the preparation of ion-exchange materials. The main results of this work were recently published in a review [170]. In general, the

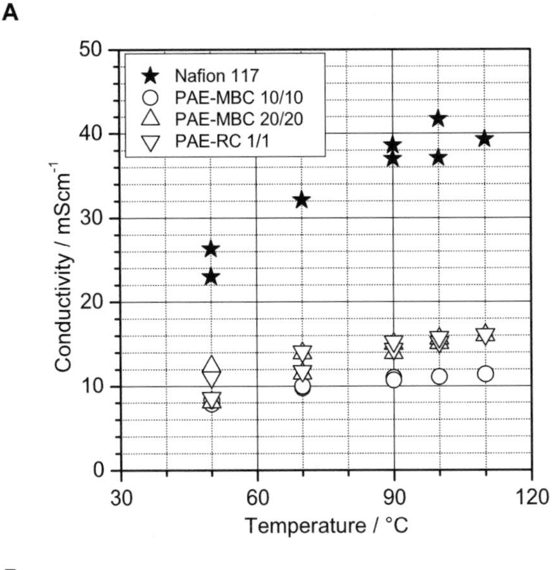

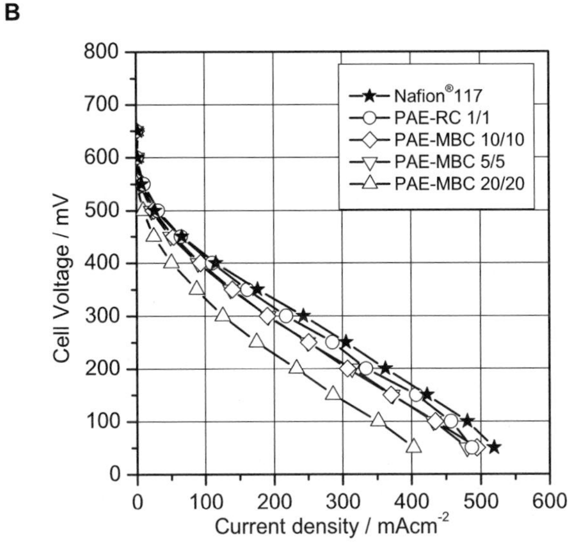

Fig. 28 Electrical properties of poly(ether sulfone) membranes compared to Nafion® 117 membrane; **A** conductivity at 100% relative humidity; **B** polarization curves from DMFC tests at 110 °C (1.5 mol/l MeOH; 2.5 bar air) (taken from [58])

free bisphenols were reacted with the dichloro compounds in NMP at 190 °C, in the presence of K_2CO_3. The prepared materials showed thermo stabilities up to temperatures of 400 °C (weight loss 5%) in nitrogen atmosphere and up to 220 °C in air (30 min). Polymers (BPSH) based on biphenol and DCDPS with up to 60 mol % of the sulfonated monomer (sDCDPS) gave stable membranes, while those with 100 mol% sulfonated monomer (IEC 3.6 mmol/g) were water soluble. The sodium salt form of the sulfonated polymers swelled less in water than in the acid form. Furthermore, the thermo stability of the sodium form was much better than that of protonated membranes. High proton conductivities were achieved with biphenol as the phenolic component. For example a polymer with 40 mol% of the sulfonated monomer (IEC = 1.72 mmol/g) showed a proton conductivity of 110 mS/cm at 25 °C, whereas a polymer with 60% sulfonated monomer (IEC = 2.42 mmol/g) had a conductivity of 170 mS/cm [160]. The value for Nafion® 1135 (IEC = 0.91 mmol/g) given in this paper was 120 mS/cm. A phase separated morphology was reported with hydrophilic domains with a size ranging from 10 to 25 nm, depending on the degree of sulfonation. At sulfonation levels higher than 50% (50 mol% of sulfonated monomer) a phase inversion was observed using AFM. These findings correlated very well with an immense increase of water uptake. Furthermore, samples with more than 50 mol% sDCDPS showed two Tgs in the DSC curves, indicating a well organized phase separation in these systems [165]. Such poly(arylene ether sulfone) membranes showed similar or even better performance in both hydrogen/air and especially direct methanol fuel cell tests than Nafion® 117 membranes [170]. The long-term tests in a hydrogen/air fuel cell (800 h at 80 °C, 0.5 V, and 100% r.h.) revealed a high stability of the poly(arylene ether sulfone) under these conditions.

In a further study, Kim et al. [164] reported on the properties of sulfonated poly(arylene ether sulfone) membranes (BPSH) blended with heteropolyacid (HPA). The blend membrane composed of 70 wt.% of an sulfonated poly(arylene ether sulfone), having an IEC of 1.72 mmol/g and 30 wt.% HPA, showed much higher conductivities than non-blended membranes and even Nafion® 1135 (Fig. 29). After treatment, the weight loss of the blend membranes increased with increasing degree of sulfonation of the polymer matrix, but was only 2% after 48 h in liquid water at 30 °C and ca. 1% after 15 h at 100 °C in water vapor. In a recent paper, Kim et al. reported on poly(arylene ether sulfone)s based on sDCDPS, bisphenol 6F, and 2,6-difluorobenzonitrile [169]. These partly fluorinated membranes with an IEC of 1.32 mmol/g exhibited a 33% higher current density at 0.5 V than that of the above discussed poly(arylene ether sulfone) membranes (IEC 1.72 mmol/g) and a recasted Nafion® 1100 membrane with comparable thicknesses. Similar proton conductivities were detected for all membranes discussed in this paper. Although having a slightly higher water uptake, the fluorinated membrane exhibited an approximately 2-fold lower methanol permeability when compared to Nafion®.

In more fundamental works, Kim et al. studied the influence of (hydro)-thermal pre-treatment and the state of water in sulfonated polymers on the membrane performance [166–168]. Again, BPSH samples with different degree of sulfonation were used for the investigations. Kim indicated three

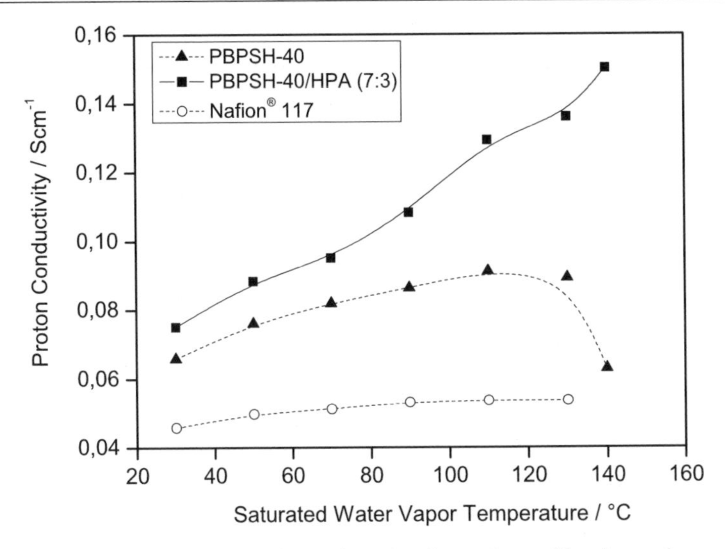

Fig. 29 Proton conductivities of sulfonated poy(arylene ether sulfone) membrane and its blend with a heteropolyacid in comparison with Nafion® 117 (data taken from [164])

irreversible morphological states (regimes, borders indicated by arrows in Fig. 30), which can be controlled by the copolymer composition, and the pretreatment parameters (temperatures). Regime 1 describes the morphology of the as-casted membrane with isolated hydrophilic domains. Within this regime the water absorption increases only slightly with the temperature. In regime 2, where interconnection between hydrophilic domains occurs, the water absorption increases steadily (linear) with treatment temperature, while in regime 3 a rapid water absorption with temperature is observed (Fig. 30). In this regime, the hydrophilic/hydrophobic domain structure was no longer well defined. But not only the water uptake is influenced by the morphological changes. The mechanical properties (Fig. 31) and the proton conductivities (Fig. 32) also show dependencies on hydro-thermal treatment.

The transition temperatures of poly(arylene ether sulfone)s and Nafion® 1135 are given in Table 8. The formation of destinct morphological regimes on thermal treatment allows to adjust the membrane properties in terms of proton conductivity, water uptake, and mechanical strength at an optimum, by treatment at an appropriate temperature. On the other hand, the knowledge of the transition temperatures could be used to predict the upper operation temperature of ion-exchange membranes in fuel cells. These findings also demonstrated that the performance loss of ion-exchange membranes at elevated temperatures might not only be based on dehydration but also on morphological changes as indicated by comparison of the ion conductivities at elevated temperature and the regime transition temperatures of the respective polymer membranes. The morphological changes can be

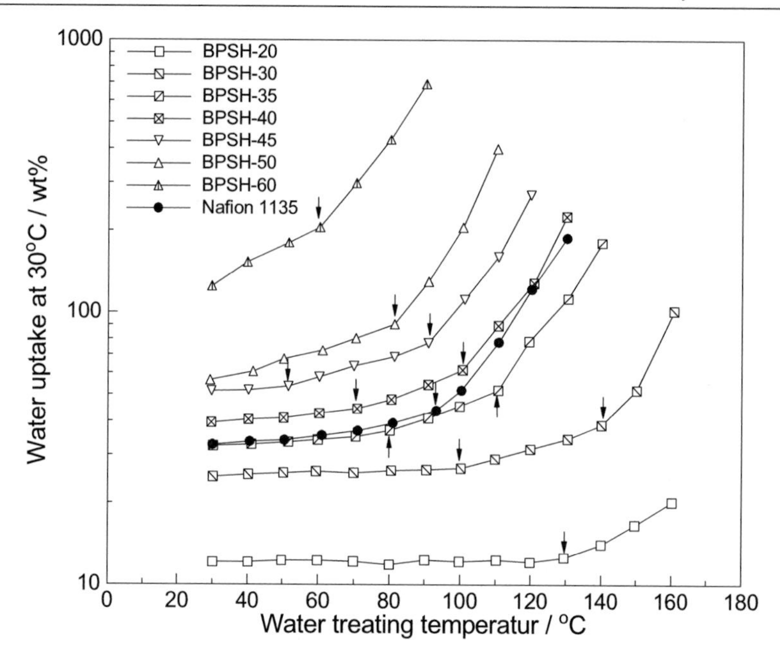

Fig. 30 Water absorption of sulfonated poy(arylene ether sulfone) membranes as a function of water treatment temperature (data taken from [166]). The regime transition temperatures are indicated by *arrows*

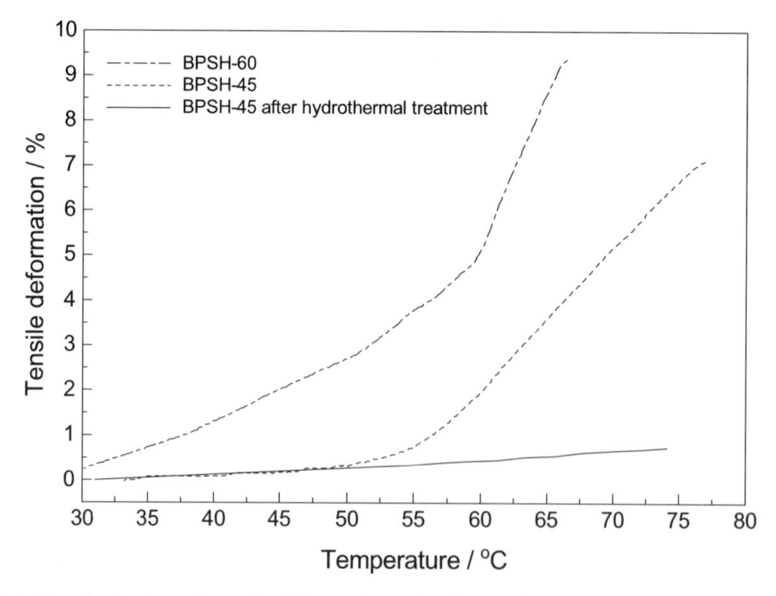

Fig. 31 Tensile deformation of sulfonated poy(arylene ether sulfone) membranes (data taken from [166])

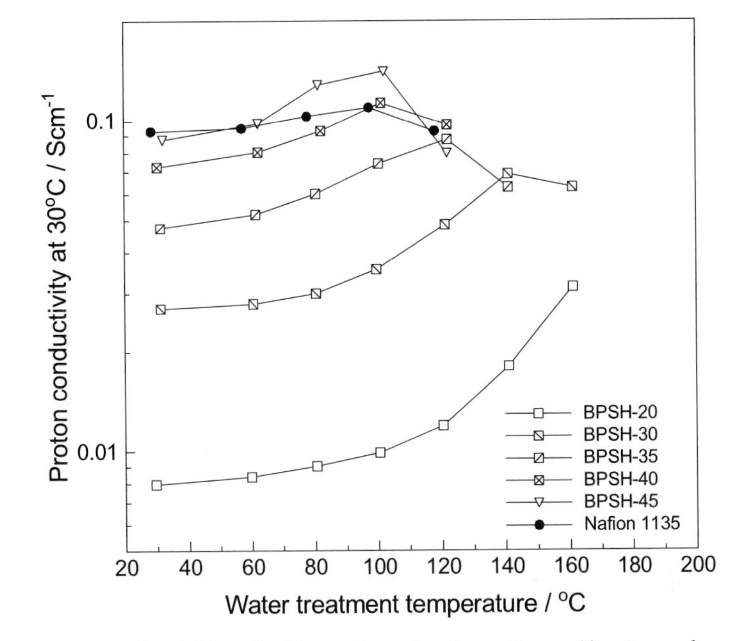

Fig. 32 Proton conductivities of sulfonated poy(arylene ether sulfone) membranes (data taken from [166])

Table 8 Transition temperatures of sulfonated poly(arylene ether sulfone)s and Nafion® (data taken from [166])

Sample	Transition temperature (°C) Regime 1/Regime 2	Transition temperature (°C) Regime 2/Regime 3
BPSH-20	130 [a]	n.d.
BPSH-30	100	140
BPSH-35	80	110
BPSH-40	70	100
BPSH-45	50 (53 [b])	90
BPSH-50	30	80
BPSH-60	n.d. [c]	60 (59 [c])
Nafion® 1135	n.d.	100

[a] Detected by swelling measurements
[b] Detected by dynamic mechanical measurements
[c] n.d. not detected

attributed to the T_g depression in the presence of absorbed water. For the poly(arylene ether sulfone) membranes, the lowering of T_g with the water content was much more pronounced than for Nafion® 1135 [168].

Similar results were reported by the same author while performing acidification treatment under different conditions (variation of time and temperature) [167]. Contrary to the findings for poly(arylene ether sulfone) membranes, the conductivity and water sorption were invariant with treatment temperature.

5.4 Poly(Arylene Thioether Sulfone)s

Poly(phenylene sulfide)s (PPS) are an interesting class of high performance engineering thermoplastics. They possess high melting temperatures, good mechanical properties, as well as good chemical resistances, which is on one hand of certain interest for fuel cell membranes, for example. On the other hand, some of these properties are restricting the processability of poly(phenylene sulfide)s, in particular to membranes, due to their very limited solubility in common organic solvents. Schauer and Brozova reported on the sulfonation of PPS with chlorosulfonic acid in 1,2-dichloroethane [180]. Even samples having an IEC of 1.6 mmol/g were insoluble in organic solvents. Liu et al. [181] reported on the effect of preparation conditions on the properties of poly(phenylene sulfide sulfone), which has been described as an amorphous polymer with a T_g around 217 °C (see ref. 4 and 5 in [181]) and which is now soluble in dipolar aprotic solvents, such as NMP or DMAc. In a paper of the McGrath group, Wiles et al. [182] reported on poly(arylene thio ether sulfone)s based on 4,4′-difluorodiphenylsulfone, sulfonated 4,4′-difluorodiphenylsulfone, and 4,4′-thiobisbenzenethiol (for further references on the synthesis of (sulfonated) poly(arylene thio ether)s the reader is referred to ref [183] and the literature cited in there). High-proton conductivities up to 160 mS/cm at 30 °C in water with an IEC of 1.65 mmol/g were observed. AFM-images revealed the build-up of a continuous hydrophilic matrix when the content of the sulfonated monomer in the polymer backbone exceeds 40 mol %. Since then, severals groups reported on the synthesis of sulfonated poly(arylene tio ether sulfone) and/or sulfonated poly(arylene thio ether ketone)s as potential ion-exchange materials for fuel cell applications [184–190]. In all cases, 4,4′-thiobisbenzenethiol was used as thiol-component, which was reacted with various sulfonated and non-sulfonated dihalogeno aromatics such as difluorobenzophenon, difluorodiphenylsulfone, and others. On treatment with water at elevated temperatures, all these sulfonated poly(sulfide sulfone)s and poly(sulfide ketone)s showed excellent dimensional stability, even at high ion-exchange capacities (up to 3.2 mmol/g). These high IECs are responsible for proton conductivities higher than those observed for Nafion under similar test conditions. Furthermore, a very high oxidative stability was observed. Shen et al. [184] reported on the formation of sulfone groups as a result of the oxidation of the thio ether groups. In order to further improve the mechanical properties and the dimensional stability of these materials, Bai et al. [187] and Lee and Kerres [189] prepared blend membranes with polybenzimidazole as blend partner. The proton conductivities were slightly lowered due to the formation of ion-pairs (SO_3H-imidazole), but the effect on water uptake (reduction) was much more pronounced. As a side-effect, the oxidative stability was also improved. Another approach towards the improvement of mechanical properties and reduction of methanol permeability was described by Zhang et al. [188]. They prepared chemically crosslinked membranes (compared to ionically crosslinked membranes discussed before) by the treatment of sulfonated poly(sulfide sulfone) membrane with polyphosphoric acid at 180 °C for 1.5 h. Since the crosslinking occurs by the intermolecular reaction of sulfonic acid groups within non-sulfonated aromatic rings of neighbor chains, a loss of ion-exchange capacity (10–20%) and, therefore, proton conductivity is unavoidable (up to 50%). Simultaneously, the water uptake and the methanol crossover was reduced by 50% and 50 to 80%, respectively. Recently, Schönberger and Kerres reported on the synthesis of multiblock copolymers, including poly(sulfide sulfone)s, consisting of hydrophilic sulfonated segments and fluorinated hydrophobic segments. These authors reported on the effect of different building blocks (monomers) linking elements on the stability of the resulting polymer. It turned out, that thio ether groups linked to assymetrically substituted rings are less stable than thio ether groups between symmetrically substituted rings. The properties of the latter were similar to those ionomers with sulfone linkages although having a slightly lower oxidative stability.

As the hydrolytic stability of aromatic sulfonic acids depends strongly on the electron densitiy and, therefore, on the substitutes at the aromatic ring, electron-withdrawing groups should protect the sulfonic acid groups from hydrolytic cleavage. The electron-withdrawing groups should further increase the acidic strength of the sulfonic acid group when attached to the same ring. Shen et al. [184] and other researchers already mentioned that the thio ether linkage in poly(arylene thio ether)s is oxidized to sulfoxides or sulfones on treatment with Fenton's reagent, for example. Taking this fact into account, Schuster et al. synthesized highly sulfonated poly(*p*-phenylen sulfide)s and converted these polymers into the corresponding sulfonated poly(*p*-phenylen sulfones) [183, 191]. This synthetic route is a smart way to get sulfonated poly(*p*-phenylene sulfone)s by converting the electron-donating thio ether groups into electron-withdrawing groups, which are not accessable by the direct sulfonation of poly(phenylene sulfone). Secondly, the hydrolytic stability and the acidity of the sulfonic acid groups is very much improved [183]. Although many of the properties for a highly effective fuel cell membrane are improved, it should be mentioned that low IEC materials are insoluble in common organic solvents and that high IEC materials are very brittle.

References

1. Roziere J, Jones DJ (2003) Annual Reviews of Materials Sci 33:503
2. Yu J, Yi B, Xing D, Liu F, Shao Z, Fu Y, Zhang H (2003) Phys Chem Chem Phys 5:611
3. Mathias MF, Makharia R, Gasteiger HA, Conley JJ, Fuller TJ, Gittleman CJ, Kocha SS, Miller DP, Mittelsteadt CK, Xie T, Yan SG, Yu PT (2005) Interface 14:24
4. Kerres J, Van Zyl AJ (1999) J Appl Polym Sci 74:428
5. Kerres J, Cui W, Reichle S (1996) J Polym Sci A Polym Chem 34:2421
6. Rikukawa M, Sanui K (2000) Progress Polym Sci 25:1463
7. Kreuer K (2003) Hydrocarbon membranes In: Vielstich W, Lamm A, Gasteiger HA (eds) Handbook of fuel Cells. John Wiley & Sons Ltd., p 420
8. Li Q, He R, Jensen JO, Bjerrum NJ (2003) Chem Mater 15:4896
9. Jannasch P (2003) Curr Opin Colloid Interface 8:9
10. Yang Y, Holdcroft S (2005) Fuel Cells 5:171
11. Smitha B, Sridhar S, Khan A (2005) J Membr Sci 259:10
12. Souzy R, Ameduri B (2005) Prog Polym Sci 30:644
13. Bredas JL, Chance RR, Silbey R (1982) Phys Rev B 26:5843
14. Lee J, Marvel CS (1984) J Polym Sci Polym Chem Ed 22:295
15. Jin X, Bishop MT, Ellis TS, Karasz FE (1985) Br Polym J 4:4
16. Litter MI, Marvel CS (1985) J Polym Sci Polym Chem Ed 23:2205
17. Ogawa TMCS (1985) J Polym Sci Polym Chem Ed 23:1231
18. Bishop MT, Karasz FE, Russo PS, Langley KH (1985) Macromolecules 18:86
19. Bailly C, Williams DJ, Karasz FE, MacKnight WJ (1987) Polym 28:1009
20. Shibuya N, Porter RS (1992) Macromolecules 25:6495
21. Wang F, Roovers J (1993) Macromolecules 26:5295
22. Kobayashi H, Tomita H, Moriyama H (1994) J Am Chem Soc 116:3153
23. Kreuer K, Dippel T, Maier J (1995) Proc Electrochem Soc 95-23:241
24. Kreuer KD (1997) Solid State Ionics 97:1
25. Kerres J, Cui W, Disson R, Neubrand W (1998) J Membr Sci 139:211
26. Kobayashi T, Rikukawa M, Sanui K, Ogata N (1998) Solid State Ionics 106:219
27. Lufrano F, Squadrito G, Patti A, Passalacqua E (2000) J Appl Polym Sci 77:1250
28. Bauer B, Jones DJ, Roziere J, Tchicaya L, Alberti G, Casciola M, Massinelli L, Peraio A, Besse S, Ramunni E (2000) J New Mater Electrochem Sys 3:93
29. Zaidi SMJ, Mikhailenko SD, Robertson GP, Guiver MD, Kaliaguine S (2000) J Membr Sci 173:17
30. Genova-Dimitrova P, Baradie B, Foscallo D, Poinsignon C, Sanchez JY (2001) J Membr Sci 185:59
31. Kreuer KD (2001) J Membr Sci 185:29
32. Kerres J (2001) J Membr Sci 185:3
33. Linkous CA, Anderson HR, Kopitzke RW, Nelson GL (1998) Int J Hydrogen Energy 23:525
34. Alberti G, Casciola M, Massinelli L, Bauer B (2001) J Membr Sci 185:73
35. Li L, Zhang J, Wang Y (2003) J Mat Sci Letters 22:1595
36. Robertson GP, Mikhailenko SD, Wang K, Xing P, Guiver MD, Kaliaguine S (2003) J Membr Sci 219:113
37. Elias H (1990) Makromoleküle. Hüthig & Wepf, Basel
38. Bai Z, Williams LD, Durstock MF, Dang TD (2004) Polym Prepr 45:60
39. Paddison SJ (2001) J New Mater Electrochem Systems 4:197–207
40. Sakaguchi Y, Kaji A, Nagahara S, Kitamura K, Takase S, Omote K, Asako Y, Kimura K (2004) Polym Prepr 45:20

41. Gao Y, Robertson GP, Guiver MD, Mikhailenko SD, Li X, Kaliaguine S (2004) Macromolecules 37:6748
42. Ponce ML, Prado L, Ruffmann B, Richau K, Mohr R, Nunes SP (2003) J Membr Sci 217:5
43. Nunes SP, Ruffmann B, Rikowski E, Vetter S, Richau K (2002) J Membr Sci 203:215
44. Chang J, Park JH, Park G, Kim C, Park Ok O (2003) J Power Sources 124:18
45. Lu X, Steckle WP, Weiss RA (1993) Macromolecules 26:5876
46. Lu X, Steckle Jr WP, Weiss RA (1993) Macromolecules 26:6525
47. Gan D, Lu S, Wang Z (2001) Polym Int 50:812
48. Shin CK (2002) PhD thesis, Technical University of Munich
49. Shin CK, Maier G, Andreaus B, Scherer Günther G (2004) J Membr Sci 245:147
50. Yang Y, Shi Z, Holdcroft S (2004) Macromolecules 37:1678
51. Yang Y, Shi Z, Holdcroft S (2004) Europ Polym J 40:531
52. Zhang X, Liu S, Liu L, Yin J (2005) Polym 46:1719
53. Zhang X, Liu S, Yin J (2005) J Membr Sci 258:78
54. Yu X, Roy A, McGrath J E (2005) Prepr Symp 50:577
55. Ghassemi H, Ndip G, McGrath JE (2003) Polym Prepr 44:814
56. Ghassemi H, Harrison W, Zawodzinski TAJ, McGrath JE (2004) Polym Prepr 45:68
57. Lee H, Einsla B, McGrath JE (2005) Prepr Symp 50:579
58. Taeger A, Vogel C, Lehmann D, Lenk W, Schlenstedt K, Meier-Haack J (2004) Macromol Symp 210:175
59. Guo X, Fang J, Watari T, Tanaka K, Kita H, Okamoto K (2002) Macromolecules 35:6707
60. Heitner-Wirguin C (1996) J Membr Sci 120:1
61. Cornet N, Diat O, Gebel G, Jousse F, Marsacq D, Mercier R, Pineri M (2000) J New Mater Electrochem Systems 3:33
62. Cornet N, Beaudoing G, Gebel G (2001) Sep Pur Technol 22–23:681
63. Piroux F, Espuche E, Mercier R, Pineri M (2003) J Membr Sci 223:127
64. Genies C, Mercier R, Sillion B, Cornet N, Gebel G, Pineri M (2001) Polym 42:359
65. Essafi W, Gebel G, Mercier R (2004) Macromolecules 37:1431
66. Wang H, Harrison W, Yang J, McGrath JE (2004) Prepr Symp 49:586
67. Shin CK, Maier G, Scherer GG (2004) J Membr Sci 245:163
68. Linares A, Acosta JL, Rodriguez S (2006) J Appl Polym Sci 100:3474
69. Wootthikanokkhan J, Seeponkai N (2006) J Appl Polym Sci 102:5941
70. Xue S, Yin G (2006) Polym 47:5044
71. Wang J, Yue Z, Economy J (2007) J Membr Sci 291:210
72. Bouzek K, Cilova Z, Podubecke P, Paidar M, Schauer J (2006) Desalination 200:650
73. Schauer J, Brozova L, Pientka Z, Bouzek K (2006) Desalination 200:632
74. Deb PC, Rajput LD, Hande VR, Sasane S, Kumar A (2007) Polym Adv Technol 18:419
75. Bae J, Honma I, Murata M, Yamamoto T, Rikukawa M, Ogata N (2002) Solid State Ionics 147:189
76. Le Ninivin C, Balland-Iongeau A, Demattei D, Coutanceau C, Lamy C, Leger JM (2004) J Appl Electrochem 34:1159
77. Fujimoto CH, Hickner MA, Cornelius CK, Loy DA (2005) Macromolecules 38:5010
78. Ghassemi H, Ndip G, McGrath JE (2004) Polymer 45:5855
79. Ghassemi H, McGrath JE (2004) Polymer 45:5847
80. Litt MH (2005) United States US 20050239994A1
81. Le Ninivin C, Balland-Longeau A, Demattai D, Palmas P, Saillard J, Coutanceau C, Lamy C, Leger JM (2006) J Appl Polym Sci 101:944
82. Hickner MA, Fujimoto CH, Cornelius CJ (2006) Polymer 47:4238

83. Stille JK, Harris FW, Rakutis RO, Mukamal H (1966) J Polym Sci B Polym Lett 4:791
84. Mukamal H, Harris FW, Stille JK (1967) J Polym Sci A-1 Polym Chem 5:2721
85. VanKerckhoven HF, Gillians YK, Stille KK (1972) Macromolecules 5:541
86. Poppe D, Frey H, Kreuer KD, Heinzel A, Mülhaupt R (2002) Macromolecules 35:7936
87. Miyatake K, Fukushima Kazuaki;Takeoka, Takeoka S, Tsuchida E (1999) Chem Mater 11:1171
88. Miyatake K, Iyotani H, Yamamoto K, Tsuchida E (1996) Macromolecules 29:6969
89. Miyatake K, Shouji E, Yamamoto K, Tsuchida E (1997) Macromolecules 30:2941
90. Jennings BE, Jones MEB, Rose JB (1967) J Polym Sci C 16:715
91. Rose JB (1974) Chimia 28:561
92. Johnson RN, Farnham AG (1964) GB Patent 1 078 234
93. Rose JB (1979) US Patent 4 268 650
94. Kricheldorf HR, Bier G (1983) J Polym Sci Polym Chem Ed 21:2283
95. Kricheldorf HR, Meier J, Schwarz G (1987) Makromolek Chem Rapid Commun 8:529
96. Kricheldorf HR (1996) Silicon in Polymer Synthesis. Springer Verlag, Berlin Heidelberg
97. Kim IC, Choi JG, Tak TM (1999) J Appl Polym Sci 74:2046
98. Chen M, Chiao T, Tseng T (1996) J Appl Polym Sci 61:1205
99. Schauer J, Albrecht W, Weigel T, Kudela V, Pientka Z (2001) J Appl Polym Sci 81:134
100. Kang M, Choi Y, Choi I, Yoon T, Moon S (2003) J Membr Sci 216:39
101. Noshay A, Robeson LM (1976) J Appl Polym Sci 20:1885
102. Lafitte B, Jannasch P (2005) J Polym Sci A Polym Chem 43:273
103. Miyatake K, Hay AS (2001) J Polym Sci A Polym Chem 39:3770
104. Jakoby K, Peinemann K, Nunes SP (2003) Macromol Chem Phys 204:61
105. Iojoiu C, Marechal M, Chabert F, Sanchez J (2005) Fuel Cells 5:344
106. Li L, Wang Y (2005) J Membr Sci 246:167
107. Hajatdoost S, Yarwood J (1997) J Chem Soc Faraday Trans 93:1613
108. Al-Omran A, Rose JB (1996) Polymer 37:1735
109. Chikashige Y, Chikyu Y, Miyatake K, Watanabe M (2005) Macromolecules 38:7121
110. Miyatake K, Chikashige Y, Watanabe M (2003) Macromolecules 36:9691
111. Tavares AC, Pedicini R, Gatto I, Dubitsky YA, Zaopo A, Passalacqua E (2003) J New Mater Electrochem Systems 6:211
112. Wang L, Meng YZ, Wang SJ, Li XH, Xiao M (2005) J Polym Sci A Polym Chem 43:6411
113. Helmer-Metzmann F, Schleicher A, Schneller A, Witteler H (1995) DE Patent 19 527 435A1
114. Nagarale RK, Gohil GS, Shahi VK, Rangarajan R (2005) J Appl Polym Sci 96:2344
115. Deimede V, Kandilioti G, Kallitsis JK, Gregoriou VG (2005) Macromolecular Symposia 230:33
116. Blanco JF, Nguyen QT, Schaetzel P (2002) J Appl Polym Sci 84:2461
117. Manea C, Mulder M (2002) J Membr Sci 206:443
118. Huang RYM, Shao P, Burns CM, Feng X (2001) J Appl Polym Sci 82:2651
119. Baradie B, Poinsignon C, Sanchez JY, Piffard Y, Vitter G, Bestaoui N, Foscallo D, Denoyelle A, Delabouglise D, Vaujany M (1998) J Power Sources 74:8
120. Dyck A, Fritsch D, Nunes SP (2002) J Appl Polym Sci 86:2820
121. Ehrenberg SG, Serpico J, Wnek GE, Rider JN (1994) US Patent 5 468 574A1
122. Byun IS, Kim IC, Seo JW (2000) J Appl Polym Sci 76:787–798
123. Hwang G, Ohya H, Nagai T (1999) J Membr Sci 156:61–65
124. Park HB, Shin H, Lee YM, Rhim J (2005) J Membr Sci 247:103–110
125. Zhang X, Liu S, Yin J (2006) J Polym Sci B Polym Phys 44:665–672

126. Meier-Haack J, Vogel C, Butwilowski W, Lehmann D (2006) Desalination 200:299–301
127. Meier-Haack J, Vogel C, Butwilowski W, Lehmann D (2007) Pure and Appl Chem 79:2083–2093
128. Meier-Haack J, Komber H, Vogel C, Butwilowski W, Schlenstedt K, Lehmann D (2007) Macromolecular Symposia 254:322–328
129. Liu B, Robertson GP, Kim D, Guiver MD, Hu W, Jiang Z (2007) Macromolecules 40:1934–1944
130. Vogel C (2007) Personal communication
131. Yamagishi H, Crivello JV, Belfort G (1995) J Membr Sci 105:249
132. Crivello JV, Belfort G, Yamagishi H (1995) US Patent 5 468 390
133. Guiver MD, Kutowy O, ApSimon JW (1989) Polym 30:1137
134. Guiver MD, Croteau S, Hazlett JD, Kutowy O (1990) British Polym J 23:29
135. Guiver MD, Robertson GP, Yoshikawa M, Tam CM (2000) Functionalized polysulfones: Methods for chemical modification and membrane applications In: Pinnau I, Freeman BD (eds) Functionalized polysulfones: Methods for chemical modification and membrane applications. ACS Symposium Series 744, American Chemical Society, Washington, DC, p 137
136. Kerres J, Zhang W, Cui W (1998) J Polym Sci A Polym Chem 36:1441
137. Van Zyl AJ, Kerres J (1999) J Appl Polym Sci 74:422
138. Kerres J, Ullrich A, Meier F, Häring T (1999) Solid State Ionics 125:243
139. Kerres J, Ullrich A, Hein M (2001) J Polym Sci A Polym Chem 39:2874
140. Jörissen L, Gogel V, Kerres J, Garche J (2002) J Power Sources 105:267
141. Kerres J, Zhang W, Jörisson L, Gogel V (2002) J New Mater Electrochem Sys 5:97
142. Lafitte B, Karlsson LE, Jannasch P (2002) Macromol Rapid Commun 23:896
143. Karlsson LE, Jannasch P (2004) J Membr Sci 230:61
144. Lafitte B, Puchner M, Jannasch P (2005) Macromol Rapid Commun 26:1464
145. Karlsson LE, Jannasch P (2005) Electrochimica Acta 50:1939
146. Jannasch P (2005) Fuel Cells 5:248
147. Lafitte B, Jannasch P (2006) J Polym Sci A Polym Chem 45:269
148. Gieselman MB, Reynolds JR (1992) Macromolecules 25:4832
149. Gieselman MB, Reynolds JR (1993) Macromolecules 26:5633
150. Kawahara M, Rikukawa M, Sanui K (2000) Polym Adv Technol 11:544
151. Kawahara M, Rikukawa M, Sanui K, Ogata N (2000) Solid State Ionics 136–137:1193
152. Kricheldorf HR, Vakhtangishvili L, Fritsch D (2002) J Polym Sci A Polym Chem 40:2967
153. Oguri M, Ikeda R, Okisaki F (2002) DE Patent 60 200 877 T2
154. Oguri M, Ikeda R, Okisaki F (2002) EU Patent 1 314 751 B1
155. Lafitte BJP (2007) Adv Funct Mater 17:2823
156. Ekström H, Lafitte B, Ihonen J, Markusson H, Jacobsson P, Lundblad A, Jannasch P, Lindbergh G (2007) Solid State Ionics 178:959
157. Ueda M, Toyota H, Ouchi T, Sugiyama J, Yonetake K, Masuko T, Teramoto T (1993) J Polym Sci A Polym Chem 31:853
158. Zhang L, Hampel C, Mukerjee S (2005) J Electrochem Soc 152:A1208
159. Wang F, Ji Q, Harrison W, Mecham J, Formato R, Kovar R, Osenar P, McGrath JE (2000) Polym Prepr 41:237
160. Wang F, Hickner M, Ji Q, Harrison W, Mecham J, Zawodzinski TA, McGrath JE (2001) Macromol Symp 175:387
161. Mecham JB, Wang F, Glass TE, Xu J, Wilkes GL, McGrath JE (2001) Polym Mater Sci Eng 84:105

162. Harrison W, Wang F, Mecham J, Glass T, Hickner M, McGrath J (2001) Polym Mater Sci Eng 84:688
163. Wang F, Kim Y, Hickner M, Zawodzinski TA, McGrath JE (2001) Polym Mater Sci Eng 85:517
164. Kim YS, Wang F, Hickner M, Zawodzinski TA, McGrath JE (2001) Polym Mater Sci Eng 85:520
165. Wang F, Hickner M, Kim YS, Zawodzinski TA, McGrath JE (2002) J Membr Sci 197:231
166. Kim YS, Dong L, Hickner MA, Pivovar BS, McGrath JE (2003) Polymer 44:5729
167. Kim YS, Wang F, Hickner M, McCartney S, Hong YT, Harrison W, Zawodzinski TA, McGrath JE (2003) J Polym Sci B Polym Phys 41:2816
168. Kim YS, Dong L, Hickner MA, Glass TE, Webb V, McGrath JE (2003) Macromolecules 36:6281
169. Kim YS, Sumner MJ, Harrison WL, Riffle JS, McGrath JE, Pivovar BS (2004) J Electrochem Soc 151:A2150
170. Harrison WL, Hickner MA, Kim YS, McGrath JE (2005) Fuel Cells 5:201
171. Every HA, Hickner MA, McGrath JE, Zawodzinski Jr TA (2005) J Membr Sci 250:183
172. Meier-Haack J, Taeger A, Vogel C, Schlenstedt K, Lenk W (2005) Sep Pur Technol 41:207
173. Lakshmi RTS M, Meier-Haack J, Schlenstedt K, Komber H, Choudhary V, Varma I (2006) React Funct Polym 66:634
174. Kozlowski JH (1975) US Patent 4 046 806
175. Vogel C, Meier-Haack J, Taeger A, Lehmann D (2004) Fuel Cells 4:320
176. Takeuchi M, Jikei M, Kakimoto M (2003) Chem Lett 32:242
177. Xiao G, Sun G, Yan D (2002) Polym Bull 48:309
178. Serpico JM, Ehrenberg SG, Fontanella JJ, Jiao X, Perahia D, McGrady KA, Sanders EH, Kellogg GE, Wnek GE (2002) Macromolecules 35:5916
179. Taeger A, Vogel C, Lehmann D, Jehnichen D, Komber H, Meier-Haack J, Ochoa NA, Nunes SP, Peinemann K (2003) React Funct Polym 57:77
180. Schauer J, Brozova L (2005) J Membr Sci 250:151
181. Liu Y, Bhatnagar A, Ji Q, Riffle JS, McGrath JE, Geibel JF, Kashiwagi T (2000) Polymer 41:5137
182. Wiles KB, Wang F, McGrath JE (2005) J Polym Sci A Polym Chem 43:2964
183. Schuster M, Kreuer K, Andersen HT, Maier J (2007) Macromolecules 40:598
184. Shen L, Xiao G, Yan D, Sun G (2005) e-Polymers 31:1
185. Bai Z, Dang TD (2006) Macromol Rapid Commun 27:1271
186. Bai Z, Houtz MD, Mirau PA, Dang TD (2007) Polymer 48:6598
187. Bai Z, Price GE, Yoonessi M, Juhl SB, Durstock MF, Dang TD (2007) J Membr Sci 305:69
188. Zhang C, Guo X, Fang J, Xu H, Yuan M, Chen B (2007) J Power Sources 170:42
189. Lee KK, Kerres J (2007) J Membr Sci 294:75
190. Schönberger F, Kerres J (2007) J Polym Sci A Polym Chem 45:5237
191. Schuster M, Kreuer K, Thalbitzer AH, Maier J (2005) DE Patent 102 005 010 411 A1
192. Tang H, Pintauro PN (2001) J Appl Polym Sci 79:49

Adv Polym Sci (2008) 216: 63–124
DOI 10.1007/12_2007_129

Published online: 22 February 2008

Polybenzimidazole/Acid Complexes as High-Temperature Membranes

Jordan Mader[1] · Lixiang Xiao[2] · Thomas J. Schmidt[3] · Brian C. Benicewicz[1] (✉)

[1]NYS Center for Polymer Synthesis, Department of Chemistry and Chemical Biology, Rensselaer Polytechnic Institute, Troy, NY 12180, USA
benice@rpi.edu

[2]BASF Fuel Cells Inc., 39 Veronica Ave, Somerset, NJ 08873, USA

[3]BASF Fuel Cells GmbH, Industriepark Hochst, G865, 65926 Frankfurt am Main, Germany

Abstract This chapter reviews the progress towards applying acid-doped polybenzimidazoles (PBIs) as polymer electrolyte membrane (PEM) fuel cell membranes over approximately the last ten years. The major focus of the first part of the chapter is on three main systems: (1) the well-developed *meta*-PBI (poly(2,2′-*m*-phenylene-5,5′-bibenzimidazole)); (2) the various derivatives and filled systems based on *meta*-PBI; and (3) poly(2,5-benzimidazole) (AB-PBI). The polymer membrane properties, such as thermal and chemical stability, ionic conductivity, mechanical properties, and ability to be manufactured into a membrane and electrode assembly (MEA), are discussed in detail. Preliminary fuel cell performance is reported for a number of PBI chemistries. The second section of the chapter highlights recent work on developing a novel process to produce phosphoric acid (PA)-doped PBI membranes for use in high-temperature PEMFCs. This novel sol-gel process, termed the polyphosphoric acid (PPA) process, allows production of a gel membrane that exhibits properties not observed with the "traditionally" prepared PBIs, such as improved ionic conductivity, mechanical properties, fuel cell performance, and long-term stability. The final section of the chapter focuses on the possible degradation modes of the commercially available products from BASF Fuel Cells.

Keywords Acid-doped membranes · High-temperature · PEMFC · Polybenzimidazole · Polyphosphoric Acid Process

Abbreviations

PBI	Polybenzimidazole
m-PBI	*meta*-polybenzimidazole
p-PBI	*para*-polybenzimidazole
IV	Inherent viscosity
DMAc	Dimethylacetamide
PA	Phosphoric acid
moles PA/PRU	Moles phosphoric acid per moles polymer repeat unit
PEM(FC)	Polymer electrolyte membrane (fuel cell)
MEA	Membrane and electrode assembly
AB-PBI	Poly(2,5-benzimidazole)
moles PA/BI	moles phosphoric acid per moles benzimidazole unit
RH	Relative humidity
ZrP	Zirconium phosphate
PWA	Phosphotungstic acid
SiWA	Silicotungstic acid
ZrPBTC	Zirconium tricarboxybutylphosphonate
SA	Sulfuric acid
BP	Boron phosphate
PBI-PrS	Propylsulfonated PBI
SD	Degree of sulfonation
PBI-BS	Butylsulfonated PBI
DMFC	Direct methanol fuel cell
sPS	sulfonated polysulfone
P4VP	Poly(4-vinylpyridine)
DABA	3,4-Diaminobenzoic acid
PMA	Phosphomolybdic acid
sAB-PBI	sulfonated poly(2,5-benzimidazole)
MSA	Methanesulfonic acid
TMM	Trimethoxymethane
PABI	Poly(amide-benzimidazoles)

PPA	Polyphosphoric acid
TAB	3,3′,4,4′-Tetraaminobiphenyl
TPA	Terephthalic acid
PDA	Pyridine dicarboxylic acid
PPBI	Pyridine-based PBI
PTFE	Polytetrafluoroethylene
GDE	Gas diffusion electrode
mtx	mass transport
ECSA	Electrochemical surface area
RMFC	Reformed methanol fuel cell
MW	Molecular Weight

1 Introduction to Polybenzimidazoles

Polybenzimidazoles (PBIs) are a class of well-known polymers, which have applications as thermally stable and nonflammable textile fibers, high-temperature matrix resins, adhesives, and foams. The wholly aromatic PBIs were developed for high-performance fiber applications in the early 1960s by the United States Air Force Materials Laboratory in conjunction with Dupont and the Celanese Research Company. The first wholly aromatic PBI was synthesized in 1961 by Vogel and Marvel [1]. Fibers and textiles made from the *m*-PBI shown in Fig. 1 display excellent properties, such as high temperature stability, nonflammability, and high chemical resistance. Because of these properties, PBI fiber has traditionally been used in firefighter's turnout coats, astronaut space suits, and gloves used in metalworking industries.

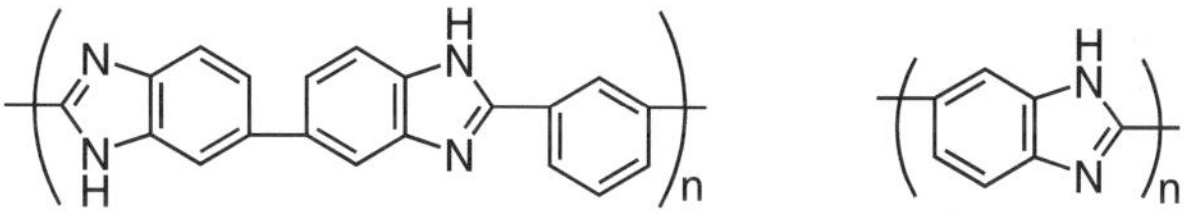

m-PBI AB-PBI

Fig. 1 Chemical structure of poly(2,2′-*m*-phenylene-5,5′-bibenzimidazole) (*m*-PBI) and poly(2,5-benzimidazole) (AB-PBI)

The commercial PBI polymer synthesis and fiber formation is a multi-step process. The polymer is made from 3,3′,4,4′-tetraaminobiphenyl and diphenyl isophthalate in a two-step, melt/solid polymerization process that produces PBI powder and byproducts of phenol and water (see Scheme 1). This process produces polymer with inherent viscosities (IV) of between 0.5 and 0.8 $dL\,g^{-1}$, which corresponds to low to moderate molecular weights. The polymer is then dissolved under high pressure in DMAc/LiCl, filtered, dry spun into fibers, washed, dried, drawn, acid treated, and wound up for subsequent textile processing. A similar process is used to produce films. For films

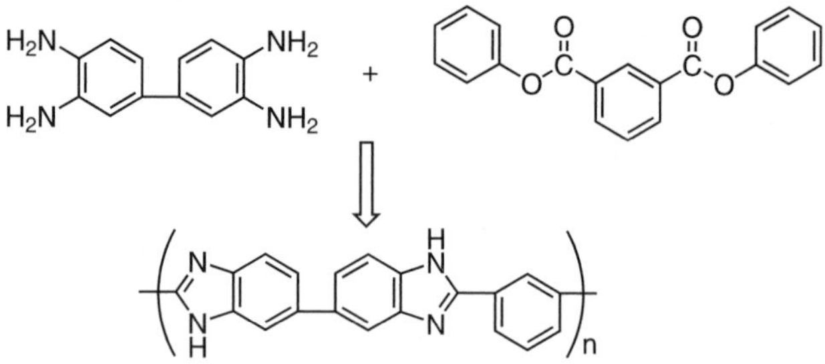

Scheme 1 Polymerization of *m*-PBI from 3,3′,4,4′-tetraaminobiphenyl and diphenyl isophthalate

to be used in fuel cells, additional processing in phosphoric acid is required to produce an acid imbibed film.

In the past several years, there has been a major emphasis on the development of high-temperature (>100 °C) polymer-based proton-exchange membrane fuel cells. The benefits of operating at higher temperatures include: the reduction or elimination of humidification requirements, increased tolerance to fuel impurities (e.g., CO), wider fuel choices, lower fuel reforming costs, improved electrode kinetics, higher conductivities, and smaller heat exchangers or radiators. Traditional polymer fuel cell membranes that rely on water for proton conduction require complicated or expensive water management systems for operation at 80 °C or higher. Initial work on PBI-phosphoric acid based membranes using the commercially available PBI polymer has shown that many of the requirements for high-temperature operation could be satisfied by this membrane system [2]. Since these early reports, much work has been done to more fully evaluate and develop fuel cell membranes based on PBI polymers.

Typically, PBI-based fuel cells use phosphoric acid (PA) as an electrolyte [2–6, 8–10], because of its high conductivity and thermal stability. It has been reported that these membranes exhibit high ionic conductivities at high temperatures, low gas permeability, excellent chemical and thermal stability in the fuel cell environment, and nearly zero water drag coefficient. Furthermore, PBI polymer is commercially available; it is well-characterized and methods of synthesis have been developed thoroughly. However, some of the perceived problems with using PBI for fuel cell membranes include: the low molecular weights (IVs of 0.5–0.8 dL g^{-1}), low phosphoric acid loading (6–10 moles of phosphoric acid/moles polymer repeat unit [moles PA/PRU]), phosphoric acid retention, and membrane durability. Improvements in these properties are the focus of much research, which should lead to improved membranes that satisfy the extensive needs of a commercially viable fuel cell membrane.

In this chapter, the early work on PBI-phosphoric acid systems that demonstrates the general applicability of this polymer-acid membrane to high-temperature PEM operation will be reviewed. Two different PBI polymer systems, based on the commercially available *meta*-PBI and AB-PBI, have been investigated for this application as acid-imbibed systems. Additionally, a new sol-gel process was discovered and applied to a much wider variety of PBI chemical structures. The sol-gel process produces an acid-imbibed membrane directly upon casting with a morphology and set of properties not attainable from the conventional imbibing process. The second part of the chapter reviews these recent advancements, which have been used to further develop an acid-imbibed membrane that forms the basis for a commercially available MEA. In the last section of this chapter, the general properties, performance, and durability of the commercial membrane and MEA will be reviewed with an outlook on possible degradation modes.

2 Various Polybenzimidazole Membranes Produced via Conventional Processes

2.1 Introduction to the Polybenzimidazole/Phosphoric Acid Complex

In 1995, Wainright et al. first described a polybenzimidazole (PBI)-phopshoric acid (PA) complex for use in high-temperature fuel cells [2]. Due to its commercial availability, a large amount of research has focused on *m*-PBI, poly(2,2′-*m*-phenylene-5,5′-bibenzimidazole) (Fig. 1), commonly referred to as PBI. In this chapter, this specific structure will be referred to as *m*-PBI to identify the orientation of the phenyl ring. Research with this polymer has expanded to include functionalization of *m*-PBI, inorganic additives, polymer blends, and doping with different electrolytes. Another PBI structure that has been widely investigated is poly(2,5-benzimidazole), or AB-PBI (Fig. 1). It is interesting to note that PBIs have been studied for use in both hydrogen and direct methanol fuel cells. These membranes have shown an incredible potential for PEM fuel cell use as alternatives to traditionally investigated perfluorinated sulfonic acid type membranes.

2.2 *Meta*-PBI

The earliest work describing a PBI/PA complex for fuel cell use was reported by Wainright et al. in 1995 [2]. This research detailed various synthesis methods and characterization of the resulting polymer, and was further reviewed by Kim and Lim [3]. Because PBI is thermally stable [4–7], mechanically robust, chemically stable, and exhibits high CO tolerance [8], it was shown to be a promising high-temperature fuel cell membrane candidate. Since its introduction, the majority of research has focused on increasing the low acid-doping levels and, consequently, the conductivity of the *m*-PBI/PA complex, as well as improving the mechanical properties of the polymer. A variety of approaches have been utilized, but most include sulfonation, inorganic fillers, or polymer blends.

Wainright's work with *m*-PBI demonstrated that this polymer could be a viable membrane candidate for fuel cells. Almost all previous PEM work had focused on perfluorosulfonic acid electrolytes, such as Nafion. This early work showed that *m*-PBI could be cast into films from dimethylacetamide solutions, doped with acid (~5 moles PA/PRU), could retain conductivity even at high temperatures ($0.025\ \mathrm{S\,cm^{-1}}$ at 150 °C), and function in a fuel cell, all without loss of polymer properties ($\mathrm{IV} = 1.2\ \mathrm{dL\,g^{-1}}$). Furthermore, this work reported a methanol permeability of $15 \times 10^{-16}\ \mathrm{m^3\,(STP)m\,m^{-2}\,s^{-1}\,Pa^{-1}}$ and methanol crossover current of $\sim 10\ \mathrm{mA\,cm^{-2}}$ for *m*-PBI films. Typically, the crossover current for Nafion is $\sim 100\ \mathrm{mA\,cm^{-2}}$.

In 1996, in order to determine if *m*-PBI was a truly acceptable candidate for high-temperature use, Samms et al. confirmed the thermal stability of the polymer ($M_{\mathrm{W}} = 25\,000$) by simulating fuel cell operating conditions (swollen with PA and loaded with Pt) and performing thermal gravimetric analysis [4]. They showed that both the dry polymer and the acid-doped polymer were stable up to 600 °C in pure nitrogen, 5% hydrogen (nitrogen balance), and air, concluding that the PBI/PA complex was very stable even under simulated fuel cell conditions.

Many researchers have worked on characterizing the proton conductivity of *m*-PBI/PA [9–13], as well as other PBI structures. Table 1 shows a comprehensive collection of polymer structures, molecular weights, doping levels, and proton conductivities (with experimental conditions, when available) reported to date. Values are widely reported throughout the literature, with typical ranges between $0.04\text{–}0.08\ \mathrm{S\,cm^{-1}}$ at 150 °C, and, as generally agreed, the conductivity is dependent on acid doping level, humidity, temperature, and pressure. Unfortunately, data on polymer molecular weight, doping levels, or details of the test conditions were not included in some reports.

Specifically, the conductivity of *m*-PBI was investigated as a function of doping electrolyte [9–11, 15–18], and was nicely discussed by Schuster and Meyer [19]. Overall, it was found that *m*-PBI with a doping level between 2–8 moles PA/PRU typically has a conductivity between 10^{-1} and $10^{-4}\ \mathrm{S\,cm^{-1}}$ in low humidification or nonhumidified conditions at high temperatures (>120 °C). In general, for the *m*-PBI/sulfuric acid complex, conductivities were similar to the *m*-PBI/PA complex or slightly lower [10, 11, 16, 17]. The higher values are comparable to the current state of the art perfluorinated sulfonic acid membrane (Nafion) at atmospheric pressure and full hydration. However, the *m*-PBI/PA complex is the most widely studied, because of its conductivity and thermal stability.

Wainright et al. [2] and Wang et al. [20] described some early results of fuel cell testing in direct methanol and hydrogen/oxygen cells, respectively.

Table 1 Polymer Structure, Characterization, and Membrane Conductivities for PBI Polymers

	Common structure	Monomers, Additives, or Misc. Info	IV ($dL\,g^{-1}$)	Acid doping (PA/PRU)	Conductivity $S\,cm^{-1}$/°C/RH NR = not reported	Refs.
m-PBI from conventional processes	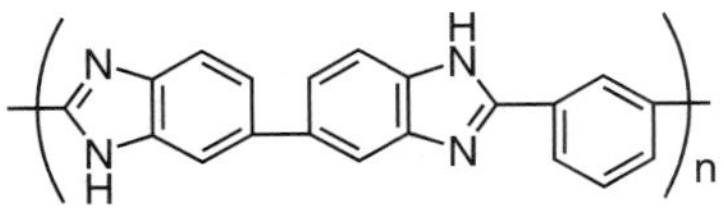	Tetraamines; difunctional phenyl carboxylic acids	0.6	5.01	2×10^{-2}/130/NR	[2, 11]
			0.6	5	2.5×10^{-2}/150/NR	[2, 11]
			0.6	3.38	5×10^{-3}/130/NR	[2]
			1.2	6.3	5×10^{-2}/140/30	[9]
			1.2	6.3	2×10^{-2}/140/5	[9]
			1.2	6.3	5.9×10^{-2}/150/30	[9]
			1.2	6.3	4.7×10^{-3}/150/5	[9]
			0.6	3.05	7×10^{-6}/30/NR	[10]
			NR	5.01	3.5×10^{-2}/190/NR	[11]
			NR	0.8	5×10^{-5}/25/NR	[11]
			NR	4.2	4×10^{-3}/25/NR	[11]
			NR	3.38	2.5×10^{-3}/130/dry	[16]
			NR	6.0	4.5×10^{-5}/25/dry	[16, 18]
			NR	6.13	10^{-4}/20–160/NR	[16]
			1.0	16	1.3×10^{-1}/160/NR	[21]
			$M_w = 21\,900$	6.6	6.1×10^{-2}/140/20	[22]
			$M_w = 25\,100$	6.2	5.7×10^{-2}/140/20	[22]
			$M_w = 55\,000$	6.6	6.3×10^{-2}/140/20	[22]
			NR	1.9	10^{-5}/160/anhydrous	[23]
			NR	5.6	6.8×10^{-2}/200/5	[31]
			NR	5.7	7.9×10^{-2}/200/5	[31]

Table 1 (continued)

	Common structure	Monomers, Additives, or Misc. Info	IV (dL g^{-1})	Acid doping (PA/PRU)	Conductivity S cm^{-1}/°C/RH NR = not reported	Refs.
		With PWA	NR	3.8	2.5×10^{-3}/130/NR	[107]
		With ZrP	NR	4.5	4.6×10^{-2}/165/NR	[108]
		With ZrPBTC	NR	NR	3.0×10^{-3}/100/100	[33]
		With ZrPBTC	NR	5.6	9.6×10^{-2}/200/5	[34]
			NR	None	3.82×10^{-3}/200/NR	[37]
			NR	NR	5.24×10^{-3}/200/NR	[37]
m-PBI PPA process		Tetraamines; difunctional phenyl carboxylic acids	1.49	14.4	5.16×10^{-2}/dry/25	[123]
			1.49	14.4	5.28×10^{-2}/dry/40	[123]
			1.49	14.4	6.23×10^{-2}/dry/60	[123]
			1.49	14.4	7.99×10^{-2}/dry/80	[123]
			1.49	14.4	9.52×10^{-2}/dry/100	[123]
			1.49	14.4	1.1×10^{-1}/dry/120	[123]
			1.49	14.4	1.2×10^{-1}/dry/140	[123]
			1.49	14.4	1.27×10^{-1}/dry/160	[123]

Table 1 (continued)

	Common structure	Monomers, Additives, or Misc. Info	IV (dL g^{-1})	Acid doping (PA/PRU)	Conductivity S cm^{-1}/°C/RH NR = not reported	Refs.
sPBI		Commercially available	NR	Not doped	3×10^{-6}/40/100	[53]
		m-PBI, thermal treatment in sulfuric acid	NR	Not doped	7.5×10^{-5}/160/100	[53]
sPBI with aryl group	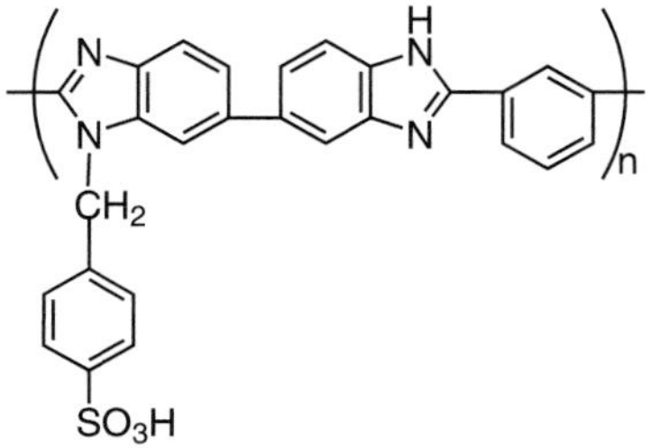	SD: 75%	NR	0.1 M	6×10^{-3}/25/100	[55]
		75%	NR	0.1 M	2×10^{-3}/25/100	[55]
		75%	NR	6 M	1×10^{-2}/50/100	[55]
		0%	NR	6 M	2×10^{-4}/50/100	[55]
PBI-PrS	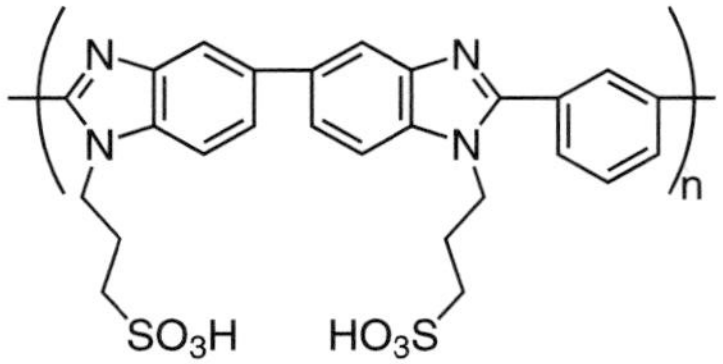	Commercially available *m*-PBI with 1,3 propanesulfone in LiH/DMAc	$M_w = 230\,000$	Not doped	$\sim 4 \times 10^{-4}$/90/100	[60]

Table 1 (continued)

	Common structure	Monomers, Additives, or Misc. Info		IV (dL g^{-1})	Acid doping (PA/PRU)	Conductivity S cm^{-1}/°C/RH NR = not reported	Refs.
PBI-BS	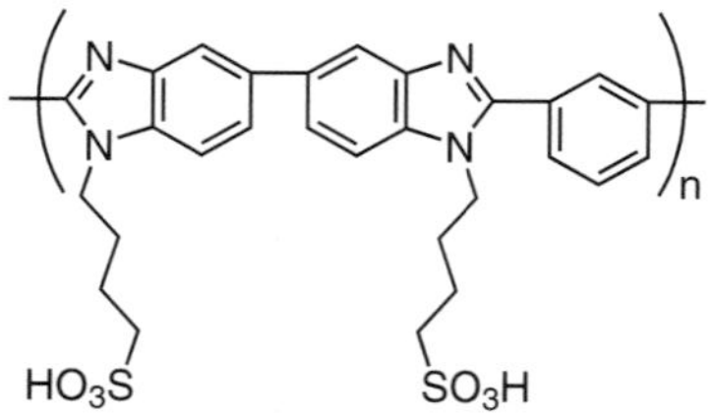	Commercially available *m*-PBI with 1,4 butanesulfone in LiH/DMAc		$M_w = 230\,000$	Not doped	$\sim 3 \times 10^{-3}$/90/100	[60]
sPS Blend with *m*-PBI	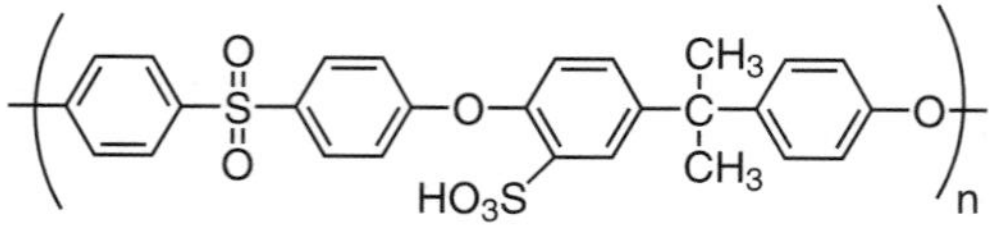	Commercially available *m*-PBI, usually sulfonated via chlorosulfonic acid (next column: wt % PBI/SD)	20/20	NR	5	1.5×10^{-2}/25/80	[50]
			75/20	NR	5	2.7×10^{-2}/25/80	[50]
			75/36	NR	5	3.9×10^{-2}/25/80	[50]
			75/70	NR	5	4.5×10^{-2}/25/80	[50]
			20/20	NR	5	5.5×10^{-2}/160/80	[50]
			75/20	NR	5	7×10^{-2}/160/80	[50]
			75/36	NR	5	7.8×10^{-2}/160/80	[50]
			75/70	NR	5	1×10^{-1}/160/80	[50]
			75/36	NR	11	7×10^{-2}/25/80	[50]
			75/36	NR	11	2.1×10^{-1}/160/80	[50]

Table 1 (continued)

	Common structure	Monomers, Additives, or Misc. Info	IV ($dL\,g^{-1}$)	Acid doping (PA/PRU)	Conductivity $S\,cm^{-1}/°C/RH$ NR = not reported	Refs.
sPhosphazene Blend with *m*-PBI	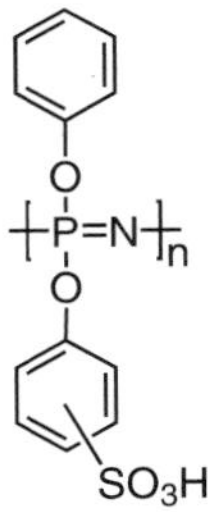	next column: wt % PBI/SD	3/NR NR	NR	6×10^{-2}/60/in water	[68]
sPPO Blend with *m*-PBI	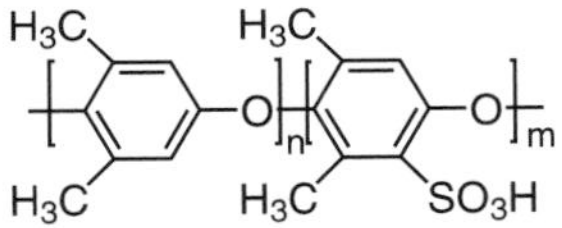	Commercially available *m*-PBI, sulfonated via sulfuric acid (SD: 25.4%)	NR NR	NR NR	~9×10^{-3}/25/0.1 M KCl (SD: 42.4%) ~1×10^{-2}/25/0.1 M KCl	[42] [42]
P4VP Blend with *m*-PBI	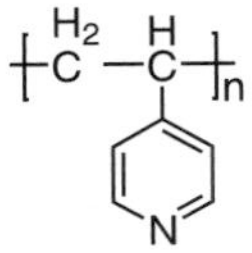	Blend compositions are 70/30 and 50/50 for doping levels 2.1 and 3.2 respectively	NR NR	2.1 3.2	10^{-3}/200/NR 10^{-2}/170/NR	[66] [66]

Table 1 (continued)

	Common structure	Monomers, Additives, or Misc. Info	IV (dL g^{-1})	Acid doping (PA/PRU)	Conductivity S cm^{-1}/°C/RH NR = not reported	Refs.
Phenylpyridine-co-sPS Blend with *m*-PBI	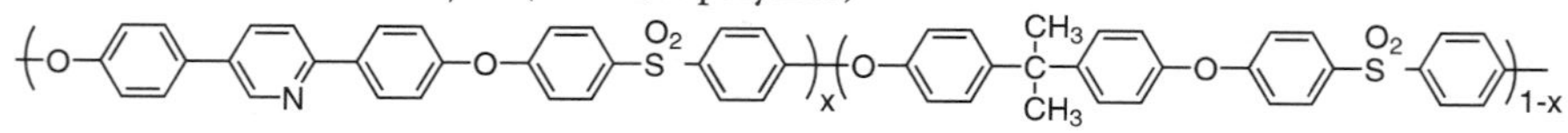	Blend composition is 50/50 (*m*-PBI/copolymer)	NR	220 wt %	7×10^{-2}/150/30	[67]
AB-PBI	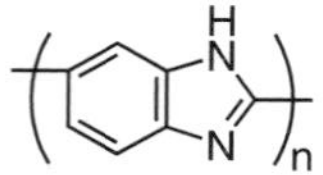		NR	5	10^{-4}/25/NR	[12]
		3,4-diaminobenzoic acid; diaminobenzoic acid salts	2.4	3	6.2×10^{-2}/150/30	[70]
			2.4	3	3.9×10^{-2}/180/5	[70]
			2.4	3	1.5×10^{-2}/180/dry	[71]
			2.4	2.7	2.5×10^{-2}/180/dry	[71, 74]
			1.5–1.8	1.6–3.7	2.6×10^{-2}–6×10^{-2}/110/dry	[72]
		With PMA and PA	2.7	NR	3×10^{-2}/185/dry	[13, 74]
sAB-PBI	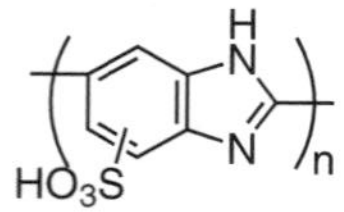	3,4-diaminobenzoic acid, sulfonation is sulfuric acid with heat treatment	NR	4.6	~3×10^{-2}/185/dry	[76]
		SD = 28% (3×10^{-2}) and 41% (3.5×10^{-2})	NR	4.6	3.5×10^{-2}/185/dry	[13]

Table 1 (continued)

	Common structure	Monomers, Additives, or Misc. Info	IV ($dL\,g^{-1}$)	Acid doping (PA/PRU)	Conductivity $S\,cm^{-1}/°C/RH$ NR = not reported	Refs.
p-PBI PPA Process		Tetra-amines; difunctional carboxylic acids	3.0 3.0	32 32	1.0×10^{-2}/25/dry 2.6×10^{-1}/200/dry	[90] [90]
2,5-PPBI PPA Process		2,5-pyridine dicarboxylic acid, 3,4-tetraaminobiphenyl	2.5–3.1 2.5–3.1	20.4 20.4	1.8×10^{-2}/25/dry 2×10^{-1}/160–200/dry	[110] [110]
2,6-PPBI PPA Process		2,6-pyridine dicarboxylic acid, 3,4-tetraaminobiphenyl	1.3 1.3	8.5 8.5	1×10^{-2}/25/dry 1×10^{-1}/160–200/dry	[110] [110]

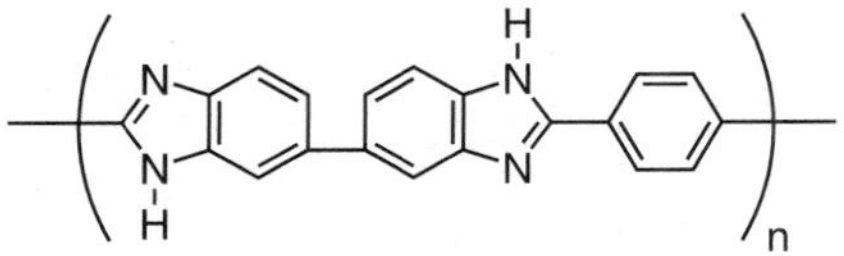

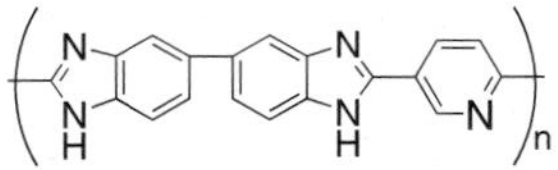

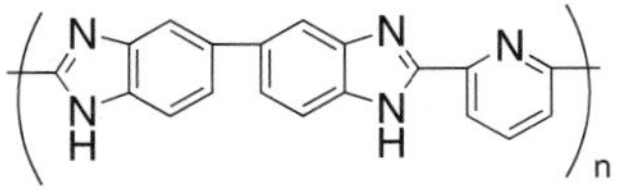

The performance of a methanol cell tested by Wainright et al. was comparable to a Nafion-based cell. The polarization curve showed a voltage of ~0.45 V at a current density of 0.20 A cm^{-2}. These results were considered promising, because of *m*-PBI's resistance to methanol crossover and the early stage of development of *m*-PBI membranes. Wang et al. operated a *m*-PBI/PA fuel cell on humidified hydrogen/oxygen and reported a voltage of ~0.6 V at a current density of 0.20 A cm^{-2}. The *m*-PBI/PA complex was also tested in a humidified hydrogen/air cell with lower performance (~0.5 V at 0.20 A cm^{-2}), as expected based on the Nernst Equation. While these results were lower than those of a Nafion cell, they are quite impressive when cell temperature is taken into consideration. All of Wang's cells operated at 150 °C and showed excellent stability for 200 hours. Nafion-based cells typically cannot operate above 80 °C without humidification or pressurized feed gases.

In an effort to increase the amount of acid held by *m*-PBI, many experimental methods have been investigated. Generally, it is believed that higher acid-doping levels lead to increased proton conductivity and, hopefully, improved fuel cell performance. This was highlighted by Li et al. [21] for a *m*-PBI/PA complex with 16 moles PA/PRU and a conductivity of 0.13 S cm^{-1} at 160 °C. This is comparable to a fully hydrated Nafion system running at 80 °C. However, at this high doping level, the membrane was mechanically unstable and could not be made into an MEA. This research also showed a correlation between conductivity and doping level; i.e., the conductivity increased with higher doping levels. Unfortunately, the mechanical strength decreased proportionally with the increase in doping level, leading to a loss of mechanical properties at higher doping levels. Preliminary hydrogen/oxygen fuel cell tests using membranes with 620 mol % doping level showed promise at varying temperatures and atmospheric pressure with no humidification (~0.6 V at 0.7 A cm^{-2}, 190 °C).

For a more in-depth look at physicochemical properties of the *m*-PBI/PA complex, He et al. [22] conducted a study on gas permeability, volume swelling, mechanical integrity, and conductivity. They separated out by fractionation PBI samples with molecular weight (MW) of approximately 25 000. They found that an increased doping level led to an increase in volume swelling (0.3 mol PA/PRU had 22% swelling, while 5 mol PA/PRU corresponded to 188% swelling), seen mostly in the thickness of the polymer sample due to separation of the PBI backbone by acid molecules. Interestingly, at 125 °C with a doping level of less than 2 mol PA/PRU, mechanical strength increased slightly due to H-bonding, then decreased sharply as acid loading increased. At 180 °C, a linear decrease in mechanical integrity was seen with increasing doping level, and values were overall lower than those at 125 °C: for PBI with doping level of 2.3, stress at break was 160 MPa at 125 °C and 48 MPa at 180 °C. Elongation at break increased with higher doping levels, because the membrane became more plastic at high acid doping levels and could more easily rearrange under load. Mechanical properties were also improved with higher molecular weights, as shown by the 3.5 MPa stress at break for PBI with MW of 17 800 and the 6 MPa stress at break for PBI with MW of 25 000. Gas permeability studies found that both the hydrogen and oxygen crossover increased as temperature and/or doping level increased, though it was more prominent for oxygen. It was found that there was no significant effect on the conductivity of membranes with different MW and similar doping levels.

Kawahara et al. also tried to increase acid doping levels by using various inorganic acids, such as phosphoric acid, sulfuric acid, and methane(or ethane)sulfonic acid [23]. They found that the *m*-PBI/PA complex was stable up to 500 °C, with an anhydrous proton conductivity of 10^{-5} S cm^{-1}. The doped films were prepared by immersing *m*-PBI films into strong acid/methanol solutions. Characterization by Fourier transform infrared (FTIR) spectroscopy, thermogravimetric analysis (TGA), and electrical impedance was then performed. Acid absorption level was found to increase with the increased concentration of the strong acid. A maximum doping level of 2.9 moles PA/PRU was achieved for the *m*-PBI/PA complex. FTIR showed that sulfuric acid, methanesulfonic acid, and ethanesulfonic acid protonated the basic N moiety on the imidazole, but phosphoric acid interacted through hydrogen bonding of the OH and NH groups instead. The thermal stability of *m*-PBI/strong acid complex was highest for phosphoric acid, and decreased in the order of sulfuric acid, methanesulfonic acid, and ethanesulfonic acid. The higher thermal stability of the PBI/PA suggests that the hydrogen bonding interaction imparts stability to the polymer/acid complex. The decomposition of the various polymer/acid complexes was believed to be due to elimination of the acid molecules from the complexes. The highest conductivity measured with PA at a doping level of 1.9 was 10^{-5} S cm^{-1} at 160 °C. The conductivity of all other polymer/acid complexes decreased at temperatures greater than 80 °C.

Li et al. [24] have examined the relationships between phosphoric acid doping level and water uptake. They showed that the water uptake of *m*-PBI was comparable to or even exceeded that of the commercially available Nafion, both when membranes were immersed in water or when placed in varying relative humidities, especially at higher acid doping levels (~6 moles PA/PRU). An in-depth discussion can be found in Li's review [25] of PBI-based membranes.

Kim et al. [26] synthesized *m*-PBI in a mixture of P_2O_5, CH_3SO_3H, and CF_3SO_3H, but the membrane did not show significant conductivity or performance improvements over other methods, even at comparable polymer inherent viscosities. Schechter and Savinell [27] studied the proton conduction pathway in *m*-PBI/PA complexes, as well as the use of imidazole or methyl imidazole additives, which did not result in any improvements in conductivity over previous work.

Hu et al. performed a five hundred hour long term performance test on PA-doped PBI. They also developed a one dimensional model to predict degradation over time [28]. The one dimensional model was developed for high-temperature operation and took into account the measured internal cell resistance and cathode area exchange current. However, in order to reduce complexity, many parameters were kept constant (such as electrochemical surface area (ESA)) and assumed simplifications, such as steady state cell operation, even temperature distribution across the electrode, ideal gases, etc., which do not reflect operational circumstances. The model was shown to have a good agreement with experimental data, with the largest deviations thought to occur because the modeling equations hold ESA constant, while it decreases with time in operational fuel cells. The MEA was built using PA-doped PBI (level unspecified) and electrodes made with some PBI ionomer in them. The cell was operated for 500 hours at 150 °C, using hydrogen and oxygen with a constant load of 640 mA cm^{-2} and measurements taken every 24 hours. The cell was assumed to be in activation phase for the first 100 hours, with the voltage increasing from 0.5 V to 0.58 V over this time period. Linear scanning voltammetry showed an increase in ESA up to 100 hours. Both the fuel cell performance and ESA decreased over the following 400 hours. The cell performance decreased linearly at a rate of ~150 mV h^{-1}. The decrease in ESA is thought to come from Pt sintering; TEM imaging confirmed an increase in average particle size from an initial value of 3.8 nm to 6.9 nm at test completion. Scanning electron microscopy (SEM) imaging was performed on the MEA before and after the long-term test to study the effect of delamination on cell failure. It was determined that this did not contribute to performance losses, as there was no increased separation after 500 h of testing. The researchers concluded that the major loss of performance was due to the large decrease in ESA.

The temperature effects on cell performance and catalyst stability were investigated by Lobato et al. [29]. Measurements were carried out on a single 5 cm^2 cell from 100 to 175 °C. The MEA was made with electrodes with acid-doped PBI ionomer and a PBI membrane with doping level of 6.5 mol PA/PRU and the fuel and oxidant were hydrogen and oxygen, respectively. Each cell was conditioned at a certain temperature for 24 h and then polarization curves were taken. All results were related to the initial conditioning temperature, rather than the temperature at which measurements were taken. Cyclic voltammetry (CV) was used to determine catalyst stability. X-ray diffraction was used to record any change in the structure of the catalyst during CV measurements. It was found that at conditioning temperatures of 100 and 125 °C, a stable current value was reached after ~5 hours, while currents for the conditioning temperatures of 150 and 175 °C continued to drop at a constant rate (1 mA cm^{-1} h^{-1} and 2.8 mA cm^{-1} h^{-1}, respectively) even after 24 hours. This may be due to the rapid loss of absorbed water at the higher temperatures and, therefore, decreased proton conductivity. Also playing a large role was the oligomerization of phosphoric acid to pyrophosphoric acid, causing a drop in proton conductivity. By collecting the Nyquist plots of these membranes at various temperatures, the various resistances could be separated. It was determined that at 100 and 125 °C, cell performance was mainly affected by the reduction of ohmic and polarization resistance, accounting for better performance. However, at 150 and 175 °C ohmic resistance increased greatly and the polarization resistance decrease was not large enough to overcome the difference. After conditioning the cells, the short-term response of the system was studied by changing the temperature from the conditioned value and immediately recording a polarization curve. It was found that the short-term response was governed by the faster electrode kinetics and increased electrolyte conductivity. For long-term effects, higher conditioning temperatures led to lower overall fuel cell performance, as the temperature was changed. This was especially true for those cells conditioned at 150 and 175 °C, which was thought to be a function of the progressive and constant dehydration of phosphoric acid during conditioning. It was thought that the dehydration was the major cause of cell-performance degradation in these experiments. Water loss (free and from PA dehydration) under the conditioning/operating temperatures was confirmed by TGA.

CV studies showed a progressive loss of active area of Pt when subjected to a harsh acid environment similar to an operating fuel cell. These losses may be explained by Pt migration and/or Pt dissolution-redeposition throughout the electrode. Particle redistribution after the CV study was confirmed by X-ray diffraction with the result of Pt agglomeration.

Zhai et al. further studied the degradation mechanisms of the MEA in PA/PBI high-temperature fuel cells [30] by performing a 550 h long-term test. The first 500 h was continuous operation at 640 mA cm^{-2}, while the last 50 h was intermittent operation with shutoffs every 12 hours. The tests were performed at 150 °C with unhumidified hydrogen and oxygen. It was found that there were three main regions in the long-term performance curve. The first region, up to ~90 hours, was considered the activation period, and the voltage increased from 0.57 to 0.66 V. The next 450 h of operation showed a continual steady decrease of ~18 mV h^{-1} in performance, with an overall change from 0.66 to 0.58 V. The last 10 h showed a rapid decrease in performance, due to severe membrane damage. The best performance was recorded at 96 h, with a power density of 0.95 W cm^{-2}. At the end of the 500 hours of continuous operation, the power density was 0.70 W cm^{-2}, which corresponds to an overall loss of ~26%. The major causes for loss of performance were concluded to be Pt agglomeration, leaching of phosphoric acid, and hydrogen crossover. Pt agglomeration was shown by CV and SEM. After about 90 h, when the active area of Pt had reached the maximum, a steady decrease in active area was observed. This was confirmed by SEM at 480 h, which showed an increase in average particle size from 4.02 to 8.88 nm and a distribution shift to larger particle sizes. It was concluded that this was the major cause of performance loss, and that it represents the main stability issue to be overcome in the future. The leaching of PA was confirmed by EIS and showed as a slight increase in internal resistance during operation. The elemental distribution of phosphorus across the MEA was analyzed using energy dispersive X-ray analysis and showed no obvious variation in membrane phosphorus during the lifetime of the test. However, the phosphorus levels in the catalyst layer decreased severely. Interestingly, it was found by linear scanning voltammetry that hydrogen crossover was steady during the continuous operation portion of the test. When the cell was switched into intermittent operation, the hydrogen crossover increased drastically, because the electrode potentials were higher. This led to cracks in the membrane (as seen by SEM), which allowed for increased hydrogen crossover. These cracks were surmised to be a result of oxidative degradation during operation.

Kongstein et al. formulated a method to make PBI-based electrode materials and tested them in a high-temperature acid-doped PBI fuel cell [31]. It is well known that the choice of electrode can significantly impact the fuel cell performance. This is especially important with PBI fuel cells, because at high voltages, carbon can be oxidized in acidic environments. To help improve PBI fuel cell performance, a dual layer electrode was formulated. A microporous layer made of carbon fiber paper treated with PTFE was fabricated, and then, the catalytic layer was sprayed from a dispersion of platinum on carbon in PBI/DMAc. First, the portion of the electrode that would be in contact with the membrane was sprayed with 50 wt % Pt/C, and then the outer part was sprayed with 20 wt % Pt/C and the DMAc was removed by evaporation at 190 °C. The MEA was made by hot-pressing of the electrodes onto a PBI membrane with a doping level of 5.6 mol PA/PRU. Polarization curves were collected in a 2×2 cm^2 cell using hydrogen and oxygen. The highest performance was found with electrodes containing both PBI and PTFE ionomers and with Pt loading of 0.4 mg cm^{-2} on the anode and 0.6 mg cm^{-2} on the cathode. The amount of PBI ionomer in the electrode was of crucial importance. Too high a concentration led to a coating of electrically insulating PBI over the Pt surfaces, while too little PBI content led to lower ionic conductivity. It was found that the best performance was with electrodes containing between 0.2 and 0.4 mg PBI cm^{-2}. The fuel cell performance was lower than Nafion, but still impressive for high-temperature operation, with a maximum of 0.6 A cm^{-2} at 0.6 V. The maximum power density achieved was 0.83 W cm^{-2} at 0.4 V.

2.2.1 Acid-Doped Polybenzimidazole/Inorganic Fillers

Inorganic fillers are typically added to increase the proton conductivity and/or acid uptake of PBI films. A number of methods for filling, blending, and sulfonating PBI have been investigated and were reviewed by Kerres [32]. The following is a more detailed look at specific research on inorganic fillers. Staiti et al. [33] used phosphotungstic acid adsorbed on silicon dioxide to increase the PA doping levels. It was found that the conductivity increased with higher loadings of phosphotungstic acid. The maximum conductivity at 100 °C and

100% RH was 3.0×10^{-3} S cm^{-1}, which was comparable with the lower values reported for a *m*-PBI/PA complexes. The membrane exhibited promising mechanical and thermal stability, which could lead to potential use in PEMFC's.

He et al. [34] synthesized a number of PA-doped *m*-PBI/inorganic filler blends, including zirconium phosphate, phosphotungstic acid, and silicotungstic acid as fillers. The conductivity of the PA-doped *m*-PBI and *m*-PBI composite membranes was found to depend on PA doping level, relative humidity, and temperature. The conductivity for a PA-doped PBI (5.6 moles PA/PRU) at 200 °C and 5% RH was 6.8×10^{-2} S cm^{-1}. The addition of 15 wt % zirconium phosphate to the membrane increased the conductivity to 9.6×10^{-2} S cm^{-1} when tested under the same conditions. *m*-PBI/PA membranes containing 20–30 wt % phosphotungstic acid and 20–30 wt % silicotungstic acid exhibited conductivities similar to the unfilled PA-doped *m*-PBI membrane at temperatures up to 110 °C.

When tested in a hydrogen atmosphere or humidified atmosphere, it was found that the conductivity of unfilled PA-doped *m*-PBI membranes increased dramatically with temperature. A humidity control experiment demonstrated that the conductivity of PA-doped *m*-PBI membranes was dependent on the relative humidity present at a specific temperature, partic-

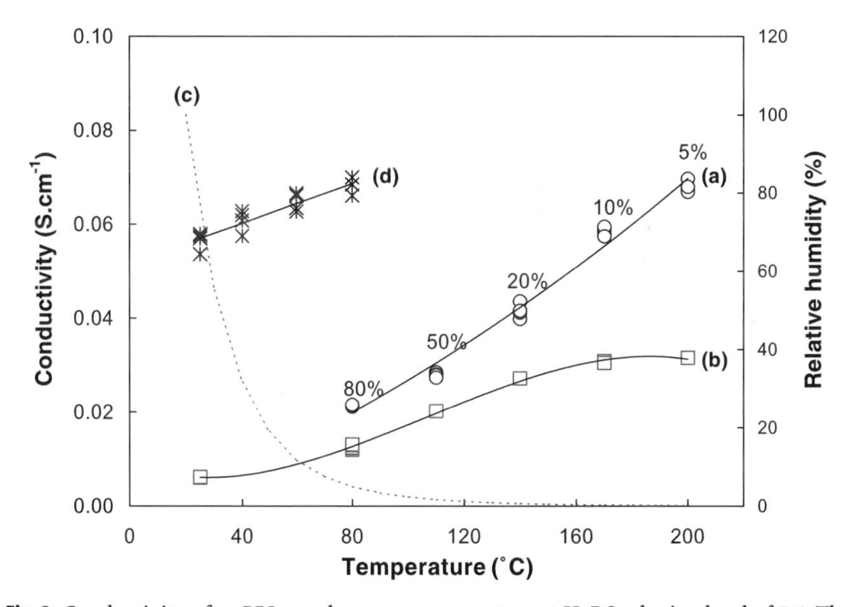

Fig. 2 Conductivity of *m*-PBI membrane vs. temperature at H_3PO_4 doping level of 5.6. The relative humidity (RH) at each temperature is indicated in the figure with (*a*) humidity control; (*b*) under hydrogen atmosphere, saturated with water vapor at room temperature; (*c*) RH of hydrogen atmosphere saturated with water vapor at room temperature vs. temperature; (*d*) conductivity of Nafion 117 at 80% RH and 25–80 °C. Reprinted from [34], with kind permission from Elsevier

ularly when dealing with very low relative humidities (RH). However, the complex's RH dependence was not as drastic as Nafion 117's conductivity dependence on RH. For example, when the RH was increased from 0.15% to 5% at 200 °C, the PBI conductivity increased from 0.032 S cm^{-1} to 0.068 S cm^{-1}. The conductivity dependency of acid-doped PBI and Nafion 117 membranes on temperature is shown in Fig. 2. Figure 3 shows the conductivity dependence of PA-doped *m*-PBI membranes at different temperatures and relative humidities. Clearly, the relative humidity dependence is much greater for Nafion than for PBI. As temperature increased, conductivity for all levels of acid doping and relative humidity also increased. The highest performance (0.079 S cm^{-1}) was measured on the membrane with a doping level of 5.7 moles PA/RPU and relative humidity of 5% at 200 °C.

Because very high acid doping can lead to deterioration of the mechanical properties of PBI films, inorganic fillers were subsequently introduced to increase film strength, as well as water uptake, thermal stability, and conductivity. Zirconium phosphate (ZrP) was the first inorganic filler tested with acid-doped *m*-PBI. The conductivity of the *m*-PBI/PA complexes increased with filler loading level, and it dramatically increased with temperature, as shown in Fig. 4. This is consistent with other literature reports of acid-doped polymer/ZrP blends.

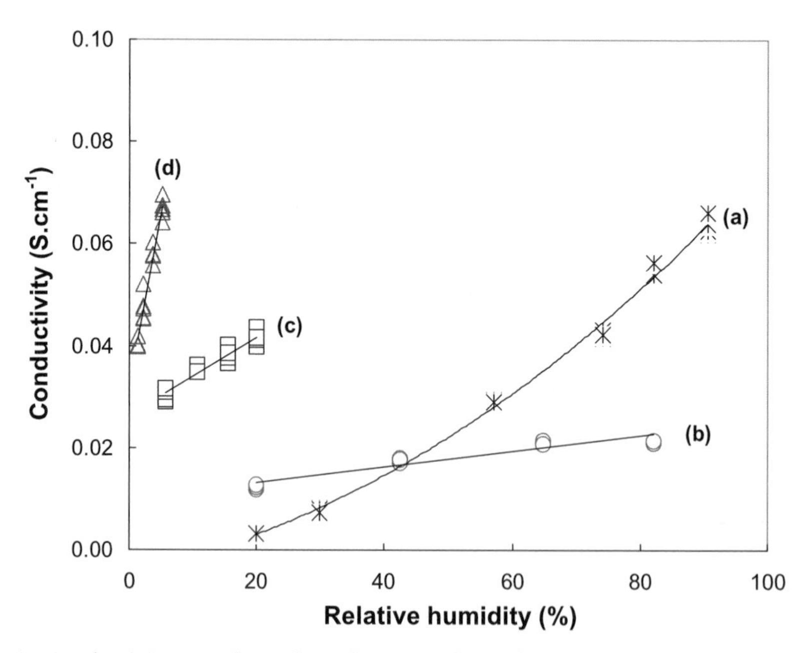

Fig. 3 Conductivity vs. relative humidity (RH) for Nafion 117 and acid doped *m*-PBI membranes at a H_3PO_4 doping level of 5.6. (*a*) Nafion 117, 50 °C; (*b*) *m*-PBI, 80 °C; (*c*) *m*-PBI, 140 °C; (*d*) *m*-PBI, 200 °C. Reprinted from [34], with kind permission from Elsevier

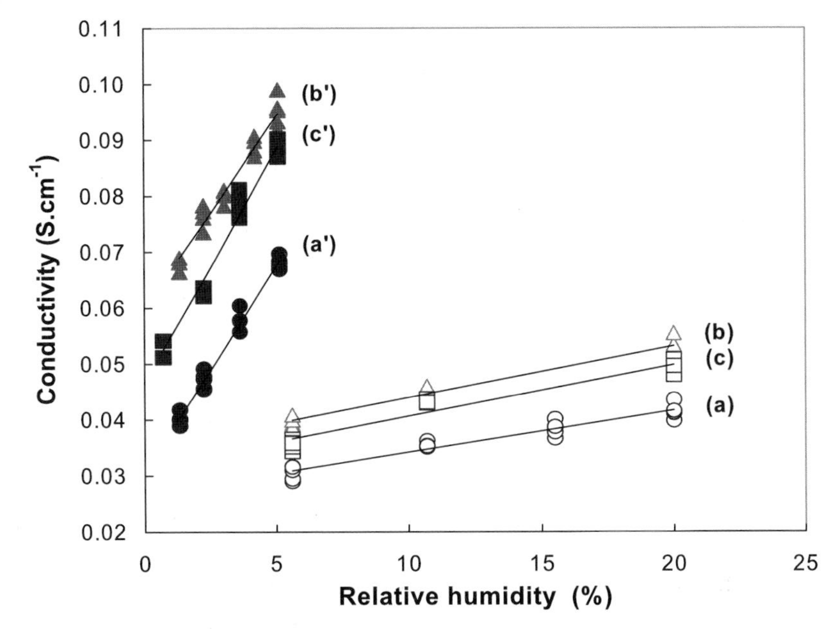

Fig. 4 Conductivity of ZrP/*m*-PBI composite membranes vs. relative humidity (RH) at H_3PO_4 doping level of 5.6. The temperatures were 140 °C for: (*a*–*c*) and 200 °C for: (*a*′–*c*′). (*a*) and (*a*′) *m*-PBI; (*b*) and (*b*′) 15 wt % ZrP in *m*-PBI; (*c*) and (*c*′) 20 wt % ZrP in *m*-PBI. Reprinted from [34], with kind permission from Elsevier

The effect of phosphotungstic (PWA) and silicotungstic acid (SiWA) on conductivity was also investigated. These additives increased the proton conductivity (Fig. 5), but they were extremely sensitive to relative humidity and temperature (Fig. 6). Because of this sensitivity, it may be difficult to use these membranes in practical fuel cell applications. Although the conductivities of filled membranes were lower than the unfilled *m*-PBI/PA complex, they may still be useful as conductivity enhancers for polymer membranes.

Other groups have investigated additional inorganic additives, such as zirconium tricarboxybutylphosphonate (ZrPBTC), for use in a direct methanol fuel cell [35–37]. They also applied a post-sulfonation thermal treatment to the *m*-PBI/ZrPBTC membrane to increase the conductivity. ZrPBTC was introduced into the polymer by dispersing ZrPBTC powder in a DMAc solution of *m*-PBI [37]. The solvent was evaporated, leaving behind a 50 wt % ZrPBTC/*m*-PBI composite membrane. The membrane was then soaked in hydrochloric acid to introduce protons. Further immersion in either phosphoric acid or sulfuric acid produced a doped membrane. The sulfuric acid (SA)-doped membrane was then thermally treated at 480 °C for 60 s.

Under fully humidified conditions, the conductivity of these membranes was improved over the native *m*-PBI/PA complex and varied between 10^{-3} and nearly 10^{-2} S cm^{-1}, with conductivities increasing with temperature. The

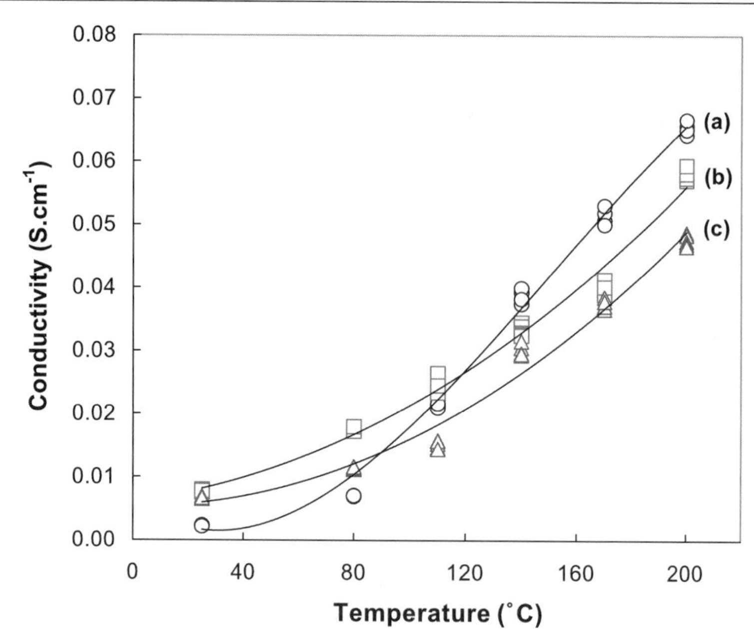

Fig. 5 Conductivity of *m*-PBI and phosphotungstic acid (PWA) *m*-PBI and silicotungtic acid (SWA)/*m*-PBI composite membranes. The relative humidity was 95% for 25 °C, 20% for 80 °C and 110 °C, 10% for 140 °C, and 5% for 170 °C and 200 °C. (*a*) *m*-PBI, PA doping level was 4.4; (*b*) 30 wt % PWA in *m*-PBI, PA doping 4.4; (*c*) 30 wt % SWA in *m*-PBI, PA doping 5.1. Reprinted from [34], with kind permission from Elsevier

m-PBI/ZrPBTC membrane showed a conductivity of 3.82×10^{-3} S cm^{-1} at 200 °C, while the *m*-PBI/ZrPBTC/PA and *m*-PBI/ZrPCTC/SA showed conductivities of 5.24×10^{-3} and 8.21×10^{-3} S cm^{-1}, respectively, at the same temperature. It was believed that the high conductivity of the thermally treated SA complex was due to a conductive network of sulfonic acid groups strongly associated with the imidazole groups.

Zaidi [38] investigated a blend of sulfonated poly(etheretherketone) (*s*-PEEK), *m*-PBI, and boron phosphate (BP). The solid boron phosphate was blended into the composite membrane at 10–40 wt %. The conductivities of the composite films increased with increasing BP content to a maximum value of 6×10^{-3} S cm^{-1}. Since these membranes were not acid doped, the conductivity was significantly lower than doped membranes, but still significantly higher than native PBI. Even though the water uptake decreased with incorporation of BP compared to the *s*-PEEK/PBI blend, the conductivity was still higher. The author concluded that the BP may have increased the acidity of the sulfonic acid groups of the PEEK component in the blend. These blended membranes still exhibited good thermal stability in the temperature range desirable for PEM fuel cell use.

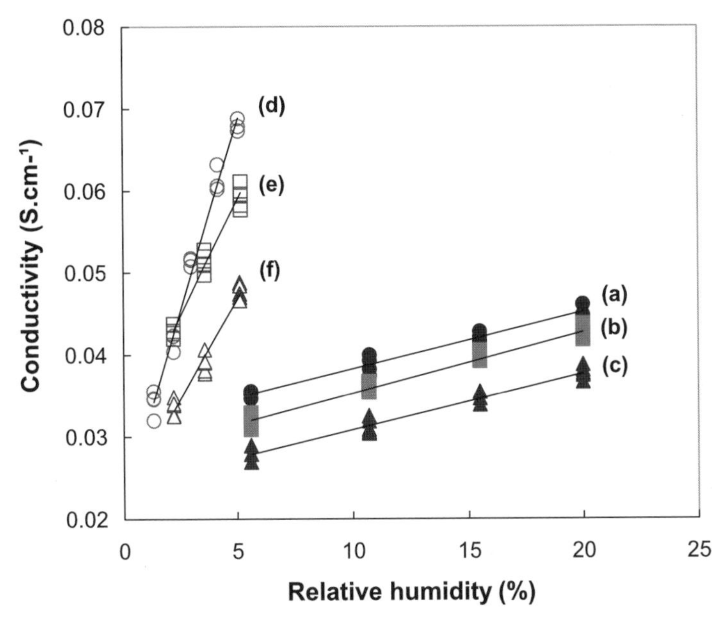

Fig. 6 Conductivity vs. relative humidity (RH) for *m*-PBI and phosphotungstic acid (PWA)/*m*-PBI and silicotungstic acid (SWA)/*m*-PBI composite membranes. (*a*) *m*-PBI doping level 4.4, 140 °C; (*b*) 20 wt % PWA in *m*-PBI, doping level 4.4, 140 °C; (*c*) 30 wt % PWA in *m*-PBI, doping level 4.4, 140 °C; (*d*) *m*-PBI, doping level 5.1, 200 °C; (*e*) 20 wt % SWA in *m*-PBI, doping level 5.1, 200 °C; (*f*) 30 wt % SWA in *m*-PBI, doping level 5.1, 200 °C. Reprinted from [34], with kind permission from Elsevier

On the whole, commercially produced *m*-PBI has shown great promise for development as a high-temperature membrane. The polymer and film are commercially available, relatively inexpensive, have outstanding chemical and thermal stability, mechanical robustness, high CO tolerance, acceptable conductivity, and can be doped with a variety of inorganic acids. The major focus of current research is to increase conductivity, generally through increasing the acid doping levels. The results so far are very encouraging and indicate that *m*-PBI is an excellent candidate for high-temperature fuel cell use. However, additional challenges still remain, such as phosphoric acid retention, membrane durability at higher doping levels, and improvement of the fuel cell performance under practical operating conditions.

2.3 Sulfonated Polybenzimidazole and Its Derivatives

Nafion, a perfluorinated sulfonic acid membrane (Fig. 7) is often considered the state of the art membrane for fuel cells operating at temperatures below about 80 °C. Because the acid functionality in the presence of water is ne-

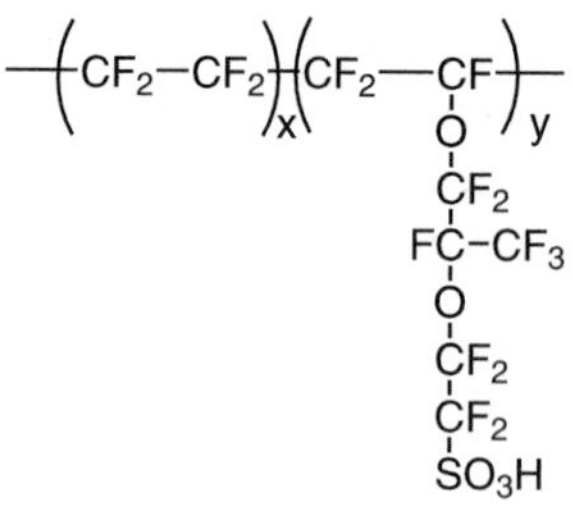

Fig. 7 Chemical structure of the perfluorosulfonic acid polymer used in Nafion membranes

cessary for the efficient transport of protons, most explorations of Nafion alternatives have included this group in the chemical structure. This adherence to water-based proton conduction is commonly reported as degree of sulfonation (SD), the equivalent weight (EW), or lambda (λ), the number of waters per sulfonic acid group. The sulfonated alternatives to Nafion are often blended with other polymers [39–42] or contain inorganic fillers [33, 43] in an effort to increase conductivity. Recent reviews by Li et al. [44] and Hickner et al. [45] provided a thorough overview of problems remaining with PEM fuel cells and some of the alternative membranes and structure variations that have been developed. Similarly, Jannasch reviewed new ionomer and hybrid membranes [46]. A significant amount of work has been done on phosphonated poly(arylene ether)s [47], sulfonated polysulfones [48], and sulfonated polyimides [49]. Recently, *m*-PBI has been used as a blend component with some of these sulfonated alternatives [40, 42, 50, 51], such as *m*-PBI/sulfonated poly(etheretherketone) or *m*-PBI/sulfonated poly(ethersulfone).

Acid-doped sulfonated PBIs (sPBI) have been synthesized by multiple research groups [12, 15, 40–42, 50–54]. Typical approaches to sulfonation include direct sulfonation of the PBI backbone, chemical grafting of functionalized monomers onto the chain, or copolycondensation of sulfonated monomers. The last approach is highly favored, because side reactions can be avoided and degree of sulfonation easily controlled. A recent review by Rikukawa and Sanui [54] thoroughly describes the preparation of sulfonated hydrocarbon polymers and the properties of the sulfonated membranes.

Qing et al. [52] synthesized sPBIs with varying sulfonation degrees by controlling the stoichiometric ratios of sulfonated to nonsulfonated monomers. These polymers were solution cast from polar aprotic solvents into tough, flexible, transparent films with good thermal stability and mechanical properties. Furthermore these sulfonated polymers retained good mechanical properties at high temperatures (>300 °C), making them potential candidates for high-temperature fuel cells. The extreme hygroscopicity of this type of membrane also makes *s*-PBIs attractive Nafion alternatives.

Glipa et al. [55] successfully grafted sulfonated aryl groups onto *m*-PBI, leading to a proton conducting polymer with various degrees of sulfonation (see Fig. 8). Room temperature conductivity increased from $\sim 10^{-4}$ S cm^{-1} to $>10^{-2}$ S cm^{-1} for highly sulfonated samples that were thermally stable in their dry state up to 350 °C. Mechanical properties of the hydrated films were preserved by balancing the degree of sulfonation with the swelling behavior of the membranes. Overall, the degree of post-sulfonation was easily controlled by reaction time.

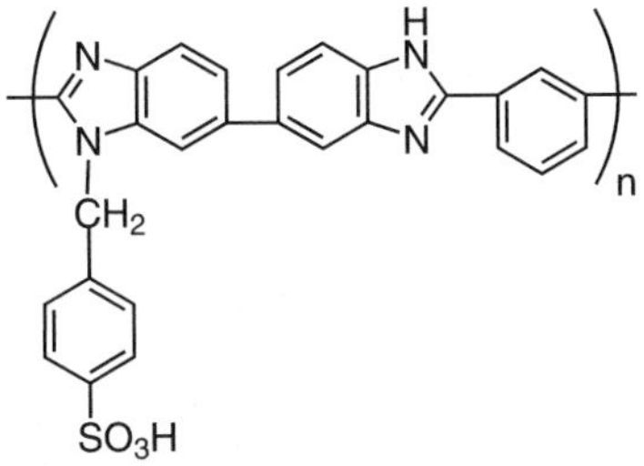

Fig. 8 *m*-PBI grafted with a methyl benzenesulfonic acid side chain

Although these initial results appear promising, additional work is needed to understand the relationships between degree of sulfonation, water uptake, proton conductivity, swelling behavior, and fuel cell performance. Earlier work by Gieselman and Reynolds [56] showed that these polymers were fully soluble in water at high degrees of sulfonation.

Ariza et al. [57] applied a post-sulfonation thermal treatment to an already synthesized PBI backbone to attach sulfonic acid groups to *m*-PBI. The attachment was achieved by soaking "a pre-formed PBI film in dilute sulfuric acid solution, then heating the acid-complexed membrane in an inert atmosphere at a high temperature during a determined period of time." While this method worked, there was not a clear definition of the effects of phosphoric acid concentration, heating time, and temperature on the membrane properties important for fuel cell applications. Characterization performed by FTIR showed evidence of sulfonation on the *m*-PBI backbone via an ionic interaction between the inorganic acid and polymer. Additional evidence of covalent attachment and possible cross-linking was obtained from combined FTIR analysis, solubility studies, and elemental analysis after thorough water washing treatments.

The membrane conductivity remained low (2.4×10^{-5} S cm^{-1}) due to the low degree of sulfonation, but it was improved by one order of magnitude over the native *m*-PBI. The membrane was stable up to at least 300 °C, but degraded at temperatures lower than that of native *m*-PBI, as shown by TGA. Future development of these membranes appears dependent on the ability to achieve higher proton conductivities.

Kawahara et al. [58] produced a propylsulfonated-PBI (PBI-PrS) via a ring-opening reaction of 1,3-propanesulfone with the N – H groups of the imidazole functionality, as previously described by Gieselman and Reynolds [59], and investigated the conductivity of the resulting membranes. The conductivity increased with both temperature and SD, and a maximum of $\sim 10^{-3}$ S cm^{-1} at 140 °C was reported for a SD of 73.1 mole percent. Conductivity also increased with increasing water uptake per sulfonic acid group, with a maximum of $\sim 10^{-3}$ S cm^{-1} at 140 °C and 11.1 H_2O/SO_3H. A comparison of Nafion 115 (7.6 H_2O/SO_3H) to PBI-PrS (SD of 73.1, 11.1 H_2O/SO_3H) showed similar conductivities up to about 90 °C. At higher temperatures, Nafion's conductivity dropped sharply and continued to decrease up to 140 °C, while the conductivity of PBI-PrS gradually increased over this temperature range.

Similarly, Bae et al. [60] compared the propylsulfonated and butylsulfonated PBIs, shown in Fig. 9, in conductivity and fuel cell tests. In conductivity tests, the butylsulfonated-PBI exhibited a higher conductivity ($\sim 3 \times 10^{-3}$ S cm^{-1}, 90 °C, 100% RH) than propylsulfonated-PBI ($\sim 10^{-3}$ S cm^{-1}, 90 °C, 100% RH) and was able to maintain acceptable conductivities up to 160 °C. The propylsulfonated-PBI had relatively low overall conductivity, perhaps because of the rigidity of the shorter alkyl spacer. Fuel cell performance (ambient pressure, H_2/O_2) of the butylsulfonated-PBI was moderate compared to Nafion systems. Under testing conditions of 80 °C and 100% RH, at a current density of 0.20 A cm^{-2}, the voltage was $\sim$0.45 V. The maximum power density of 200 mW cm^{-2} was achieved at 0.7 A cm^{-2} and 0.3 V.

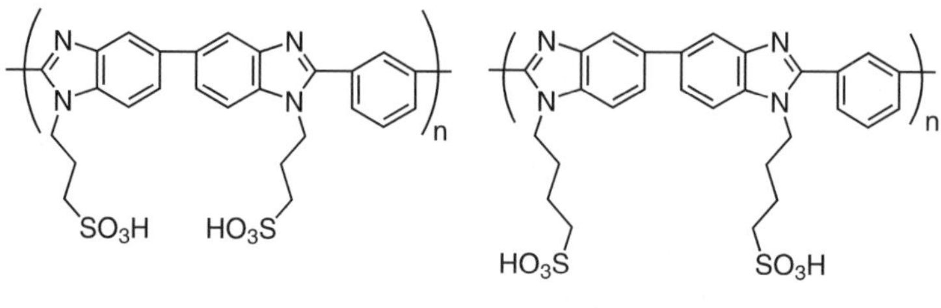

Fig. 9 Chemical structures of propylsulfonated and butylsulfonated *m*-PBI

Using the Eaton reagent, Jouanneau et al. [61] synthesized a novel sulfonated PBI based on a sulfonated tetraamine monomer, bis-3-amino-4-[3-(triethylammoniumsulfonato) phenylamino]phenyl sulfone (BASPAPS). A nonsulfonated amine (bis-3,4-diaminophenyl sulfone, BDAPS) was used as a comonomer in copolymer synthesis to vary the IEC. Before polymerization, model compounds containing only one carboxylic acid were made to help refine the polymerization conditions needed to achieve high molecular weight polymer. Both random and sequenced copolymers were made to control the variation of IEC. Blends of the sulfonated PBI were also made. Polymer structures were confirmed by FTIR, ^{13}C, and ^{1}H NMR. It was found that by varying the composition of nonsulfonated tetraamine, the IEC could be controlled between 0 and 2.57 meq g^{-1} (0–100% BASPAPS). Using the novel monomer, the architecture, MW, and IEC could be controlled by using various types of polymer synthesis. However, because of the low lambda (5–8 water molecules per sulfonic acid), only low conductivities were achieved. While the performance is not yet as high as Nafion-based polymers, these membranes show that there are promising hydrocarbon alternatives that need further optimization.

2.4 Blends of Polybenzimidazole and Sulfonated Polymers

Many different sulfonated polymers have been investigated as single components or as blends with *m*-PBI to improve different aspects of membrane properties (Fig. 10) with widely varying results [40, 42, 44, 50, 62–65]. Some polymer blends swelled to such an extent that mechanical properties were lost, while others were immiscible and did not form usable blends. However, some polymer blends showed improvements in conductivity at temperatures greater than 100 °C. Efforts have also been made to reduce the swelling problems through cross-linking and to increase conductivity via inorganic fillers. However, limited fuel cell performance testing was reported and, therefore, limited conclusions can be made on the significance of the improved conductivities. As mentioned earlier, a recent review by Li et al. [44] discusses and compares the fundamental membrane properties of these exploratory polymers and blends.

A blend of *m*-PBI and sulfonated polysulfone (*m*-PBI/sPS) was investigated for direct methanol fuel cell use [62]. The ion exchange capacity of one membrane of particular interest was 1.01 meq per gram dry polymer. The methanol permeation was approximately one order of magnitude less than that of Nafion 117 for methanol weight concentrations of 20–75%, while the swelling was similar.

Silva et al. [63] developed a mixture of sulfonated-poly(etheretherketone) (*s*-PEEK) with a 42 or 68% degree of sulfonation, *m*-PBI, and zirconium phosphate (ZrP) for use in a direct methanol fuel cell. This blend was originally designed to increase the chemical and thermal stability of the s-PEEK. The proton conductivity and DMFC performance were also tested. Although the membrane swelling and methanol permeability decreased, the membrane conductivity also decreased. In general terms, however, the addition of the *m*-PBI and ZrP imparted chemical stability and increased DMFC efficiency at temperatures up to 130 °C.

Sulfonated PBIs, other sulfonated polymers, and their blends show great potential for use as membranes in high-temperature fuel cells. The synthesis, conductivity, mechanical properties, and performance still require further development, but results so far are promising. Further investigation remains to determine whether these problems can be overcome and useful chemistries developed to meet the needs of high-temperature membranes with performance characteristics comparable to lower-temperature membranes.

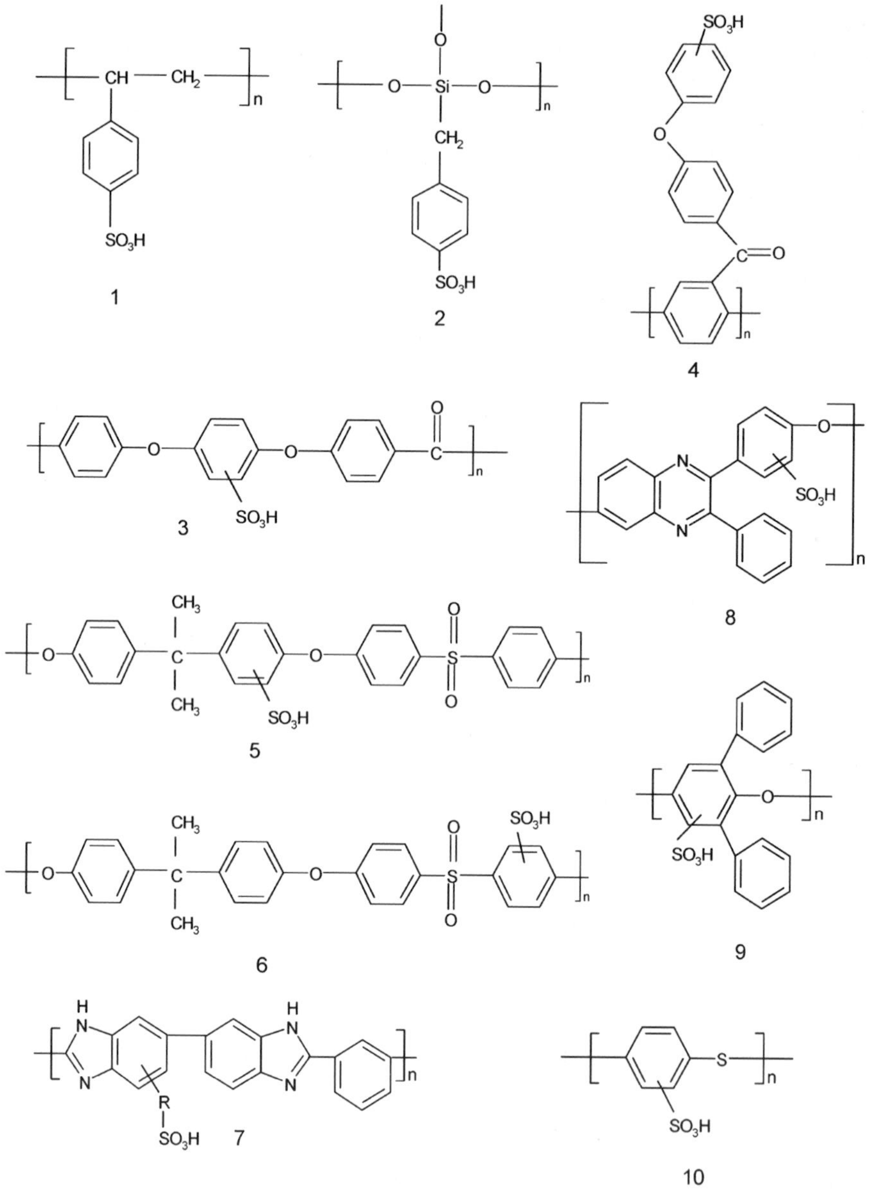

Fig. 10 Some sulfonated hydrocarbons used alone or as blends with *m*-PBI. *1*: sulfonated polystyrene; *2*: poly(benzylsulfonic acid siloxane); *3*: sulfonated poly(etheretherketone); *4*: sulfonated poly(4-phenoxybenzoyl-1,4-phenylene); *5*: sulfonated polysulfone; *6*: sulfonated polysulfone; *7*: sulfonated *m*-PBI; *8*: sulfonated poly(phenylquinoxalines); *9*: sulfonated poly(2,6-diphenyl-4-phenylene oxide); *10*: sulfonated polyphenylenesulfide. Reprinted with kind permission from [44]

2.5
Acid-Based Blends of Polybenzimidazole and Other Polymers

To increase the thermal stability, acid uptake, water uptake, and conductivity of *m*-PBI, many blended membranes have been investigated. Pu investigated *m*-PBI/poly(4-vinylpyridine) (P4VP) blends and their proton conductivity after acid doping [66]. The polymers were chosen because, due to strong intermolecular hydrogen bonding, they form miscible blends and can be used as proton conductors. The thermal stability of the blends was lower than that of *m*-PBI, but significantly higher than that of P4VP. The thermal stability decreased with increasing P4VP content, with decomposition beginning around 350 °C. The conductivity for a PBI/P4VP blend (70/30) reached a maximum of approximately 10^{-3} S cm^{-1} at 200 °C with a doping level of 2.1 moles PA/PRU (RH not specified). The maximum conductivity was reported for a 50/50 blend (doping level of 3.2 moles PA/PRU) and was on the order of $\sim 10^{-2}$ S cm^{-1}.

Daletou et al. blended *m*-PBI with aromatic polyethers containing pyridine units [67] to improve acid uptake and conductivity. The oxidative stability of each membrane was tested by immersion in hydrogen peroxide and iron (II) chloride (Fenton Test), followed by TGA and DMA. The copolymers made were based on bisphenol A combined with 2,5-bis(4-hydroxyphenyl)pyridine and, in some cases, blended with *m*-PBI. DMA was also used to assess the miscibility of the polymer blends. The doping level, specified as weight percent of PA per gram of copolymer or per gram copolymer/blend, reached a maximum of 450 wt % with a PBI/(50/50 phenylpyridine-co-sPS) blend. The conductivity of a PBI/(50/50 phenylpyridine-co-sPS) blend with doping level of 220 wt % was ~0.07 S cm^{-1} at 150 °C and 30% RH. It was found that there was still relatively high conductivity at elevated temperatures. Furthermore, the thermal stability of the blends remained high even after oxidative degradation, as did the mechanical properties, making these miscible blends viable membrane candidates for fuel cells.

Wycisk et al. studied sulfonated polyphosphazene/*m*-PBI blended membranes for direct methanol fuel cell use [68], where *m*-PBI provided stabilization via strong ionic interactions. The water swelling and conductivity were measured at room temperature, while the methanol permeability and DMFC performance were studied at 60 °C. Sulfonated polyphosphazene was prepared and blended with *m*-PBI (3, 5, 8, 10, and 12 wt %) to control swelling and mechanical properties. The ion exchange capacity, swelling, and conductivity decreased as the amount of PBI increased, presumably due to the ionic interactions of the sulfonic acid groups with the basic nitrogens of *m*-PBI. The methanol permeability also decreased with increased percentage of PBI (3–20 times lower than Nafion 117 at 60 °C). Fuel cell performance at 60 °C of blends with 3 or 5 wt % *m*-PBI was comparable to that of Nafion 117. A maximum conductivity of 0.06 S cm^{-1} was measured for the 3 wt % *m*-PBI blend, which is a promising result. These membranes showed good thermal stability and mechanical properties as well as excellent fuel cell performance, warranting further study for use in DMFCs.

Jeong et al. [69] synthesized acid-doped sulfonated poly(aryl ether benzimidazole) (s-PAEBI) copolymers for use in high-temperature fuel cells. The polymer was made in a direct polymerization (structure confirmed with ^{1}H NMR) and doped with phosphoric acid to levels of 0.7–5.7. The degree of sulfonation was varied from 0–60%. The copolymer's physicochemical properties were studied using AFM, TGA, and conductivity measurements. TGA runs showed good stability of the nonsulfonated, sulfonated, and acid-doped sulfonated membranes up to $\sim$450 °C, with a slow decline above this temperature. The conductivity depended on the doping level of the polymer. At 130 °C with no humidification, a polymer with a doping level of 5.7 had a conductivity of 7.3×10^{-2} S cm^{-1}.

A number of other blends have been studied for use in fuel cells, but are similar in approach and properties to those presented here. Clearly, derivatives or blends of PBI may provide property and performance improvements over the simple homopolymer. Ongoing investigations will explore further functionalization and extend our understanding of the relationships between polymer properties and fuel cell performance.

2.6 AB-PBI: Poly(2,5-benzimidazole)

Poly(2,5-benzimidazole), or AB-PBI, (Fig. 1) is another polybenzimidazole derivative that has been investigated as an alternative fuel cell membrane material. AB-PBI is synthesized from 3,4-diaminobenzoic acid (DABA), a relatively inexpensive and widely available monomer. Discussions on AB-PBI and comparisons to other PBIs must be done with some caution. The repeat unit of AB-PBI contains a single benzimidazole moiety, while the repeat unit of PBIs made from TAB contain two benzimidazoles. If one assume the acid-base interactions are important for membrane properties, then this difference is important. For clarification in this chapter, the polymer acid ratio for AB-PBI will be expressed as moles phosphoric acid per moles benzimidazole moiety (moles PA/BI). Doubling this value will provide an estimate for the TAB-based PBI "equivalent" loading levels.

AB-PBI was doped with phosphoric acid (up to 5 moles PA/BI) and remained thermally stable at temperatures well above those needed for PEM fuel cells [12]. Conductivities as high as 10^{-4} S cm^{-1} were reported in this study. Interestingly, although AB-PBI absorbed more acid than the sulfonated and *m*-PBI tested (above 3 moles PA/BI), conductivity improvements were not observed.

Asensio et al. [13] incorporated different polyanions, such as phosphomolybic acid (PMA), into the AB-PBI system, as well as some sulfonated

PBI derivatives. Sulfonation was performed by immersing pre-cast AB-PBI membranes in sulfuric acid followed by heat treatment. The amount of PA absorption in the unfilled membranes was in the order of sAB-PBI (4.6 moles PA/BI), AB-PBI (2.7 moles PA/BI), and *m*-PBI (6.7 moles PA/PRU or 3.35 moles PA/BI). Although the molar PA doping level of *m*-PBI was higher than the AB-PBIs, the weight percent of PA was the same for all the membranes. The PMA-doped membranes also showed increased acid uptake over those without PMA. Thermogravimetric analysis of all phosphoric acid impregnated membranes showed stability up to at least 200 °C. The conductivity was tested up to 185 °C, with the conductivity order being sAB-PBI/PA > AB-PBI/PMA/PA > *m*-PBI/PA > AB-PBI/PA. The maximum conductivity for sAB-PBI/PA was 3.5×10^{-2} S cm^{-1} at 185 °C in dry conditions, while AB-PBI/PMA/PA reached 3.0×10^{-2} S cm^{-1}. The conductivities of *m*-PBI/PA and AB-PBI/PA were very similar, indicating that AB-PBI is a viable alternative membrane to *m*-PBI, as studied under these conditions.

Additional work by the same group focused on further developing and characterizing AB-PBI [70]. The membranes were prepared in a similar manner to the previous work (casting followed by PA bath immersion). Characterization included thermogravimetric analysis, conductivity, FTIR spectroscopy, X-ray diffraction, and scanning electron microscopy (SEM). It was found that the membranes had an average inherent viscosity of 2.3–2.4 dL g^{-1}, which was high enough for membrane casting. Immersion in a PA bath led to doping levels of 5 moles PA/BI. Interestingly, if the AB-PBI was immersed in a concentrated acid bath (85%), the membrane fully dissolved. X-ray diffraction showed that the polymer was amorphous in both the doped and undoped states, but developed more crystallinity on heating. TGA showed the membrane was stable up to 150 °C; above this temperature, absorbed water was lost. Between 150–210 °C, additional loss of water was detected from phosphoric acid dehydration.

Conductivity measurements (membrane with 3.0 moles PA/BI) were made in the range 50–200 °C with 5–30% RH. Conductivities as high as 6.2×10^{2} S cm^{-1} were measured at 150 °C and 30% RH. As reported earlier, conductivity increased with temperature, and exhibited lower values (3.9×10^{2} S cm^{-1} at 180 °C, 5% RH) at lower RH between 180 and 200 °C. Preliminary fuel cell tests with hydrogen and oxygen were performed from 100 to 150 °C. When the gases were switched from dry to humidified at 150 °C, a 50% increase in power density was measured. Maximum values for power densities (175 mW cm^{-2}) were obtained at 130 °C and are comparable to other reports.

Asensio et al. [71] also developed a method for producing acid-doped membranes by direct casting from an AB-PBI/phosphoric acid (PA)/methanesulfonic acid (MSA) solution. The methanesulfonic acid was evaporated to produce a very homogenous, nearly transparent film with controlled composition and up to 3 moles PA/BI. This method of preparation was much more convenient than the typical multi-step, organic solvent based process.

X-ray diffraction measurements show a much higher crystallinity than the conventionally imbibed membranes, which was increased further by heating. Unfortunately, the conductivities of the directly cast membranes were lower than the conventionally imbibed membranes, attaining a maximum of 1.5×10^{-2} S cm^{-1} and 2.5×10^{-2} S cm^{-1}, respectively (dry conditions, 180 °C), even though the conventionally imbibed membrane had a slightly lower acid doping level (2.7 vs. 3.0 moles PA/BI). The lower conductivity was believed to result from dehydration of phosphoric acid at the temperatures needed to evaporate MSA (>150 °C).

Kim et al. also developed a direct casting method from a mixture of AB-PBI, methanesulfonic acid, and P_2O_5 [72]. This solution casting method produced very fine polymer fibers that were easy to work up or could also be cast directly into a translucent membrane and assembled into an MEA. The acid-doped film was obtained by immersion in a PA bath. Conductivities ranged from 0.02 to 0.06 S cm^{-1} at 110 °C with no humidification, while inherent viscosities ranged from 1.5 to 1.8 dL g^{-1}, and doping levels ranged from 1.6–3.7 moles PA/BI. These undoped membranes also showed good mechanical properties, with tensile strengths between 88 and 121 MPa and elongation at break between 31 and 65 percent, with the higher values from polymer produced at longer polymerization times and, therefore, higher molecular weights.

Cho et al. performed a more in-depth study of the structure of the AB-PBI/PA complex [73]. The AB-PBI films were cast from a mixture of sodium hydroxide and ethanol, and then followed by immersion in a PA bath for doping. The doped films were then stretched about 450% and washed with boiling water to remove excess acid. X-ray diffraction was performed to investigate the complex structure. Crystalline ordering was apparent in the AB-PBI films before doping, but not afterwards. Molecular modeling was also performed to determine the conformation of the polymer chain based on the torsion of the bond between the repeat units. In general, it was found that the cast films had a uniplanar orientation, which was destroyed upon addition of phosphoric acid. Subsequent annealing helped plasticize the polymer and aided in the retention of the thermal and mechanical stability.

Gomez-Romero et al. [74] revisited the phosphomolybdic acid (PMA)-doped AB-PBI in a recent paper. They cast PMA impregnated films directly from a methanesulfonic acid (MSA) solution, and then doped the films with phosphoric acid. The AB-PBI had an IV of 2.3–2.4 dL g^{-1} and the films contained up to 60 wt % PMA. It was found that a 60 wt % PMA film could be doped in a bath of up to 68% PA. The AB-PBI films dissolved when placed in higher PA bath concentrations. FTIR spectroscopy showed that the PMA and phosphoric acid were interacting with the polymer. X-ray diffraction of the polymer-acid complex indicated a quasi-amorphous structure, which is consistent with previous reports. TGA showed the membranes to be stable up to 200 °C after phosphoric acid doping, which is well within the temperature range needed to operate a PEM fuel cell. The conductivity of the

AB-PBI/PMA/PA membrane was slightly higher than the only PA-doped AB-PBI (maximum of 0.03 and 0.025 S cm^{-1} at 185 °C, respectively). The work of the Gomez-Romero et al. group was recently reviewed [75].

Asensio et al. also revisited the concept of a sulfonated AB-PBI (sAB-PBI) [68]. The membranes were prepared as discussed before (i.e., sulfuric acid and thermal treatment, casting from a MSA solution, then PA doping). The degree of sulfonation, thermal stability, and nonhumidified proton conductivity were studied. At a degree of sulfonation of 41%, the imbibed membrane held 4.6 moles PA/BI. These undoped membranes were stable up to 400 °C and initial decomposition related to the sulfonic acid groups began at 490 °C. The maximum conductivity for this membrane was 3.5×10^{-2} S cm^{-1} at 185 °C with no humidification. As previously observed, conductivity increased with temperature and acid doping level, as well as sulfonation level (e.g., $\sim 3.0 \times 10^{-2}$ S cm^{-1} for sulfonation level 28%). These membranes are interesting because of their increased conductivity over native AB-PBI and PBI, good thermal stability, and excellent mechanical properties.

Overall, AB-PBI and sAB-PBI are possible alternative to Nafion-type membranes, because of the nearly zero dependence of conductivity on water. These membranes also have the requisite mechanical and thermal stability, while achieving moderate levels of proton conductivity that can be further modified with inorganic additives, such as heteropolyacids.

2.7 Other Polybenzimidazole Explorations

A number of other interesting uses have been found for polybenzimidazole membranes, including a propane fueled fuel cell, an alkaline based fuel cell, a trimethoxymethane based fuel cell, and a quasi-direct methanol fuel cell. Wang et al. investigated trimethoxymethane (TMM) as an alternative fuel for a *m*-PBI direct oxidation fuel cell [77]. The oxidation of TMM was analyzed by an online mass spectrometer and online FTIR spectroscopy. The PBI membranes used in the TMM study were doped with 5 moles PA/PRU. The TMM was hydrolyzed to form a mixture of methylformate, methanol, and formic acid. At temperatures at or above 120 °C, the TMM hydrolyzed in the presence of water without an acid catalyst. The anode performance of the different fuels increased in the order of methanol < TMM < formic acid/methanol < methylformate. The improved performance of TMM over just methanol was most likely due to the electrochemical activity of formic acid.

Xing and Savadogo investigated a hydrogen/oxygen fuel cell based on an alkaline-doped PBI instead of the traditional acid-doped film [78]. The researchers doped *m*-PBI films with potassium hydroxide, lithium hydroxide, and sodium hydroxide. The concentration of base in the film depended on the immersion time and temperature. It was found that *m*-PBI had a remarkable ability to hold potassium hydroxide and attained room temperature conductivities of up to 9×10^{-2} S cm^{-1}, which approaches maximum values reported for Nafion 117. The lowest conductivity was found with lithium hydroxide. The conductivity reported for *m*-PBI/LiOH was 1500 times lower than *m*-PBI/NaOH, and 3000 times lower than *m*-PBI/KOH. Optimum conductivities were reported at base concentrations of 4 M LiOH (2.5×10^{-5} S cm^{-1}), 15 M NaOH (3×10^{-2} S cm^{-1}), and 6 M KOH (4×10^{-2} S cm^{-1}). Conductivity increased with immersion time and the alkaline-doped *m*-PBI membrane showed comparable performance to both acid-doped *m*-PBI and Nafion membranes. Further investigation is needed to optimize immersion techniques, measure fuel cell performance and durability, determine the mechanism of conduction, and investigate the effects of water uptake, but this early research appears quite promising.

Cheng et al. investigated propane fuel cells using acid-doped *m*-PBI membranes [79]. Under anhydrous conditions, the overall reaction for the fuel cell was $2C_3H_{8(g)} + O_{2(g)} \rightarrow 2C_3H_{6(g)} + 2H_2O_{(g)}$. The cells were tested up to 250 °C with a maximum open circuit voltage of 0.9 V. For the propane-oxygen anhydrous fuel cells, performance was very poor and unsustainable (the current density decreased from 2 mA cm^{-2} to about 0.3 mA cm^{-2}, over 600 s at 200 °C). However, when humidity was introduced into the system, the cell was able to generate higher and somewhat sustainable current densities (the current density decreased from 0.6 mA cm^{-2} to 0.4 mA cm^{-2} over 1000 s at 250 °C) with only CO and CO_2 as the carbon byproducts, which were formed from an oxygen-containing partial oxidation C_3 intermediate and facilitated by water present in the humidified gas stream.

Li et al. investigated a quasi-direct or reformed methanol fuel cell (RMFC) based on a polybenzimidazole/polysulfone (*m*-PBI/PS) blend [80]. This cell operated up to 200 °C, significantly above a Nafion based system, and tolerated up to 3 volume percent CO poisoning due to the higher temperature. Since the methanol reformer operated at a similar temperature, system integration is possible. The membrane was doped at 5 moles PA/PRU of the blended polymer. Figure 11 shows the performance of the fuel cell at 200 °C (*m*-PBI/PS blend was 75/25, PS had SD 36%). At 200 °C with pure hydrogen and oxygen feed gases, the current density was 0.67 A cm^{-2} at 0.6 V. Performance decreased only slightly with 1.0 or 3.0 volume percent CO poisoning. These results demonstrate the interest in using acid-doped PBI and PBI blends for high-temperature fuel cells.

Similarly, Pan et al. integrated a high-temperature fuel cell with a methanol reformer [81]. Methanol was reformed via steam reforming to produce a hydrogen rich fuel stream that would power a high-temperature (185–260 °C) *m*-PBI based PEMFC. The MEA was made from acid-doped *m*-PBI and acid impregnated Pt–C electrodes. The performance at 205 °C at atmospheric pressure with the reformed methanol was adequate with a current density of 0.2 A cm^{-2} and corresponding voltage of ~0.7 V.

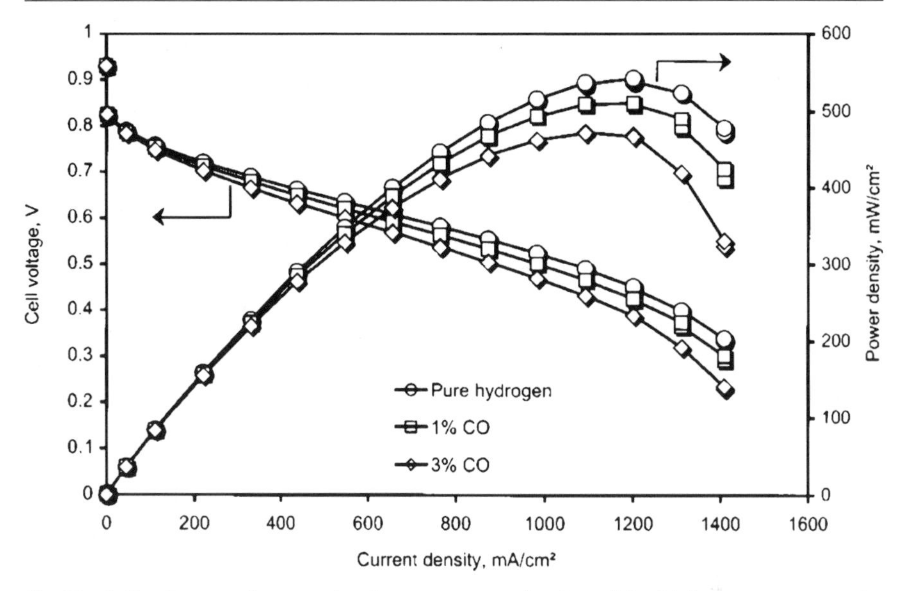

Fig. 11 Cell voltage and power density vs. current density of the high-temperature polymer membrane electrolyte fuel cell under atmospheric pressure at 200 °C. Electrodes were prepared with a platinum loading of 0.45 mg cm^{-2}. The membrane comprised 75% PBI and 25% SPSF (sulfonation degree 36%), doped with 520 mol % H_3PO_4. The fuel gas was pure hydrogen or hydrogen containing 1.0 and 3.0% CO, as indicated in the figure, and the oxidant was oxygen gas. Reprinted with kind permission from [80]

An interesting variation on the *meta*-PBI structure was explored by Banihashemi and Atabaki [82] in hopes of increasing the backbone solubility and, thus, processability. They prepared a new benzofuro[2,3-*b*]benzofuran-2,9-dicarboxyl-bis-phenylamide-4,4′-dicarboxylic acid monomer and a series of poly(amide-benzimidazoles) (PABIs) with several aromatic tetraamines in polyphosphoric acid. These model compounds contained larger aromatic functional groups between imidazole functionalities, as well as an aromatic/oxygen containing moiety in place of the normal meta phenylene ring. All of the new polymers were produced in good yield and exhibited excellent thermal stability up to at least 400 °C. Solubility studies showed that most of the various polymers were insoluble in polar aprotic solvents, such as DMSO, DMAc, NMP, and DMF. However, all of the polymers were swollen or soluble in sulfuric acid. These membranes need further characterization and development before their application in fuel cells can be demonstrated, but this research shows the current interest in developing alternative PBI chemistries for fuel cells. The data on phosphoric acid loading or characterization were not reported.

Another interesting change in PBI morphology was performed by Mecerreyes et al. [83]. They prepared porous films, which were then doped with phosphoric acid. The films were made by leaching out a low-molecular weight compound using a selective solvent to control porosity up to 75%. Initially, *m*-PBI and porogen were dissolved in DMAc. After casting and solvent evaporation, the porogen was removed by soaking in methanol, leaving a PBI film with pores. The porogens investigated were dimethyl phthalate, diethyl phthalate, dibutyl phthalate, diphenyl phthalate, and triphenyl phosphate. These films were then doped by immersion in 11 M phosphoric acid for four days. It was found that the pore size and morphology were dependent on the porogen/PBI ratio. Acid uptake increased with increasing porosity and a room temperature film conductivity of 5×10^{-2} S cm^{-1} was achieved while remaining mechanically stable. SEM micrographs clearly showed a microporous film structure with larger pores, as the weight percent porogen increased. At the lowest weight percent porogen loading (25 wt %), no pores were seen, indicating the pores were less than 100 nm (a transparent film). At approximately 50 wt % porogen loading, the pores were irregular, roughly spherical with diameters of 1–5 μm. At 70 wt % porogen loading, the pores were interconnected, with irregular shapes of 2–10 μm. By 80 wt %, the large (5–15 μm) pores formed irregular and highly interconnected structures. SEM micrograph studies of the different porogens showed a clear relationship between pore size and the aliphatic/aromatic structure of the porogen. A larger aliphatic tail on the porogen molecule produced larger pores at equivalent loading levels. The highest PA uptake measured was 439 wt % (for the 70 wt % dibutyl phthalate film), as compared to 132 wt % (nonporous *m*-PBI). This high weight percent corresponds to 1460 mol % for the 70% porous membrane. In general, a linear increase in conductivity was seen with increasing porogen content. The long-term stability and conductivity are under investigation to determine possible application in fuel cells.

Xu et al. synthesized novel hyperbranched PBIs with interesting properties via $A_2 + B_3$ monomers [84]. It was theorized that the three dimensional branched structure of these polymers might open up many cavities for the sorption of phosphoric acid. Also, by crosslinking these structures, it may be possible to "lock" the phosphoric acid in place and prevent leaching. The polymers were modified using varying amounts of the crosslinker terephthaldehyde (TPA). It was found that these membranes exhibit good mechanical properties and doping levels (5–7 mol PA/PRU) and are comparable to commercially available PBI even with the higher doping level of the hyperbranched PBIs.

Larson et al. compared the performance of PA-doped *m*-PBI with bisfluorinated acid-doped phenylene oxide benzimidazole (PBIO) [85]. PBI was purchased from Celanese and PBIO was purchased from Fumatech. The polymers were dissolved in DMAc and NMP, respectively, after IR, elemental analysis, and NMR confirmed structure. The PBIO polymer solution was combined with solutions of bisfluorinated acids (disulfonate, C1-bis-imide, or C4-bis-imide) or solutions of bisfluorinated acid and silica. The polymers were cast onto glass plates and solvent was removed to form a membrane. *m*-PBI membranes were doped with 85% phosphoric acid to a level of 600 mol percent. Fuel cell testing was done with carbon paper electrodes (0.4 mg cm^{-2} Pt) in a 5 cm^2 cell with hydrogen and air. The properties of the membrane-acid complex were tested by attempting to wash out the acid with water or squeeze it out with mechanical pressing. It was found that the addition of silica decreased the amount of acid exuded. Interestingly, the best fuel cell performances were not shown for the membranes with highest doping, but for those with the best balance between amount of acid exuded, resistance to wash out, and mechanical strength. Better performances were obtained with PBIO than *m*-PBI at both 100 and 110 °C, even though the conductivity of the PBIO was an order of magnitude lower than *m*-PBI at all temperatures tested. There was also a much greater improvement in fuel cell performance between 100 and 110 °C for the PBIO membrane than for the *m*-PBI membrane.

The effect of the transient evolution of carbon monoxide poisoning on PBI fuel cells was modeled by Wang et al. [86] and confirmed with experimental studies. A one dimensional model of hydrogen/carbon monoxide fuel streams for fuel cell performance was developed. It was found from modeling that over time, with all fuel compositions (hydrogen 40–80%, CO 1–3%, balance is a mix of CO_2 and nitrogen), CO – Pt bonding increased, while H_2 – Pt bonding decreased. The hydrogen coverage of Pt also decreased. At higher concentrations of CO, this hydrogen dilution effect becomes significantly more prominent. The modeling simulation shows good agreement with experimental procedures, especially considering that the model does not take into account important parameters, such as the three dimensional nature of the fuel cell, flow channels, or the gas diffusion layer. With further refinement, it is hoped that this model could be applied to the designing of an integrated reformer and fuel cell system with accurate prediction of performance.

A more in-depth look on the effect of electrode PBI ionomer content on fuel cell performance was undertaken by Kim et al. [87]. A cathode electrode was developed using Pt on carbon paper and *m*-PBI with a doping level of 6 moles PA/PRU as the ionomer (5–40 wt %). The 2 cm^2 cell was made using acid-doped AB-PBI, and it was tested at 150 °C without humidification and hydrogen as the fuel, and oxygen or air as the oxidant. It was found that fuel cell performance increased with ionomer loadings up to 20 wt %, and then decreased with higher *m*-PBI content. The best performance was observed at 20 wt % ionomer content, because the ohmic resistance was lowest at this loading and showed the optimum balance between ionic and electrical conductivities. The activation loss was lowest between 10 and 20 wt % loading. An increase in the concentration loss with increased ionomer content was observed, due to the increased mass transport of hydrogen and oxygen. The degree of catalyst particle interconnection increased with ionomer content and, therefore, the active area was decreased. The researchers concluded that the activation loss was the largest contributor to overall fuel cell performance loss, perhaps because PBI is applied as a liquid ionomer (unlike PTFE) and, therefore, it penetrates into all pores and decreases Pt utilization.

Recently, BASF Fuel Cells has developed a new membrane for use with liquid feed DMFCs called CeltecV [88]. CeltecV is based on a blend of polybenzimidazole and polyvinylphosphonic acid. The phosphonic acid based electrolyte was immobilized in the PBI matrix and could not be leached out during operation. Single cell performance tests of CeltecV and Nafion 117 were carried out and compared. It was found that CeltecV MEAs showed ~50% lower methanol crossover than Nafion. The electroosmotic drag of water was found to be five times lower for CeltecV which greatly reduced cathode flooding and allowed a lower cathode air flow stoichiometry. Interestingly, Nafion performed better at 90 and 110 °C at a low methanol concentration (0.5 M). However, at methanol concentrations of 1.0 and 2.0 M, CeltecV showed superior performance, especially at higher concentrations, probably due to the lower crossover of methanol and, therefore, increased stoichiometric availability of oxygen. A long-term durability test performed over 500 h showed an increase in membrane resistance of 18%, and future work will seek to elucidate whether this is an effect of a change in membrane chemistry or a deterioration of the membrane/electrode interface.

In order to improve both the mechanical properties and acid retention of *m*-PBI membranes, Li et al. [89] synthesized cross-linked polymers. *p*-Xylenedibromide was used as the cross-linking agent in varying amounts. Cross-linked membranes showed significantly lower solubility in DMAc, especially at higher cross-linker content. High doping levels were also achieved for these membranes. Linear *m*-PBI had a doping level of 15.5 mol PA/PRU, while membranes with cross-linking degrees of 1.1, 3.6, and 13.0% had 15.1, 14.1 and 8.5 mol PA/PRU, respectively. Doping level can be tailored by controlling immersion time and temperature of the phosphoric acid-doping bath. The volume percent swelling of these membranes was much lower than that of the native PBI. The cross-linked membranes showed significantly improved mechanical stability over native PBI. For example, a 13% cross-linked membrane with doping of 8.5 mol PA/PRU was comparable to linear PBI with a doping level of 6 mol PA/PRU. The conductivity of these cross-linked membranes was vastly increased over linear PBI at the same RH and temperature, due to the improved mechanical properties and higher doping levels. The Fenton test was performed on the membranes to determine the chemical stability and resistance to radical attack. The cross-linked films showed improved stability over linear PBI and a higher degree of cross-linking had improved stability, because of the reduced number of sites available for radical attack.

The alternative explorations of PBI as electrode ionomer, modified membrane, and in various types of fuel cells show great promise for the wide adaptability of PBI membranes. Further characterization and development are needed, but these initial forays give important insights into the nature of PBI interactions with electrode and reactants.

3
A New Approach: Polybenzimidazole from the Polyphosphoric Acid Process

3.1
Introduction to the Polyphosphoric Acid Process

A new process for synthesizing high molecular weight polybenzimidazoles and membrane casting was developed at Rensselaer Polytechnic Institute (RPI) in conjunction with the group that now constitutes BASF Fuel Cells. Collaboration between the two groups began in late 1998. This new method, termed "the PPA process", uses polyphosphoric acid (PPA) as the polycondensation agent, polymerization solvent, and membrane casting solvent [76]. PBIs were synthesized mostly from 3,3′,4,4′-tetraaminobiphenyl (TAB) and various dicarboxylic acids, although many combinations of tetraamines and diacids are possible. After polymerization, the PBI solutions in PPA were cast and the PPA hydrolyzed in-situ to phosphoric acid (PA). Under appropriate conditions, a sol-gel transition occurred to produce a film with a combination of desirable physicochemical properties not obtainable from conventional imbibing processes. These membranes had high PA-doping levels, good mechanical properties, excellent conductivities, and excellent long-term stabilities, even when operating at temperatures over 150 °C. A state diagram, shown in Fig. 12, was proposed to describe the multiple chemical and physical transformations that occurred during the conversion of monomer to final membrane. A critical part of the process is the sol-gel transition that occurs for many heteroaromatic polymers. The sol-gel transition is induced by a change in the nature of the solvent, when PPA (a good solvent for many PBI polymers) is converted to PA (a poor solvent) via a simple hydrolysis reaction following absorption of water during a post-casting process. In its simplest form, this is performed by exposing the cast solutions to ambient air at a set

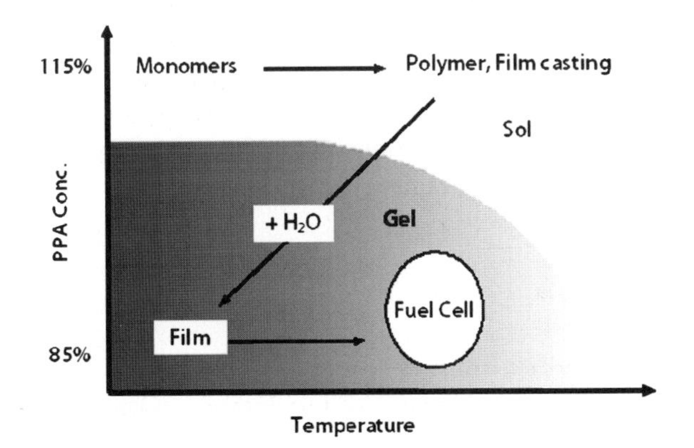

Fig. 12 State diagram of the PPA Process

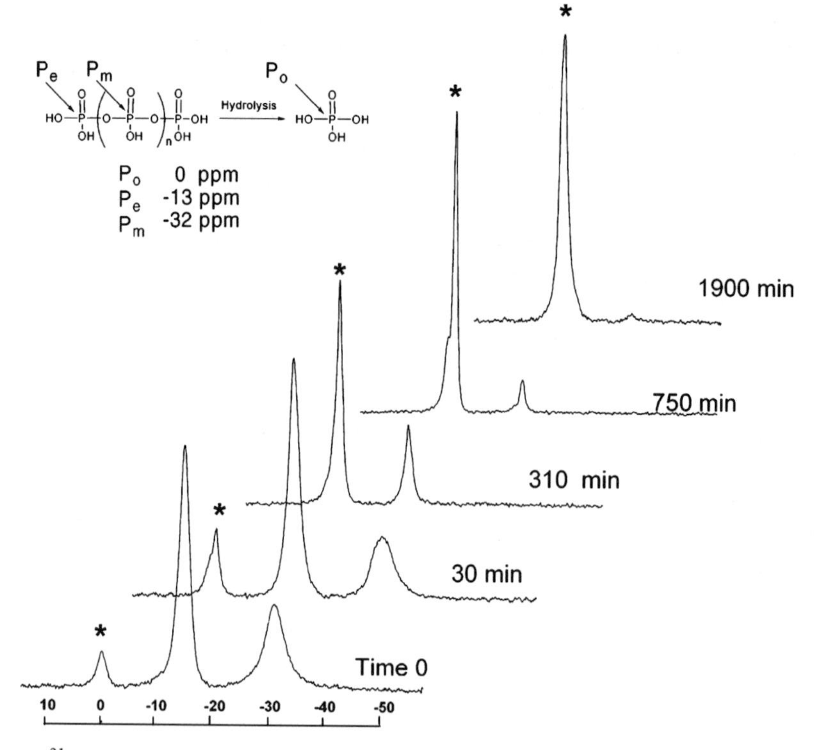

Fig. 13 ^{31}P NMR of PBI membrane cast from polyphosphoric acid and its conversion to phosphoric acid. Note that the *asterisk* identifies the 0 ppm peak

relative humidity. The in-situ PPA hydrolysis process was confirmed using ^{31}P spectroscopy [90] (Fig. 13).

As can be seen from these spectra, the amount of PPA decreased with time upon exposure to the moisture in the ambient air, while the amount of PA increased proportionately. If desired, quantitative information, such as rate of hydrolysis, may be obtained from these spectra. A number of PBI chemistries have been investigated in conjunction with the PPA process, including *m*-PBI, *p*-PBI, AB-PBI, and pyridine-based PBIs. Since the thermal conversion of PA to PPA occurs at temperatures far above 200 °C, the films remain stable for extended periods of time in an operating fuel cell.

3.2
Meta- Polybenzimidazole

One of the first PBI polymers to be synthesized by the PPA process was *meta*-PBI (*m*-PBI) (Fig. 1). Many polymerizations were performed to determine the effect of monomer concentration (2–12 wt %), PA loading, conductivity, mechanical properties, and morphology on fuel cell performance. It was found that monomer concentration had a marked effect on the polymer molecular weight, most easily seen in terms of inherent viscosity (IV). As seen in Fig. 14, maximum IV's were obtained at monomer concentrations of 8–8.5 wt %. The polymerization method (PPA process) produced high molecular weight polymers as compared to typical values reported in the literature. The highest IV's were almost a threefold increase over typical commercially available *m*-PBI polymers.

The levels of phosphoric acid doping that occur in *m*-PBI membranes produced by the PPA process are 14–26 moles PA/PRU. A membrane with 14.4 moles PA/PRU generally corresponds to about 64 wt % PA, 14.1 wt % PBI polymer, and 21.5 wt % water in the gel film [91]. The level of phosphoric acid doping for conventionally imbibed membranes has been reported to be 6–10 moles PA/PRU [2].

Proton conductivies for the PPA produced membranes were relatively high. A conductivity-temperature curve is presented in Fig. 15, showing conductivity of 0.13 S cm^{-1} at 160 °C measured under non-humidified conditions. The curve was recorded as a second heating run after an initial heating to 160 °C was conducted to remove the excess water. Literature reports of *m*-PBI/PA membranes [92] from conventional imbibing processes report conductivities of 0.04–0.08 S/cm at 150 °C at varying humidities. The conductivity was reported to be stable at these conditions for extended periods.

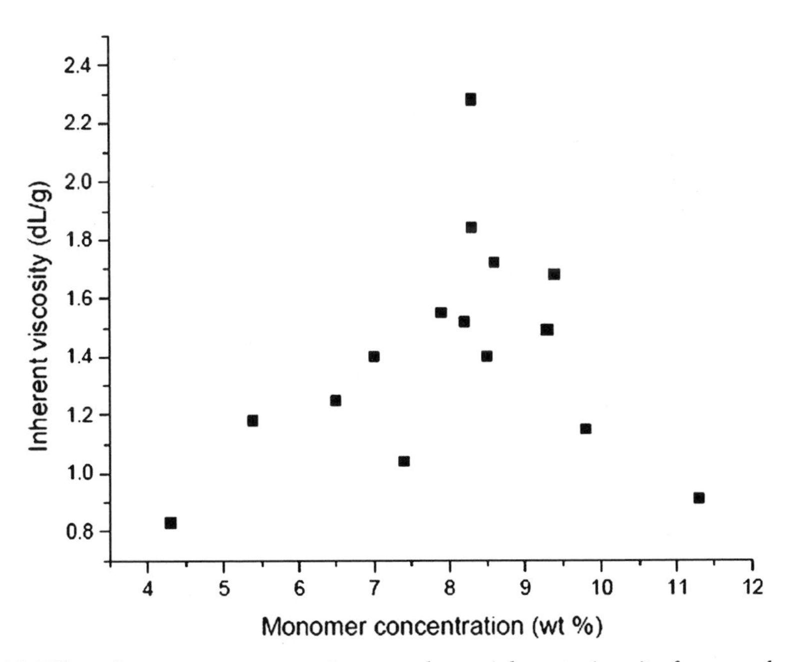

Fig. 14 Effect of monomer concentration on polymer inherent viscosity for a number of *m*-PBI polymerizations at 190 °C via the PPA Process

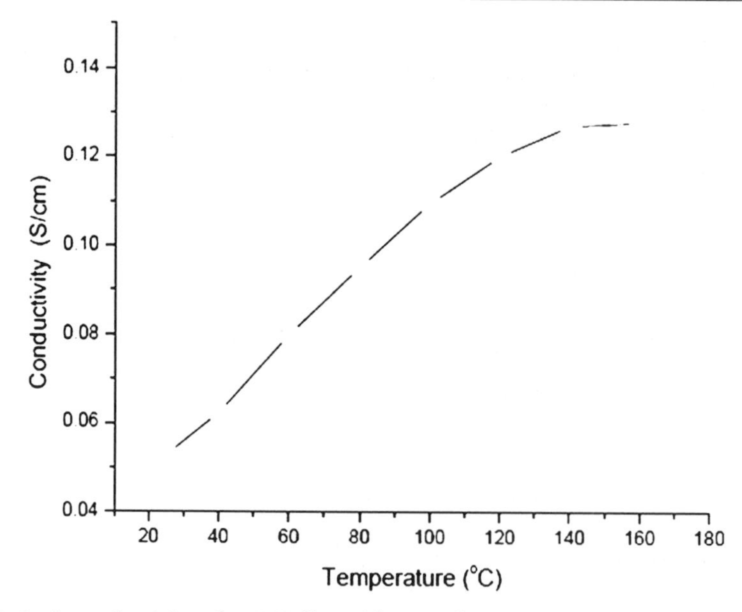

Fig. 15 Ionic conductivity of *m*-PBI film with 14 moles PA/PRU produced from the PPA Process. Recorded curve is the second run after initial heating to 160 °C to remove water

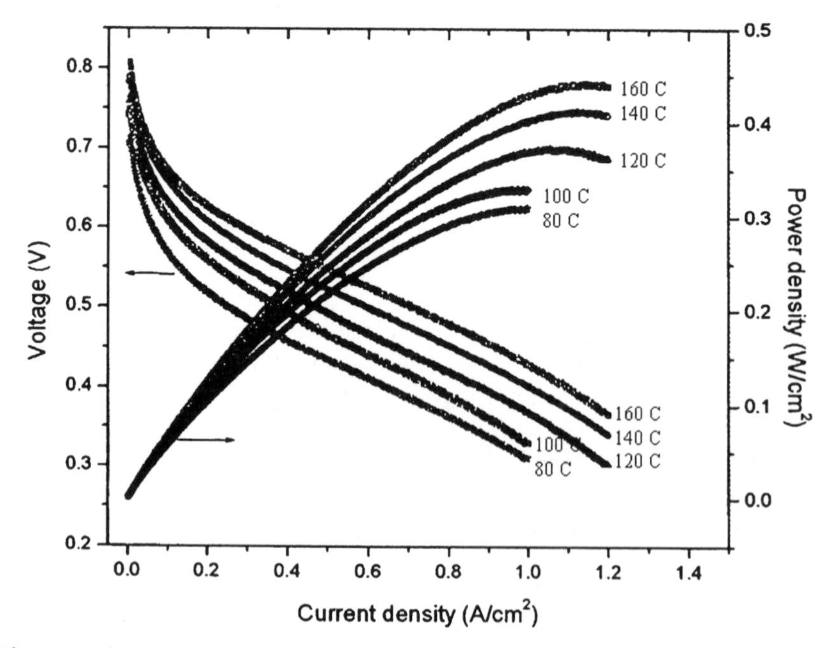

Fig. 16 Performance curves of hydrogen/air *m*-PBI/PA fuel cells at different temperatures and 1 atm (absolute) pressure. Note that the fuel cell operating conditions are as follows: constant flow rate, H_2 at 400 SCCM, air at 1300 SCCM, no humidification, 44 cm^2 active area, 1.0 mg cm^{-2} Pt catalyst loading, Pt – C 30% on each electrode

Fuel cell performance of the PPA produced *m*-PBI membranes is shown in Fig. 16. Although the fuel cell performance was measured under constant high flow-rate conditions, the membranes were operated reliably at high temperatures and dry gases. Additional work also showed that the fuel cells could run on a synthetic reformate containing 2000 ppm of carbon monoxide.

3.3
Para-Polybenzimidazole

Another chemical structure, *para*-PBI (*p*-PBI), shown in Fig. 17 has also been investigated using the PPA process. The typical polymerization process previously reported in the literature was a melt/solid polymerization of diphenyl terephthalate and 3,3′,4,4′-tetraaminobiphenyl, which produced a final polymer with an IV of $\sim$1 dL g^{-1}. Historically, little research has been reported for *p*-PBI, because it was thought to be too difficult to synthesize and process. These limitations probably result from the rigid nature of the polymer and the inability to produce fibers unless copolymerized with other monomers to prepare polybenzoxazole or polybenzthiazole copolymers [93–96]. It was not until the 1970s that high molecular weight (IV of 4.2 dL g^{-1}) *p*-PBI was synthesized [97]. However, since *p*-PBI was soluble only in strong acid solvents and at concentrations too low for fiber processing, research on *p*-PBI was not pursued [98, 99].

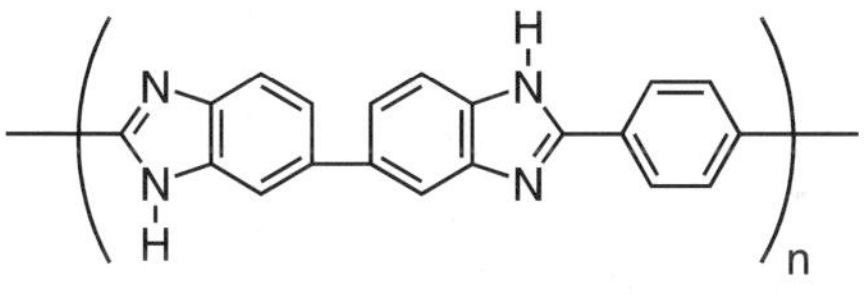

Fig. 17 Chemical structure of *p*-PBI

The synthesis of *p*-PBI in PPA was also difficult, due to the poor solubility of terephthalic acid (TPA) in PPA. The method developed by Delano [97] required several weeks of heating to produce the 4.2 dL g^{-1} IV polymer. Ultimately, the low solubility of TPA was not entirely limiting, because the solid TPA slowly dissolved in the PPA, ensuring a constant concentration throughout the polymerization.

Recently, Xiao et al. [90] reported that high molecular weight *p*-PBI could be produced using the PPA process at much shorter polymerization times and that membranes could be cast from the polymerization solutions. The mechanical properties of the doped films were critically dependent on the polymer IV and doping levels were very high, typically 20–40 moles PA/PRU. The doping levels of these mechanically stable films were much higher than all previous reports on PA/PBI films, and they resulted in conductivities that were greater than 0.2 S cm^{-1} at temperatures above approximately 140 °C.

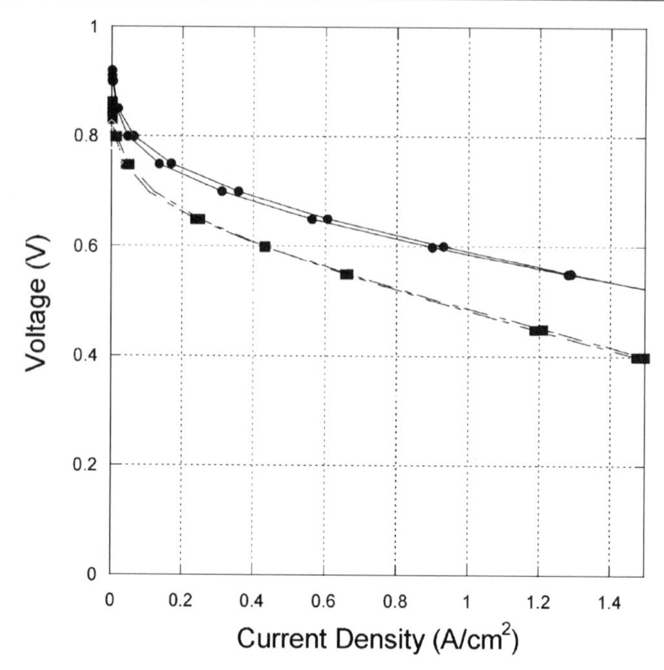

Fig. 18 Fuel cell performance for the *p*-PBI membranes from the sol-gel process. Polarization curves of fuel cells under H_2/air (*squares*) and H_2/O_2 (*circles*), without any feed gas humidification. The membrane PA doping level was approximately 32 mol PA/PRU. The catalyst loading in both electrodes was 1.0 mg cm^{-2} Pt, and the cell was operated at 160 °C at constant stoichiometry of 1.2 stoic and 2.5 stoic at the anode and the cathode, respectively

This is probably the first report of the processing of high molecular weight *p*-PBI into useful articles.

These membranes also showed excellent high-temperature performance in fuel cells when tested with dry hydrogen/air and hydrogen/oxygen gases at 160 °C. The polarization curves are shown in Fig. 18 using anode and cathode stoichiometries of 1.2 and 2.5, respectively. Exceptional long-term stability was demonstrated in cell performance tests.

3.4
Pyridine-Based Polybenzimidazole Membranes

Among a variety of PBI structures, only limited types of PBIs, which primarily included the commercially available PBI, poly[2,2′-(*m*-phenylene)-5,5′-bibenzimidazole] (i.e., *m*-PBI), sulfonated or phosphorylated *m*-PBI, as well as the poly(2,5-benzimidazole) (i.e., AB-PBI), have been explored for fuel cell applications [100]. A systematic synthesis of PBIs with different structures was initiated at RPI to study the effect of the PBI polymer molecular structure on the final film properties. The substitution of pyridine dicarboxylic acids (PDA) for the iso-/terephthalic acids is particularly interesting, since it increases the number of basic groups in the polymer backbones. Xiao et al. synthesized a series of pyridine-based polybenzimidazole (PPBI) homopolymers from 3,3′,4,4′-tetraaminobiphenyl (TAB) and 2,4-, 2,6-, 2,5-, and 3,5-pyridine dicarboxylic acids using the PPA Process, as shown in Scheme 2 [90, 101]. Different diacid monomers gave different substitution patterns on the pyridine ring. For instance, two carboxylic acid groups on 2,5-PDA are opposite to each other and give "*para*-orientation," while all the other monomers (3,5-, 2,6-, 2,4-) give "*meta*-orientation."

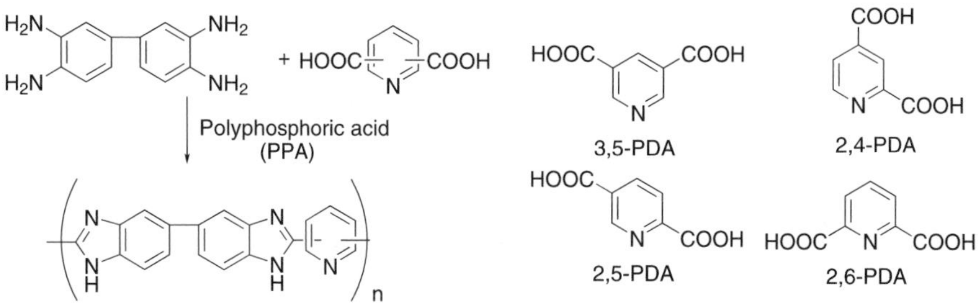

Polyphosphoric acid (PPA)

3,5-PDA

2,4-PDA

2,5-PDA

2,6-PDA

Scheme 2 Synthesis of PPBI homopolymers

There were some previous reports on the preparation of the PPBIs in PPA that resulted in low IV polymers or incomplete imidazole ring closure [102, 103]. In the present study, the initial polymerizations using diacid monomers without further purification gave PPBIs with IVs less than 1.0 $dL\,g^{-1}$, as shown in Table 2. Based on the Carother's equation, monomer purity and accurate stoichiometry are crucial to obtain high IV polycondensation polymers. Therefore, a detailed study of the monomer purity and purification method was carried out on all the diacid monomers. DSC scans were employed to monitor the relative purity of the monomers. They confirmed the improved monomer purity after purification. Table 2 shows that all diacid monomers polymerized to yield high IV PPBIs after appropriate purification of starting materials and careful manipulation of polymerization conditions.

As mentioned previously, one of the major barriers to the extensive application of the rigid rod PBI polymers has been their poor solubility and processability [104]. The polymerizations of *p*-PBI in PPA are difficult to control at a polymerization concentration higher than 4.5%, because the polymer precipitates [105, 106]. Surprisingly, the simple substitution of the pyridine ring for the phenyl ring allowed the polymerization of 2,5-PPBI in PPA at polymerization concentrations up to 18%. Clearly, the incorporation of the extra nitrogen in the polymer backbone significantly improved the solubility of the polymer in PPA and, thus, enhanced the polymer processability.

Table 2 Inherent Viscosity and Polymerization Data of PPBIs

Polymer	Monomer purity	IV ($dL\,g^{-1}$)	Polymerization concentration (wt % of the final polymer)	Film formation process
2,5-PPBI	As received	0.8	4% ~ 18%	Mechanically strong film
	Recrystallized	2.5–3.1		
3,5-PPBI	As received	0.6	4% ~ 20%	No liquid drain-off, Not film-forming, honey-like solution
	Recrystallized	1.3–1.9		
2,4-PPBI	As received	0.3	~ 7%	Mechanically strong film
	Recrystallized	1.0		
2,6-PPBI	As received	0.2	~ 7%	Liquid drain-off Mechanically weak film
	Recrystallized	1.3		

The TGA thermograms of the pyridine-based PBIs were obtained in both flowing nitrogen (20 $mL\,min^{-1}$) and flowing air (20 $mL\,min^{-1}$) at 20 $°C\,min^{-1}$. Curves for all the PPBI structures (2,4-, 2,6-) were similar and showed that the thermal stabilities of all PPBIs were quite high in both nitrogen and air, with an initial decomposition temperature of approximately 420 °C in air, as expected from the characteristic wholly aromatic structures of the PPBI polymer backbone. Thus, it was demonstrated that PPBI polymers incorporating main chain pyridine groups retained the inherently high thermo-oxidative stability of polybenzimidazoles.

The ionic conductivities of the PA-doped 2,5-PPBI and 2,6-PPBI membranes are shown in Fig. 19. The *para*-oriented 2,5-PPBI membrane with 20.4 moles of PA/PRU exhibited a conductivity of 0.018 $S\,cm^{-1}$ at room temperature and approximately 0.2 $S\,cm^{-1}$ at 160–200 °C. For the *meta*-oriented 2,6-PPBI membranes with 8.5 moles PA/PRU, the conductivity was 0.01 $S\,cm^{-1}$ at room temperature and approximately 0.1 $S\,cm^{-1}$ at 160–200 °C. It was concluded that polymer structure exerted a strong influence on membrane processing, PA doping levels, and final membrane properties.

Preliminary fuel cell evaluations were performed on the 2,5-PPBI membrane from the PPA process and are shown in Fig. 20. All of the tests were conducted on non-humidified feed gases and performed reliably under these completely dry conditions.

Among four types of PPBI homopolymers, the *para*-structured 2,5-PPBI gave mechanically strong membranes with a high PA doping level of approximately 20 moles PA/PRU, which contributed to the high proton conductivities. As expected, the fuel cell performance increased with increasing temperature. Overall, the major differences in membrane formation, PA loadings, mechanical properties, and fuel cell performance were surprisingly large, considering the seemingly minor changes in chemical structure. Fur-

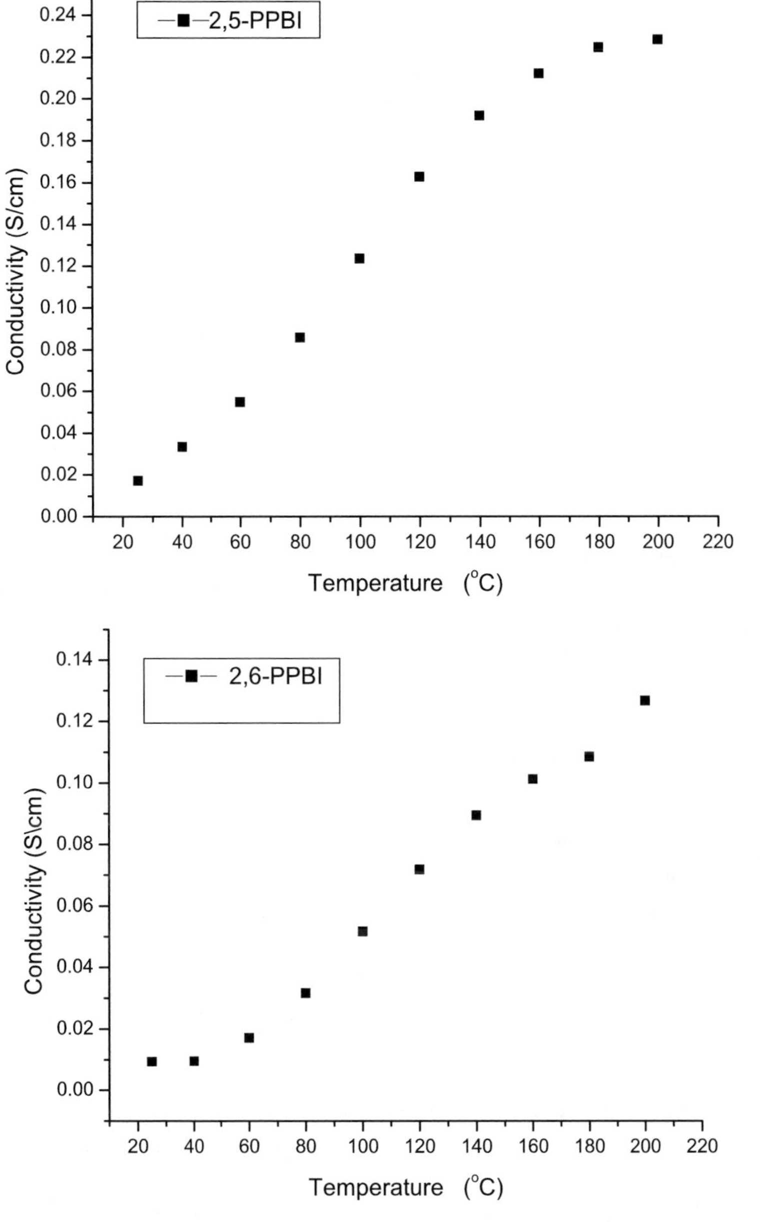

Fig. 19 Temperature dependence of ionic conductivity of PA-doped 2,5-PPBI and 2,6-PPBI membranes

ther work is being conducted to fully understand the effects of polymer structure on the fundamental membrane properties (PA content, PA retention, etc.) as well as the effects on fuel cell performance.

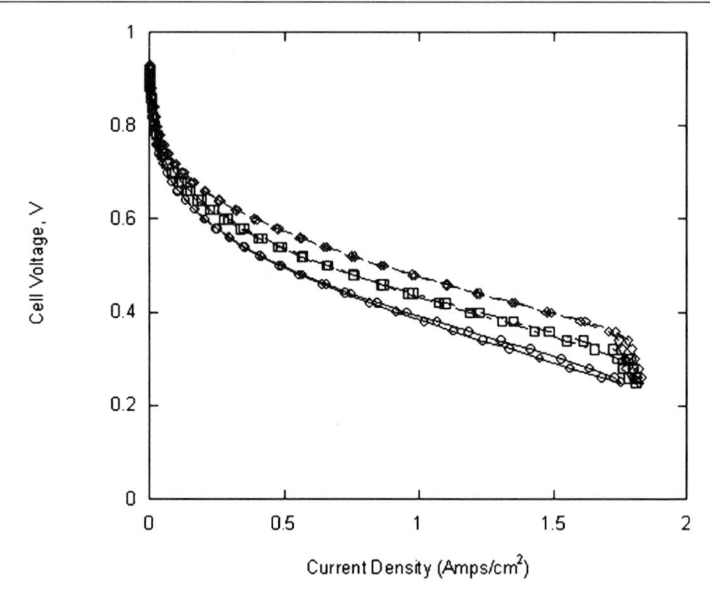

Fig. 20 Polarization curves of H_2/O_2 fuel cells with 2,5 PPBI membrane from the PPA process. Electrodes were 1.0 mg cm^{-2} Pt, 30% Pt on Vulcan XC-72 carbon. Fuel cell tests were performed on 10 cm^2 cells at ambient pressure, no external humidification, constant flow H_2 (120 mL min^{-1}) and O_2 (70 mL min^{-1}) at temperatures of 120 °C (*open circles*), 140 °C (*open squares*), and 160 °C (*open diamonds*)

4 Durability and Degradation in High-Temperature Polymer Electrolyte Membrane Fuel Cells

BASF Fuel Cells (formerly PEMEAS or Celanese Ventures) produces polybenzimidazole (PBI)-based high-temperature membrane and electrode assemblies sold under the brand name Celtec®. These MEAs operate at temperatures between 120 °C and 180 °C. One of the distinct advantages of high-temperature PEMFCs is exhibited in their high tolerance toward fuel gas impurities, such as CO (up to 3%), H_2S (up to 10 ppm), NH_3, or methanol (up to several percent), compared to low-temperature PEMFCs. Additionally, waste heat can be effectively used and, therefore, the overall system efficiency is increased.

Although several distinct advantages from an electrocatalysis perspective are present when a fuel cell operates at temperatures above 120 °C, care has to be taken when selecting the materials for MEAs. The catalyst materials must be highly active for the oxidation of realistic reformates and the oxygen reduction reaction, but, in addition, high stability towards corrosion is needed in order to ensure long fuel cell lifetimes (e.g., for stationary power applications at least 40 000 hours) at sustained high power output.

In order to get insight into some of the typical degradation modes described in the next section, specific experiments were designed [111] and are presented here.

4.1 Typical Degradation Mechanisms and Material Requirements

Table 3 summarizes the most important degradation modes observed in high-temperature MEAs operating up to 200 °C. However, it must be noted that, except for acid loss modes, which are unique to liquid acid based fuel cells, all other degradation modes are also observed in low-temperature PEMFCs.

Table 3 a: Possible Membrane Degradation Modes

Cause	Effects	Mechanisms
Pin hole formation	H_2 cross over, fuel loss, short cuts	Creep, fibers, *f(compression)*
Membrane thinning	H_2 cross over, fuel loss, short cuts	Creep, fibers, *f(compression)*
Phosphoric acid evaporation	Proton conductivity, IR-drop increase	Evaporation, $f(T,p)$

Table 3 b: Possible Electrode Degradation Modes (mtx = mass transport; ECSA = electrochemical surface area)

Cause	Effects	Mechanisms
Pt particle growth	Loss of ECSA, decrease of reaction kinetics	Migration, $f(T)$; Dissolution/recrystallization, $f(T,E)$
Pt/alloy dissolution	Loss of ECSA, decrease of reaction kinetics	Electrochemical dissolution, $f(T,E)$
Phosphoric acid evaporation from catalyst layer	Loss of ECSA, decrease of reaction kinetics	Evaporation, *f(T,p)*; *f(porosity changes)*
Carbon corrosion	Loss of ECSA, flooding, decrease of reaction kinetics, increase of mtx overpotentials, increase of IR-drop	Electrochemical oxidation, $f(T,E,p)$
GDL corrosion	Loss of structural integrity, flooding, decrease of reaction kinetics, increase of mtx overpotentials	Electrochemical corrosion $f(T,E,p)$
PTFE degradation	Loss of hydrophobicity, flooding, increase of mtx overpotentials	tbd

4.1.1
Degradation Modes Related to the Membrane

Typical membrane degradation modes include (i) the possibility of pin hole formation, due to thinning of the membrane, which leads to increased fuel crossover and loss of fuel efficiency and (ii) acid evaporation. By choosing the appropriate gasket material, the former effect can be minimized even over extended operation times. In order to effectively reduce the pressure on the membrane in the MEA, hard gasket materials are most suitable.

The acid evaporation due to the operating temperatures of 120 °C to 180 °C was found to be of no concern, due to the unique properties of the membrane. The small amounts of gaseous PA that are transported with the product water and excess gases out of the fuel cell were collected over approximately 8000 hours operation at 160 °C. The overall acid evaporation rate was found to be $\sim 0.5\,\mu g\,PA\,m^{-2}\,s^{-1}$. Based on the acid evaporation rates and the initial MEA acid content, a theoretical lifetime of more than 80 000 hours can be calculated. Additionally, during the 8000 hour test, the ohmic cell resistance changes by only $\pm 15\,m\Omega\,cm^2$, demonstrating that there is no significant impact of acid evaporation on cell lifetime.

4.1.2
Degradation Modes Related to Electrodes

The typical electrode degradation modes are: (i) corrosion of the catalyst metal (both particle growth and dissolution), (ii) corrosion of the carbon materials in electrodes (catalyst support and GDL materials), and (iii) degradation of PTFE. The first two modes are both strong functions of the electrochemical potential and the operating temperature. Pt-metal particle growth can occur through Ostwald ripening in a simple surface diffusion process or through dissolution/recrystallization process [112]. Pt dissolution concomitant with an irreversible loss of the active metal phase can become a severe issue when the fuel cell is operated for extended periods of time at potentials above 0.85 V [112]. Both lead to reduction of the electrochemical active surface area and manifest mainly as a decrease in the cathode kinetics.

Electrochemical corrosion of carbon supports was widely studied in the context of phosphoric acid fuel cell development [113, 114]. However, recently, the low-temperature fuel cell community has also given more attention to this process [115, 116]. Carbon corrosion at the fuel cell cathode in the form of surface oxidation can lead to functionalization of the carbon surface (e.g., quinone, lactones, carboxylic acids, etc.), with a concomitant change in the surface properties. This clearly results in changes to the hydrophobicity of the catalyst layer. Additionally, and even more severe, total oxidation of the carbon with the overall reaction leads to a substantial loss of the catalyst layer itself and can also result in losses of the electrical connection of the Pt particles in the electrode. The overall reaction is as follows:

$$C + 2H_2O \rightarrow CO_2 + 4H^+ + 4e^- \tag{1}$$

In addition to the electrochemical carbon corrosion rates, which are functions of temperature, electrochemical potential, and water partial pressure, the carbon morphology also plays an important role [99, 103].

In order to get some impression of the corrosion behavior of different carbon materials at the PBI/PA/carbon interface, a corrosion study of different carbon materials was performed at 180 °C and 1.0 V vs. RHE. The corrosion currents were continuously recorded for ~1000 minutes. Subsequently, from the mass specific charge at a given corrosion time, $Q(t)$, the maximum weight loss, assuming a 4-electron reaction, is calculated according to the following:

$$\Delta W(t) = 100 \cdot Q(t) \cdot M/(4F) \tag{2}$$

with M being the atomic mass of carbon and F the Faraday constant. Although the assumption of the 4-electron reaction process may not be entirely valid, especially at shorter time scales, the high potential, temperature, and low pH drive the reaction towards CO_2 formation [114]. In Table 4, the results from eight different carbons are presented. Six of these samples are typical carbon black materials (furnace or thermal blacks), whereas two samples represent synthetic high-surface area graphites. Comparing the weight loss values from the carbon black material, the well-known trend observed was

Table 4 Corrosion Behavior of Different Carbon Materials

Carbon	Weight loss [%] $t = 100$ min	Weight loss [%] $t = 800$ min	BET surface area [$m^2\,g^{-1}$]	Carbon Type
TIMCAL HSAG 100	2	12	130	Synthetic graphite
TIMCAL HSAG 300	4	13	280	Synthetic graphite
TIMCAL Ensaco 350G	13	73	770	Furnace black
Degussa HiBlack 40B	4	22	125	Furnace black
Degussa Printex L6	8	50	250	Furnace black
Vulcan XC72	9	35	250	Furnace black
Ketjen Black	14	45	800	Furnace black
Shawinigan Acetylene Black	3	12	75	Thermal black

that high surface area carbons seem more susceptible to corrosion than carbon blacks with lower surface area. However, this did not seem to be the case for the high surface area graphites (Timcal HSAG 100 and 300) with almost identical weight losses after 800 minutes, even though surface area differed by a factor of two. Based on the results from Table 4, a much higher stability of fuel cell catalysts can be expected when using one of the high surface area graphite supports, as compared to conventional carbon black materials. Even more important, the well-known loss of catalyst dispersion on graphitized carbon blacks, due to the significant reduction of the carbon surface area during graphitization, can be avoided, due to the high surface area of the synthetic graphite samples.

4.2 Impact of High Cathode Potentials

In order to get detailed insights on the impact of high cathodic operation potentials on the degradation of a Celtec®-P Series 1000 MEA, a 50 cm^2 single cell was operated at 180 °C and 0.02 A cm^{-2} for close to 2000 hours. The test was designed to determine the kinetic and mass transport changes induced by Pt particle growth and carbon (surface) corrosion.

The steady state cell potential at 0.02 A cm^{-2} as a function of runtime is illustrated in Fig. 21a (gray circles). Quite obviously, the cell potential decreased from 0.795 V to ~0.765 V. Since at this low current density, mass transport overpotentials are negligible, this difference of 30 mV can be safely attributed to a reduction of the cathode kinetics. For confirmation, $H_2 - O_2$ polarization curves were recorded at the beginning of life ($t = 72$ h) and after 500, 1000, and 1800 h of operation (see the resulting Tafel plots in Fig. 21b). The Tafel slope was found to be ~95 mV dec^{-1}, very close to the theoretical 2.3 RT/F value (90 mV dec^{-1} at 180 °C). Interestingly, the cathode kinetics decreased within the first 500 hours operation by 30 mV and then remained constant for the rest of the experiment. However, from an initial straight Tafel line with one single slope of the complete current density range, a bending of the Tafel line was observed for longer operation times (>500 h), pointing to additional changes in the mass transport properties of the cathode [118]. The reduction of the cathode kinetics by 30 mV corresponded to a reduction of the electrochemical surface area by ~52% with respect to the initial surface area, which was mainly attributed to the typical initial sintering/particle growth process in fuel cell cathodes.

In order to quantify changes in the mass transport overpotentials, additional H_2-air polarization curves (stoichiometries 1.2/1.3 for H_2/air) were recorded. In Fig. 21a (white squares), the cell potential at 0.2 A cm^{-2} is plotted, showing a total reduction of 47 mV after 1800 h, compared to beginning of life. The additional 17 mV of degradation were due to increased mass transport overpotentials, which resulted from changes in the hydrophobic-

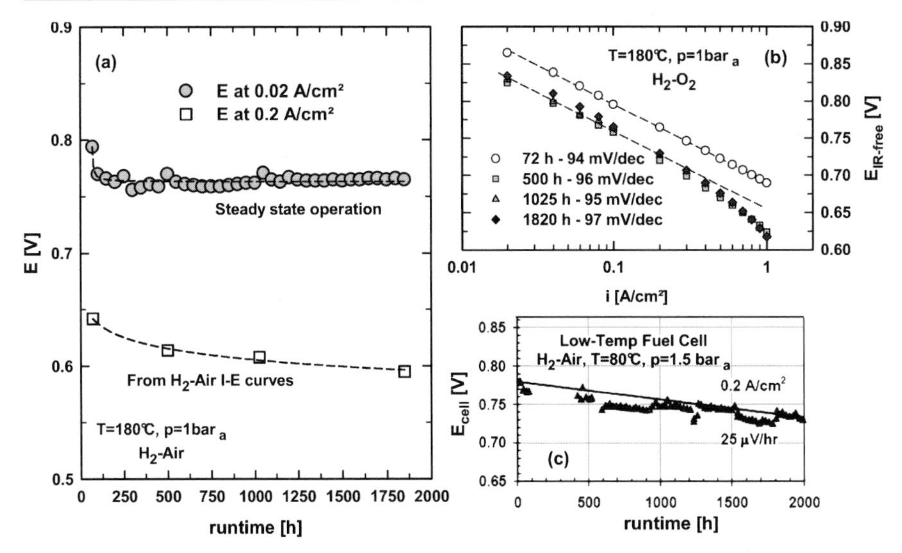

Fig. 21 **a** Steady state potential at 0.02 A cm^{-2} (*gray circles*) and cell voltage at 0.2 A cm^{-2} from H_2-Air polarization curves. $T = 180\,°C$, H_2-Air stoichiometries 1.2/1.3, $p = 1$ bar$_a$; **b** IR-corrected Tafel plots after 72, 500, 1025, and 1820 hours, respectively. $T = 180\,°C$, $H_2 - O_2$ stoichiometries 1.2/9.5, $p = 1$ bar$_a$; **c** Low temperature short stack data taken from [104]. $i = 0.2$ A cm^{-2}, $T = 80\,°C$, H_2-Air stoichiometries 2/2, $p = 1.5$ bar$_a$, fully humidified

ity of the catalyst layer due to oxidation of the carbon. Although the total degradation did not seem to be a linear function (see Fig. 21a), the calculated degradation rate was 26 μV h^{-1} under the conditions of 0.2 A cm^{-2} used in the study.

Interestingly, recently published data from a GM low-temperature short stack with Gore 5510 MEAs operated at 0.2 A cm^{-2} at 80 °C and 1.5 bar$_a$ (100% RH) at cell potentials between 0.78 V and 0.73 V showed a nearly identical degradation rate of 25 μV h^{-1} [119], shown in Fig. 21c. In this publication, the kinetic losses were attributed to a reduction of the electrochemical Pt surface area of 50% (similar to the value of 52% found with Celtec® MEAs). The similarity of the degradation rates and mechanisms of MEAs operated at 80 °C (Gore) and 180 °C (Celtec®) are quite astonishing, especially when considering the 100 °C temperature difference. One factor, which certainly plays an important role in the corrosion/oxidation of the carbon materials, is the water partial pressure. In the low-temperature case, the stack is operated with fully humidified gases, whereas the high-temperature MEA is operated with dry gases. In conclusion, although the operating temperature of a Celtec® MEA was 100 °C higher than for typical PFSA-type MEAs, no increase in degradation rate was observed.

4.3
Membrane and Electrode Assembly Durability

This section summarizes still ongoing durability experiments performed on a Celtec®-P Series 1000 MEA. The MEA was operated at a constant load of 0.2 A cm^{-2} at 160 °C using nonhumidified dry hydrogen and air (Fig. 22). The overall degradation rate under these conditions amounts to –6 μV h^{-1}.

To obtain insight into the different cathode degradation modes, a Tafel slope analysis was performed at the beginning of life and subsequently for an additional nine times. The initial IR-free cell performances using H_2-air and $H_2 - O_2$ are shown in Fig. 23. A single Tafel slope of 90 mV dec^{-1} can be fitted throughout the whole current density range (the fitted slope closely reflects the theoretical value of 2.3 RT/F (85 mV dec^{-1} at 160 °C)) and directly points to pure kinetic reaction control without interference of mass transport losses. The same Tafel slope can be fitted for the H_2-air polarization curve, although only up to 0.1 A cm^{-2}.

At higher current densities, the plot significantly deviates from linearity, pointing to the presence of the well-known mass transport resistances, when using air as the oxidant. Following a similar analysis as described in de-

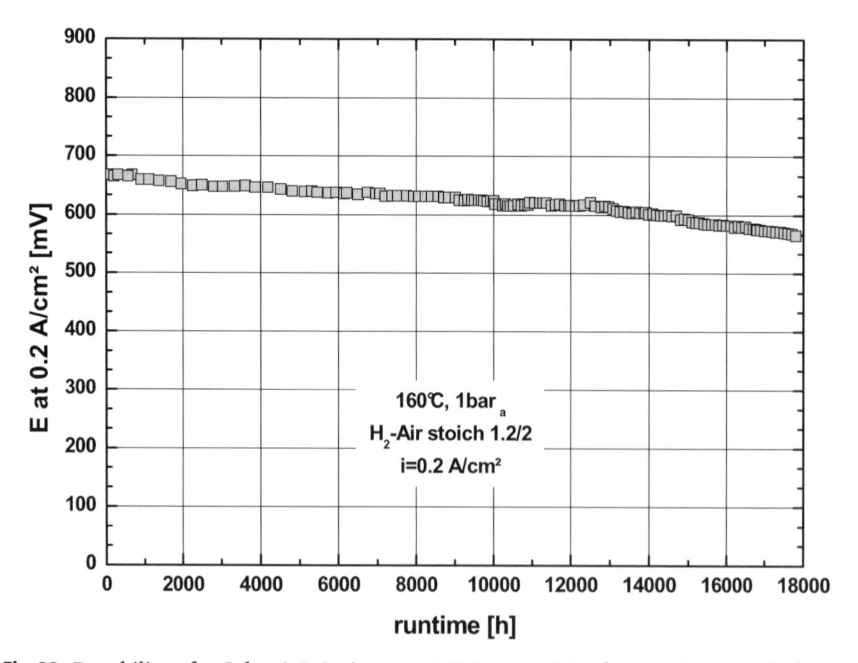

Fig. 22 Durability of a Celtec®-P Series 1000 MEA at 160 °C, 1 bar_a, using pure hydrogen and air with stoichiometries of 1.2 and 2, respectively. The test was conducted in a 50 cm^2 single cell

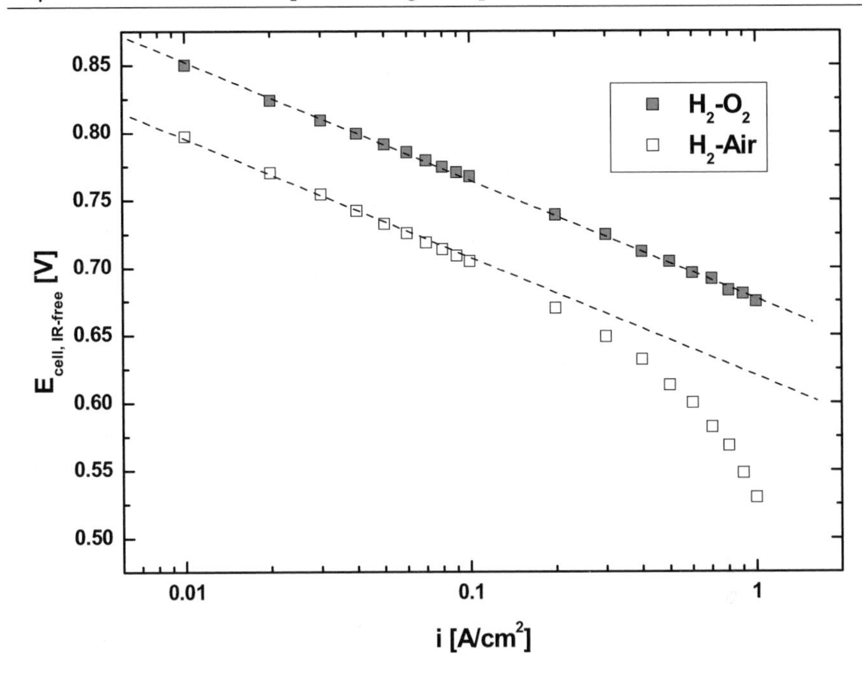

Fig. 23 Initial IR-free polarization data with H_2-Air and $H_2 - O_2$ at 160 °C. The current density is corrected for the initial H_2 crossover current of 1 mA cm^{-2}. H_2 stoichiometry 1.2, air stoichiometry 2, O_2 stoichiometry 9.5

tail [120, 121], the IR-free cell potential, $E_{\text{cell, IR-free}}$, can be given by Eq. 3:

$$E_{\text{cell, IR-free}} = E^\circ(p_{H_2/O_2/H_2O}, T) - \eta_{\text{ORR}} - \eta_{\text{cathode, mtx}} - \eta_{\text{HOR}} - \eta_{\text{anode, mtx}} , \tag{3}$$

where $E^\circ(p_x, T)$ represents the temperature and H_2, O_2, and H_2O partial pressure dependent equilibrium potential, η_{ORR} and η_{HOR} the kinetic overpotentials for the oxygen reduction and hydrogen oxidation reaction and $\eta_{x,\text{mtx}}$, the mass transport overpotentials on the cathode and anode, respectively. All anodic overpotentials are negligible under the applied conditions. This was checked by polarization measurements with increased hydrogen utilizations. Additionally, as concluded from Fig. 23, the oxygen polarization curve is under pure kinetic reaction control and no mass transport resistances are present. That is, Eq. 3 reduces to:

$$E_{\text{cell, IR-free}} = E^\circ(p_{H_2/O_2/H_2O}, T) - \eta_{\text{ORR}} . \tag{4}$$

By proper calculation of $E^\circ(p_x, T)$, the oxygen reduction overpotential can be determined using Eq. 4 from the measured oxygen polarization curve, which follows an oxygen partial pressure dependent Butler-Volmer expression [121, 122]. The theoretical Tafel line for air polarization should be just shifted to lower cell potentials by $\Delta E_{\text{O}_2\text{-Air}}$, given by:

$$\Delta E_{\text{O}_2\text{-Air}} = \Delta E^\circ + b \cdot \gamma \cdot \log(p_{\text{O}_2}/p_{\text{Air}}) \tag{5}$$

with ΔE° being the difference between the equilibrium potentials for pure oxygen and air, b and γ are the Tafel slope ($b = 2.3$ RT/F), and the kinetic reaction order ($\gamma \approx 0.6$) and p_x being the partial pressures for pure oxygen and air. Deviations from the theoretical Tafel line for air polarization to lower cell potentials can be considered to be cathodic mass transport overpotentials. The changes in oxygen reduction overpotentials and cathode mass transport overpotentials are plotted as functions of current density (Fig. 24) and runtime (Fig. 25).

All oxygen Tafel slopes were found to be in the range of 90–94 mV dec^{-1}, although at longer operation times (>10 000 h), the lines could only be fitted up to 0.1–0.2 A cm^{-2}. At higher current densities, a bending to lower cell voltages could be clearly observed, pointing to the fact that at higher current densities both kinetic and mass transport reaction control were present. However, over approximately the first 10 000 h operation, the oxygen reduction kinetics decreased by only 15% as compared to the initial kinetics, which was ascribed to initial loss of electrochemical surface area through sintering of the catalyst particles within the first 1000 hours operation. However, at

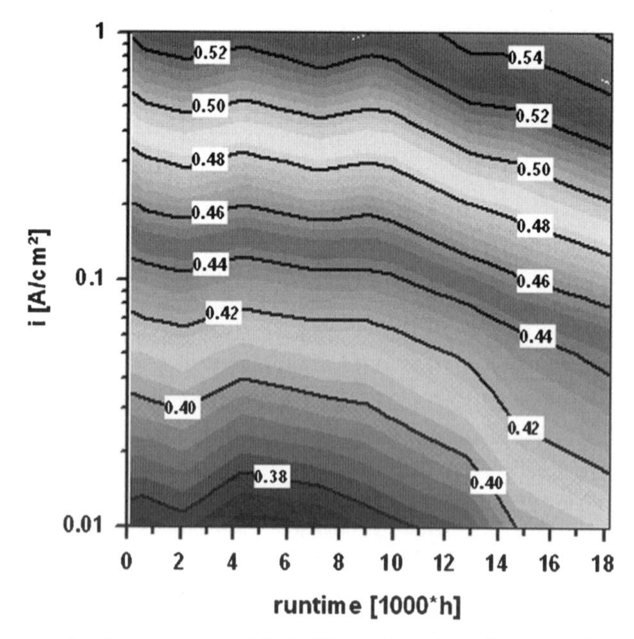

Fig. 24 Oxygen reduction overpotentials in V as a function of current density and MEA runtime

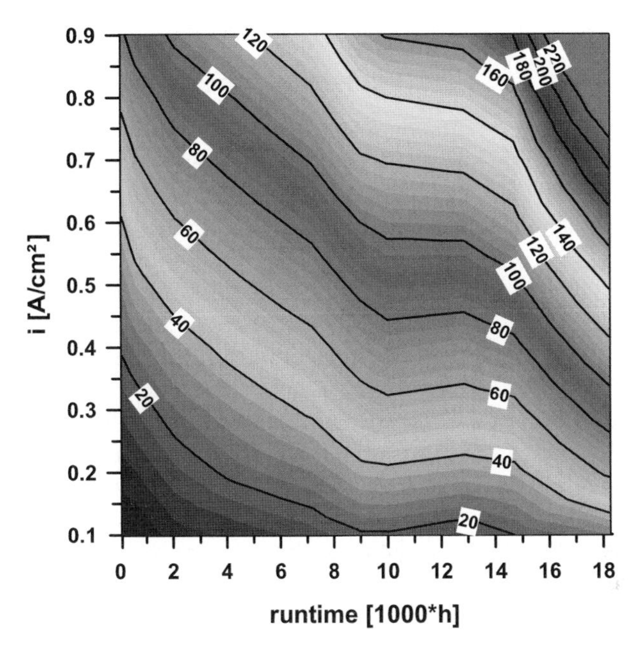

Fig. 25 Mass transport overpotentials in mV as a function of current density and MEA runtime

longer operation times, the kinetics further decreased (the oxygen reduction kinetics were reduced by ~35 mV at 0.2 A cm^{-2} after 18 000 hours). Although not entirely clear at this point in time, the major part of the kinetic losses might be due to a reduction of the electrochemical surface area, due to particle sintering. However, most likely, at least during the last 4000 to 5000 hours operation, electrical isolation of active catalyst particles due to the corrosion of the carbon perimeter can be assumed.

The mass transport overpotentials calculated from the air polarization curves were increasing during the first 9000 hours of operation from ~5 mV (0.2 A cm^{-2}) to 39 mV (0.2 A cm^{-2}). During the following 5000 h, the mass transport overpotentials remained almost constant. During the last 4000 h of operation, an additional 30 mV increase was observed. The main reason for an increase of the mass transport overpotenials on the cathode side is the slowly occurring flooding of the cathode catalyst layer structure with PA (see Table 3). It is noteworthy that only during the last 4000 hours of operation, both the kinetic and mass transport overpotentials increased significantly, which was interpreted as a concomitant corrosion of the carbon perimeter of the catalyst particles. This led to an increased flooding and an electrical disconnection of the catalyst particles. Further work (both in-situ and ex-situ), especially at the end of life, is necessary in order to completely clarify this process.

4.4 Summary and Conclusions

This summarizes the major degradation modes observed in polymer electrolyte fuel cells with a strong emphasis on the modes observed for high-temperature PEMFCs based on PBI/PA membranes. In a test over 8000 hours at 160 °C, the overall PA evaporation rate was only 0.5 $\mu g\, m^{-2}\, s^{-1}$, resulting in projected lifetimes of more than 50 000 hours, and showing that acid evaporation does not constitute an issue that affects MEA durability. Furthermore, it was shown that the degradation rate of a Celtec®-P Series 1000 MEA operated at 180 °C is very similar to a low-temperature MEA operated at 80 °C when both cells were operated in the same potential range.

Finally, a continuously ongoing durability test of a Celtec®-P Series 1000 MEA was discussed in light of cathode degradation. The MEA itself has currently been running for more than 18 000 hours at constant current density with an overall degradation rate of $-6\,\mu V\, h^{-1}$. At this current density, the main part (55%) of the degradation is due to increased mass transport overpotentials and by reduced oxygen reduction kinetics (30%). The remaining 15% can be ascribed to a slight increase in ohmic cell resistance.

5 Conclusions and Future Directions

The fuel cell systems based on acid-doped *m*-PBI and AB-PBI membranes have been well-characterized with respect to polymer properties, such as ionic conductivity, thermal and chemical stability, and in many cases, mechanical properties. In addition to these standard chemical structures, a number of modifications to the polymer backbone have been investigated. These modifications mostly include some type of sulfonated derivatives or blending with other membranes with the aim of increasing ionic conductivity via higher acid-doping levels while still retaining adequate mechanical properties. Further investigation of filled systems (nanoparticles, inorganic acid, etc.) has been done to improve mechanical strength and also to increase acid-doping levels. While a general picture that defines the properties and performance of these membranes has emerged, specific comparisons are sometimes difficult because of the minimal polymer and fuel cell characterization performed in some studies.

In comparison with the PBI membranes cast from organic solvents and subsequently imbibed with phosphoric acid, the PA-doped PBI polymer electrolyte membranes prepared via the PPA process possessed high PA loading levels with good mechanical properties and enhanced proton conductivities. It was shown that the polymer molecular structures significantly affected the properties of the polymers and the corresponding film formation process. In addition to this, improved fuel cell performance (especially at high temperatures) is seen compared to the "traditionally" prepared acid imbibed membranes.

Industrially, BASF Fuel Cells has a commercially prepared, ready made MEA (Celtec®) that shows great promise. The carbon corrosion, PA evaporation rates, and degradation losses are all minimal. The demonstrated long-term stability is excellent, with the ability to run for at least 18 000 hours, nearly halfway the United States Department of Energy target of 40 000 hours, for stationary applications with only a very small voltage loss over this time period.

In the future, for all membranes prepared and characterized, more fuel cell testing needs to be done under a variety of conditions. This is especially important because data gathered during initial studies may not accurately reflect fuel cell performance. At this time, there is still a relatively small amount of fuel cell performance data reported for PBI membranes. Also, different applications (stationary, portable, automotive) may require significantly different testing protocols and MEA requirements to fulfill their needs.

To attain widespread use of fuel cells, performance data must be available to prove to industry developers and consumers that fuel cells are indeed a viable option for replacing or supplementing conventional power systems. In order to make fuel cell systems practical, more research on preparation of MEAs, catalysts, and incorporation into stacks is needed. However, an excellent and promising start to using PBI in real world applications has been achieved and will undoubtedly continue to blossom in the future.

Acknowledgements Brian C. Benicewicz would like to personally acknowledge BASF Fuel Cell, and Dr. Gordon Calundann for the long-standing support of the work at Rensselaer Polytechnic Institute. Plug Power is also acknowledged for their support and technical assistance over the last several years. The U.S. Department of Energy (EERE, DE-FC36-03GO13097 and BES, DDE-FG02-OSER 46258), and the NSF IGERT program (DGE-0504361) have also provided support during the preparation of this chapter. D. Ott, F. Rat, and M. Jantos are greatly acknowledged for performing some of the experiments detailed in the Durability and Degradation section.

References

1. Vogel H, Marvel CS (1961) J Polym Sci 50:511
2. Wainright JS, Wang JT, Weng D, Savinell RF, Litt M (1995) J Electrochem Soc 142:L121
3. Kim H, Lim T (2004) J Ind Eng Chem 10:1081 (Seoul, Repub Korea)
4. Samms SR, Wasmus S, Savinell RF (1996) J Electrochem Soc 143:1225
5. Kerres J, Ullrich A, Meier F, Haring T (1999) Solid State Ionics 125:243
6. Cassidy PE (1980) Polybenzimidazoles. In: Thermally stable polymers: synthesis and properties. Marcel Dekker, Inc., New York, Sect. 6.10, 6.18-2
7. Hogarth WHJ, Diniz da Costa JC, Lu GQ (2005) J Power Sources 142:223
8. Li Q, He R, Gao J, Jensen JO, Bjerrum NJ (2003) J Electrochem Soc 150:A1599
9. Ma YL, Wainright JS, Litt MH, Savinell RF (2004) J Electrochem Soc 151:A8
10. Bouchet R, Siebert E (1999) Solid State Ionics 118:287

11. Glipa X, Bonnet B, Mula B, Jones DJ, Roziere J (1999) J Mater Chem 9:3045
12. Asensio JA, Borros S, Gomez-Romero P (2002) J Polym Sci, Part A: Polym Chem 40:3703
13. Asensio JA, Borros S, Gomez-Romero P (2003) Electrochem Commun 5:967
14. Xing B, Savadogo OJ (1999) New Mat Electrochem Systems 2:95
15. Pu H, Liu Q (2004) Polym Int 53:1512
16. Pu H, Meyer WH, Wegner G (2002) J Polym Sci, Part B: Polym Phys 40:663
17. Bouchet R, Miller S, Duclot M, Souquet JL (2001) Solid State Ionics 145:69
18. Fontanella JJ, Wintersgill MC, Wainright JS, Savinell RF, Litt M (1998) Electrochim Acta 43:1289
19. Schuster M, Meyer W (2003) Annu Rev Mater Res 33:233
20. Wang JT, Savinell RF, Wainright J, Litt M, Yu H (1996) Electrochim Acta 41:193
21. Li Q, Hjuler HA, Bjerrum NJ (2001) J Appl Electrochem 31:773
22. He R, Li Q, Bach A, Jensen JO, Bjerrum NJ (2006) J Membr Sci 277:38
23. Kawahara M, Morita J, Rikukawa M, Sanui K, Ogata N (2000) Electrochim Acta 45:1395
24. Li Q, He R, Berg RW, Hjuler HA, Bjerrum NJ (2004) Solid State Ionics 168:177
25. Li Q, He R, Jensen JO, Bjerrum NJ (2004) Fuel Cells 4:147 (Weinheim, Ger)
26. Kim H, An SJ, Kim J, Jin KM, Cho SY, Eun YC, Yoon H, Park Y, Kweon H, Shin E (2004) Macromol Rapid Commun 25:1410
27. Schechter A, Savinell RF (2002) Solid State Ionics 147:181
28. Hu J, Zhang H, Zhai Y, Liu G, Hu J, Yi B (2006) Electrochim Acta 52:394
29. Lobato J, Canizares P, Rodrigo M, Linares J (2007) Electrochim Acta 52:3910
30. Zhai Y, Zhang H, Liu G, Hu J, Yi B (2007) J Electrochem Soc 154:B72
31. Kongstein OE, Berning T, Borresen B, Seland F, Tunold R (2007) Energy 32:418
32. Kerres JA (2005) Fuel Cells 5:230 (Weinheim, Ger)
33. Staiti P, Minutoli M, Hocevar S (2000) J Power Sources 90:231
34. He R, Li Q, Xiao G, Bjerrum NJ (2003) J Membr Sci 226:169
35. Yamazaki Y, Jang MY, Taniyama T (2004) Sci Technol Adv Mater 5:455
36. Jang MY, Yamazaki Y (2005) J Power Sources 139:2
37. Jang MY, Yamazaki Y (2004) Solid State Ionics 167:107
38. Zaidi SMJ (2005) Electrochim Acta 50:4771
39. Jones DJ, Roziere J (2001) J Membr Sci 185:41
40. Deimede V, Voyiatzis GA, Kallitsis JK, Qingfeng L, Bjerrum NJ (2000) Macromolecules 33:7609
41. Kim S, Cameron DA, Lee Y, Reynolds JR, Savage CR (1996) J Polym Sci, Part A: Polym Chem 34:481
42. Kosmala B, Schauer J (2002) J Appl Polym Sci 85:1118
43. He R, Li Q, Xiao G, Bjerrum NJ (2003) J Membr Sci 226:169
44. Li Q, He R, Jensen JO, Bjerrum NJ (2003) Chem Mater 15:4896–4915
45. Hickner MA, Ghassemi H, Kim YS, Einsla BR, McGrath JE (2004) Chem Rev 104:4587
46. Jannasch P (2003) Curr Opin Colloid Interface Sci 8:96
47. Meng YZ, Tjong SC, Hay AS, Wang SJ (2003) Eur Polym J 39:627
48. Xiao GY, Sun GM, Yan DY, Zhu PF, Tao P (2002) Polymer 43:5335
49. Fang J, Guo X, Harada S, Watari T, Tanaka K, Kita H, Okamoto K (2002) Macromolecules 35:9022
50. Hasiotis C, Deimede V, Kontoyannis C (2001) Electrochim Acta 46:2401
51. Hasiotis C, Li Q, Deimede V, Kallitsis JK, Kontoyannis CG, Bjerrum NJ (2001) J Electrochem Soc 148:A513

52. Qing S, Huang W, Yan D (2005) Eur Polym J 41:1589
53. Staiti P, Lufrano F, Arico AS, Passalacqua E, Antonucci V (2001) J Membr Sci 188:71
54. Rikukawa M, Sanui K (2000) Prog Polym Sci 25:1463
55. Glipa X, El Haddad M, Jones DJ, Roziere J (1997) Solid State Ionics 97:323
56. Gieselman M, Reynolds J (1992) Macromolecules 25:4832
57. Ariza MJ, Jones DJ, Roziere J (2002) Desalination 147:183
58. Kawahara M, Rikukawa M, Sanui K (2000) Polym Adv Technol 11:544
59. Gieselman MB, Reynolds JR (1993) Macromolecules 26:5633
60. Bae JM, Honma I, Murata M, Yamamoto T, Rikukawa M, Ogata N (2002) Solid State Ionics 147:189
61. Jouanneau J, Mercier R, Gonon L, Gebel G (2007) Macromolecules 40:983
62. Manea C, Mulder M (2002) Desalination 147:179
63. Silva VS, Weisshaar S, Reissner R, Ruffmann B, Vetter S, Mendes A, Madeira LM, Nunes S (2005) J Power Sources 145:485
64. Kerres JA (2001) J Membr Sci 185:3
65. Kreuer KD (2001) J Membr Sci 185:29
66. Pu H (2003) Polym Int 52:1540
67. Daletou MK, Gourdoupi N, Kallitsis JK (2005) J Membr Sci 252:115
68. Wycisk R, Lee JK, Pintauro PN (2005) J Electrochem Soc 152:A892
69. Jeong J, Yoon K, Choi J, Kim YJ, Hong YT (2006) Memburein 16:276
70. Asensio JA, Borros S, Gomez-Romero P (2004) J Electrochem Soc 151:A304
71. Asensio JA, Borros S, Gomez-Romero P (2004) J Membr Sci 241:89
72. Kim H, Cho SY, An SJ, Eun YC, Kim J, Yoon H, Kweon H, Yew KH (2004) Macromol Rapid Commun 25:894
73. Cho J, Blackwell J, Chvalun SN, Litt M, Wang Y (2004) J Polym Sci, Part B: Polym Phys 42:2576
74. Gomez-Romero P, Asensio JA, Borros S (2005) Electrochim Acta 50:4715
75. Asensio JA, Gomez-Romero P (2005) Fuel Cells 5:336 (Weinheim, Ger)
76. Asensio JA, Borros S, Gomez-Romero P (2004) Electrochim Acta 49:4461
77. Wang JT, Lin WFL, Weber M, Wasmus S, Savinell RF (1998) Electrochim Acta 43:3821
78. Xing B, Savadogo O (2000) Electrochem Commun 2:697
79. Cheng CK, Luo JL, Chuang KT, Sanger AR (2005) J Phys Chem B 109:13036
80. Li Q, Hjuler HA, Hasiotis C, Kallitsis JK, Kontoyannis CG, Bjerrum NJ (2002) Electrochem Solid-State Lett 5(6):A125–A128
81. Pan C, He R, Li Q, Jensen JO, Bjerrum NJ, Hjulmand HA, Jensen AB (2005) J Power Sources 145:392
82. Banihashemi A, Atabaki F (2002) Eur Polym J 38:2119
83. Mecerreyes D, Grande H, Miguel O, Ochoteco E, Marcilla R, Cantero I (2004) Chem Mater 16:604
84. Xu H, Chem K, Guo X, Fang J, Yin J (2007) J Polym Sci, Part A: Polym Chem 45:1150
85. Larson JM, Hamrock SJ, Haugen GM, Pham P, Lamanna WM, Moss AB (2006) Prepr Symp-Am Chem Soc Div Fuel Chem 51:615
86. Wang CP, Chu HS, Yan YY, Hsueh KL (2007) J Power Sources 170:235
87. Kim JH, Kim HJ, Lim TH, Lee HI (2007) J Power Sources 170:275
88. Gubler L, Kramer D, Belack J, Unsal O, Schmidt TJ, Scherer GG (2007) J Electrochem Soc 154:B981
89. Li Q, Pan C, Jensen JO, Noye P, Bjerrum NJ (2007) Chem Mater 19:350
90. Xiao L, Zhang H, Scanlon E, Ramanathan LS, Choe EW, Rogers D, Apple T, Benicewicz BC (2005) Chem Mater 17:5328

91. Zhang H (2004) PhD thesis, Rensselaer Polytechnic Institute
92. Litt M, Ameri R, Wang Y, Savinell R, Wainright J (1999) Mater Res Soc Symp Proc 548:313
93. Wierschke SG, Shoemaker JR, Haaland PD, Pachter R, Adams WW (1992) Polymer 33:3357
94. So YH, Heeschen JP, Bell B (1998) Polym Prepr 39:280
95. So YH, Heeschen JP, Bell B, Bonk P, Briggs M, DeCaire R (1998) Macromolecules 31:5229
96. Hu X, Jenkins SE, Min BG, Polk MB, Kumar S (2003) Macromol Mater Eng 288:823
97. Delano CB, Doyle RR, Miligan RJ (1975) United States Air Force Materials Laboratory – Technical Report 1974–22
98. Chung TS (1997) Plast Eng 41:701
99. Chung TS (1997) J Macromol Sci, Rev Macromol Chem Phys C37:277
100. Choe EW, Choe DD (1996) In: Salamone JC (ed) Polymeric Materials Encyclopedia. CRC Press, New York, pp 5619–5938
101. Xiao L, Zhang H, Jana T, Scanlon E, Chen R, Choe EW, Ramanathan LS, Yu S, Benicewicz BC (2005) Fuel Cells 5:287 (Weinheim, Ger)
102. Brock T, Sherrington DC, Tang HG (1991) Polymer 32:353
103. Gerber AH (1976) US Patent 3 943 125
104. Twieg R, Matray T, Hedrick JL (1996) Macromolecules 29:7335
105. Iwakura Y, Uno K, Nume K (1973) US Patent 3 741 938
106. Arnold FE (1974) US Patent 495 452
107. Savinell RF, Litt MH (1997) WO Patent 9 737 396
108. Wang JT, Wainright J, Yu H, Litt M, Savinell RF (1995) Proc – Electrochem Soc 95:202
109. Kawabe M, Ohashi O, Yamaguchi I (1970) Bull Chem Soc Jpn 43:3705
110. Xiao L (2003) PhD thesis, Rensselaer Polytechnic Institute
111. Schmidt TJ (2006) ECS Transactions 1:19
112. Ross PN Jr (1987) Deactivation and Poisoning of fuel cell catalysts. In: Peterson EE, Bell AT (eds) Catalyst deactivation. Marcel Dekker, New York, p 197
113. Kinoshita K (1988) Carbon: electrochemical and physicochemical properties. John Wiley and Sons, New York
114. Antonucci PL, Romeo F, Minutoli M, Alderucci E, Giordano N (1988) Carbon 26:197
115. Kangasniemi KH, Condit DH, Jarvi TD (2004) J Electrochem Soc 151:E125
116. Roen LM, Paik CH, Jarvi TD (2004) Electrochem Solid-State Lett 7:A19
117. Stoneheart P (1984) Carbon 22:423
118. Perry ML, Newman J, Cairns EJ (1998) J Electrochem Soc 145:5
119. Mathias MF, Makharia R, Gasteiger HA, Conley JJ, Fuller TJ, Gittleman CJ, Kocha SS, Miller DP, Mittelstaedt CK, Xie T, Yan SG, Yu PT (2005) Electrochem Soc Interface 14:24
120. Gasteiger HA, Gu W, Makharia M, Mathias MF, Sompalli B (2003) Beginning of life MEA performance – efficiency loss contributions. In: Vielstich W, Gasteiger HA, Lamm A (eds) Handbook of fuel cells – fundamentals, technology and applications, Volume 3: fuel cell technology and applications, part 1. Wiley, Chichester, p 593
121. Neyerlin KC, Gasteiger HA, Mittelstaedt CK, Jorne J, Gun W (2005) J Electrochem Soc 152:A1073
122. Newman JS (1973) Electrochemical Systems, 1st edn. Prentice Hall, New Jersey
123. Internal Rensselaer Polytechnic Institute Data

Adv Polym Sci (2008) 216: 125–155
DOI 10.1007/12_2008_131

Published online: 20 February 2008

Proton-Conducting Polymers and Membranes Carrying Phosphonic Acid Groups

Alexandre L. Rusanov[1] (✉) · Petr V. Kostoglodov[2] · Marc J. M. Abadie[3] · Vanda Yu. Voytekunas[3] · Dmitri Yu. Likhachev[4]

[1]A.N. Nesmeyanov Institute of Organoelement Compounds, Russian Academy of Science, 28 Vavilova str., 119334 Moscow, Russia
alrus@ineos.ac.ru

[2]YRD Centre, Ltd., 55/1 b.2, Leninski pr., 119333 Moscow, Russia

[3]Laboratory of Polymer Science and Advanced Organic Materials – LEMP/MAO, Universite Montpellier 2, Science et Techniques du Languedoc, 34095 Montpellier, France

[4]Materials Research Institute, UNAM, Cirquito Exterioir s/n, CU, Apdo Postal 70-360 Coyoacan, 04510 Mexico City, Mexico

Abstract The results of the research of novel proton-conducting membranes carrying phosphonic acid groups have been analyzed and summarized with respect to their application in fuel cells. General approaches to the preparation of heterochain and heterocyclic polymers carrying phosphonic acid groups have been considered, including polycondensation of phosphonic acid functionalized monomers and phosphorylation of existing polymers. The bibliography includes 85 references.

Keywords Condensation polymers · Fuel cells · Phosphorylation · Polyelectrolytes · Polyphosphazenes · Polysulfones · Proton exchanging membranes

Abbreviations

BPA	Bis-phenol A
NMR	Nuclear magnetic resonance
^{1}H NMR	Proton nuclear magnetic resonance
^{31}P-NMR	Phosphorus-31 nuclear magnetic resonance
DCB	*o*-dichlorobenzene
DEP	Diethyl phosphite
DMA	Dimethylacetamide
DMF	Dimethylformamide

DMSO	Dimethylsulfoxide
DCP	3,5-dicarboxylbenzenephosphonic acid
GPC	Gel permeation chromatography
IEC	Ion-exchange capacity
N-MP	*N*-methyl-2-pyrrolidone
PBI	Poly(Benzimidazole)
PEMFC	Proton-exchanging membrane fuel cells
PSU	Polysulfone
STA	2-sulfoterephthalic acid
TEA	Triethylamine
TGA	Thermogravimetric analysis
THF	Tetrahydrofuran

1 Introduction

Proton-conducting ionomers for polymer electrolyte membrane fuel cell (PEMFC) is one area in which extensive research is ongoing to modify aromatic main chain polymers to tailor their properties as proton conductors [1–18]. When it comes to modifying these polymers with the aim of preparing ionomers, sulfonic acid units have been by far the acidic moiety of choice [1–19]. This is at least partly due to the wide range of reactants that are commercially available for sulfonation. Besides direct sulfonation by electrophilic substitution reactions with, for example, fuming sulfuric acid, sulfonated ionomers have been also accomplished by chemical grafting [20, 21] and by direct polymerization of sulfonated monomers [22–27].

Nowadays, research in the field of PEMFCs is mainly driven by the insufficiency of properties of Nafion membrane at temperatures above 100 °C [11, 13, 28]. Operating a fuel cell at these temperatures increases the kinetics of the electrode reactions, reduces the risk of catalyst poisoning, lowers the amount of catalyst needed, and reduces cathode flooding [7].

As a result, the overall performance and cost efficiency of the fuel cell system can be expected to be significantly improved. However, at high temperature, desulfonation, which is a loss of the sulfonic acid unit through hydrolysis, may become a critical issue [29]. This is currently motivating the search for ionomers with alternative acidic moieties, such as phosphonic acid [30, 31], which have higher chemical and thermal stabilities than sulfonic acid moieties [32–34].

Even though phosphonic acid units are less acidic than sulfonic acid units [35], they are known to have a greater ability to retain water than sulfonic acid unit [36]. This is crucial for maintaining high conductivities at elevated temperatures.

In spite of their potential usefulness, polymers containing phosphonic acid units have scarcely been studied for fuel cell applications, because the synthetic procedures available for phosphonation are limited and complex in comparison to sulfonation procedure. Furthermore, lower acidity of phosphonic acids, compared to sulfonic acids, requires higher degrees of substitution for sufficient proton exchange performance.

2 Proton-Conducting Polymers and Membranes Carrying Phosphonic Acid Groups

The first proton-conducting perfluorinated membrane carrying phosphonic acid units was prepared by Kotov et al. [34]. Several *co*- and *ter*-fluoropolymers based on tetrafluoroethylene and perfluorovinylenes, such as

$$CF_2 = CFO(CF_2)_2P(O)(OC_2H_5)_2\,,$$
$$CF_2 = CFO(CF_2)_3P(O)(OC_2H_5)_2\,,$$
$$CF_2 = CFOCF_2CF(CF_3)O(CF_2)_2P(O)(OC_2H_5)_2, \quad \text{and}$$
$$CF_2 = CFOC_3F_7\,,$$

were synthesized via redox-initiated emulsion polymerization. The polymers were subjected to acid hydrolysis and the corresponding fluorinated polymeric phosphonic acids were obtained. Both the polymeric phosphonated and phosphonic acids can be readily processed into films by compression molding. Perfluorinated phosphonic acid membranes with an ion-exchange capacity (IEC) within the range 2.5–3.5 milliequivalents per gram and a thermal stability of 300–350 °C were prepared. The characteristics of these acid membranes were explored, and these films show promising electrochemical properties, such as proton conductivity (0.006–0.076 S/cm) comparable with Nafion membranes.

Perfluorinated polymers with pendant phosphonic acid groups were also described in [37]. These polymers are analogues of Nafion and their synthesis requires demanding and expensive fluorine chemistry.

In addition, there are several investigations on proton exchange polymers, whose phosphonic acid sites are linked to aromatic ring directly or by means of alkylene spacers. These polymers can be prepared by polycondensation of phosphonic acid functionalized monomers or phosphorylation of existing aromatic condensation polymers. There are several methods reported for the preparation of arene phosphonates from aryl bromides [38–40]. The classical Arbuzow reaction [41, 42] is commonly used, in which triethylphosphite and nickel dichloride are employed as catalysts. The reaction often requires harsh conditions (160 °C) and suffers from low yields. Alternative palladium-catalyzed reaction [43–45] useful for phosphonation proceeds more smoothly.

2.1 Polycondensation of Phosphonic Acid Functionalized Monomers

Phosphonic and phosphinic acid moieties may be incorporated in the main chains of macromolecules or in the side groups.

In general, few articles on the polycondensation of phosphonic acid functionalized monomers can be found.

Miyatake and Hay [46] reported the first phosphonic acid containing polymers and copolymers from three phosphine-containing aromatic difluorides, prepared in accordance with the following Scheme 1:

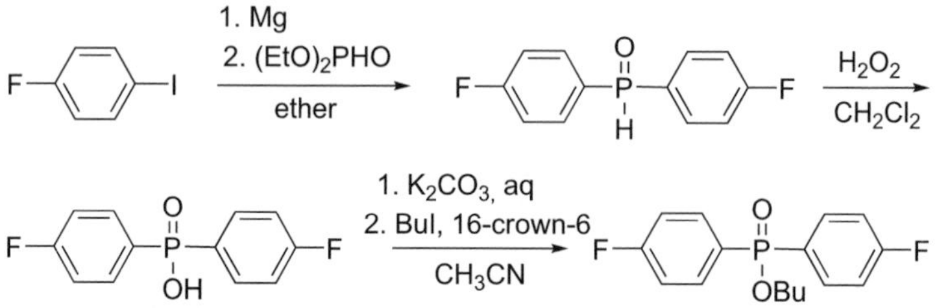

F–C6H4–I
1. Mg
2. $(EtO)_2PHO$
ether
F–C6H4–P(=O)(H)–C6H4–F
H_2O_2
CH_2Cl_2
F–C6H4–P(=O)(OH)–C6H4–F
1. K_2CO_3, aq
2. BuI, 16-crown-6
CH_3CN
F–C6H4–P(=O)(OBu)–C6H4–F

Scheme 1

The aforementioned difluoromonomers were polymerized with bisphenol A (BPA) bis-(propylcarbamate) under basic conditions at 160 °C in DMSO to produce the corresponding poly-(arylene ether)s (Scheme 2):

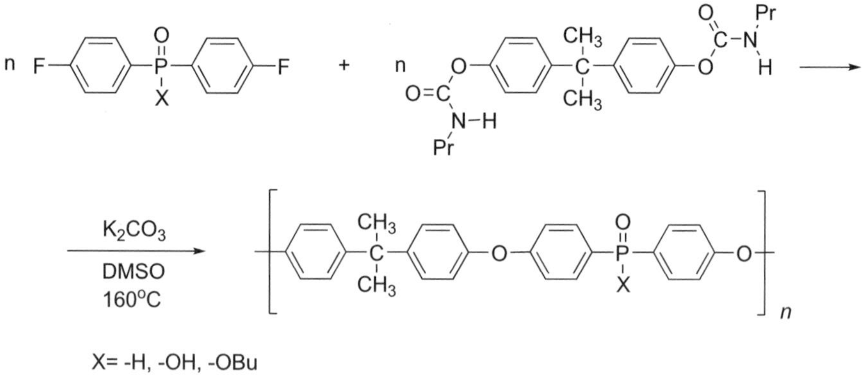

n F–C6H4–P(=O)(X)–C6H4–F + n Pr–NH–C(=O)–O–C6H4–C(CH_3)$_2$–C6H4–O–C(=O)–NH–Pr
K_2CO_3
DMSO
160°C
[–C6H4–C(CH_3)$_2$–C6H4–O–C6H4–P(=O)(X)–C6H4–O–]$_n$

X= -H, -OH, -OBu

Scheme 2

Because the carbamoyl-masking groups improve these kinds of nucleophilic-substitution polymerization reactions [47], BPA bis-(propylcarbamate) was used.

Target polyethers were prepared with 25–78% yields [47].

Their general properties are given in Table 1.

Table 1 Properties of the polyethers of general formula

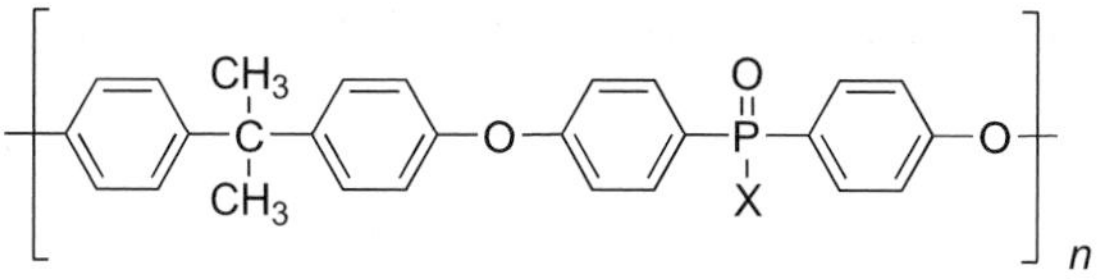

– X	Yield %	3_{inh} (dL g^{-1})[a]	*Tg* (°C)	$Td_{10\%}$[b] (°C)
– H	52	0.24	200	477
– OH	25	0.14	102	422
– OBu	78	0.30	125	325

[a] In NMP at 27 °C
[b] Temperature for 10% weight loss under nitrogen

Polyether with phosphonic acid groups in the main chain (– X = – OH) was very soluble in polar organic solvents (DMF, N-MP, DMSO), slightly soluble in chloroform, but insoluble in water. Polyether with *n*-butil phosphinate ester groups was soluble in many organic solvents like benzene, chloroform, THF, and DMF.

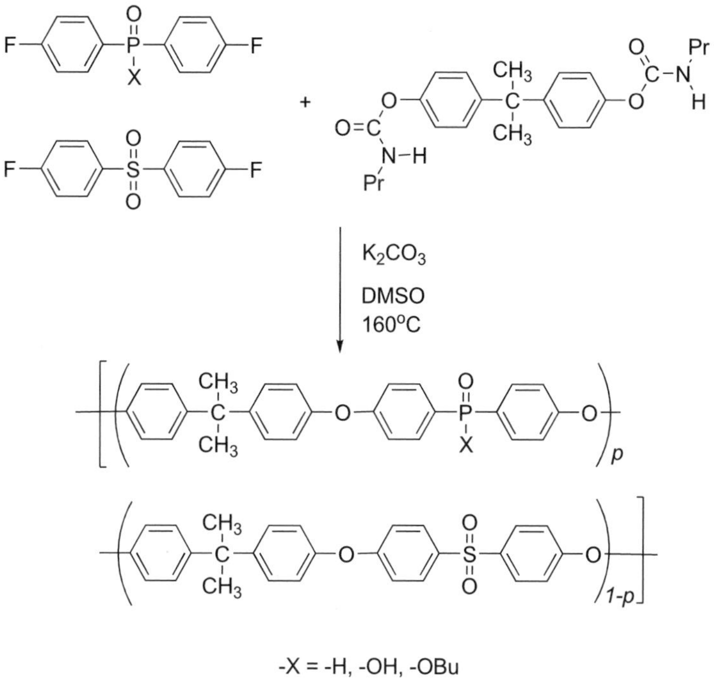

K_2CO_3
DMSO
160°C

-X = -H, -OH, -OBu

Scheme 3

Films obtained by casting from solution were brittle and self-supporting.

To produce higher molecular weight polymers, copolymerization with *bis*-(4-flourophenyl)sulfone was investigated (Scheme 3).

The copolymerization was successful and gave copolymers with a higher solution viscosities (up to 0.39 dL g^{-1} in N-MP) than that of corresponding homopolymers (Table 2).

Table 2 Properties of the copolyethers of general formula

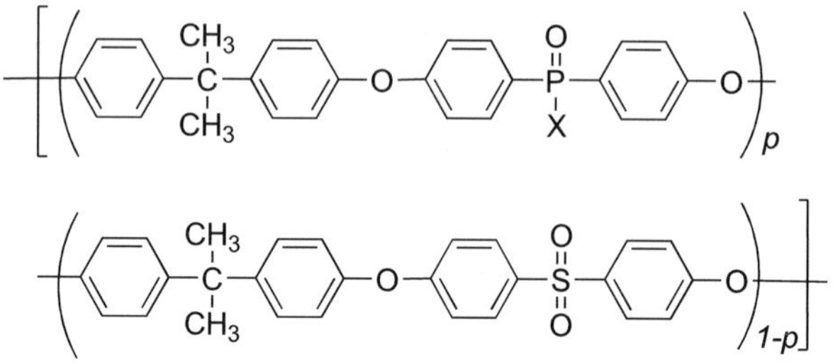

–X	*p*	Yield %	3_{inh} (dL g^{-1})[a]	*Tg* (°C)	$Td_{10\%}$[b] (°C)
–H	0.5	78	0.32	202	432
–H[c]	0.5	65	0.39	225	454
–OH	0.1	85	0.33	191	490
–OH	0.2	67	0.37	183	452
–OH	0.3	59	0.14	184	465
–OH	0.5	31	0.14	157	428
–OBu	0.2	91	0.19	175	445
–OBu	0.4	81	0.11	183	401

[a] In NMP at 27 °C
[b] Temperature for 10% weight loss under nitrogen
[c] Polymerization run in *N,N*-dimethylacetamide

Because the copolymers were more soluble than the homopolymers and even somewhat soluble in methanol, the isolated yields were not quantitative.

From the stoichiometric investigations, it was found that trace amounts of water are crucial, and that they hydrolyze fluoride monomers. A 7 mol % excess of difluoride monomer was used to obtain the highest molecular weight copolymer (Table 3).

Copolymer with the highest viscosity was soluble in many organic solvents and gave a film by casting from solution. A high thermal stability ($Tg = 178$ °C, $T_{dec10\%} = 432$ °C) was observed.

The polymers did not have high hydrophilicity; however, small amount of polar organic solvents helped them to swell in water.

Hay et al. [48, 49] have synthesized phosphonic acid containing *bis*-phenol by the interaction of 4,9-*bis*-(4-hydroxyphenyl)benzoisobenzofuran-1,3-

Table 3 Stoichiometric effect on the copolymerization of bis-(4-fluorophenyl)-phosphonic acid and bis-(4-fluorophenyl)-sulfone with BPA

Excess of bis-(4-fluoro-phenyl)-phosphinic acid	Yield %	3_{inh} (dL g^{-1})[a]	*Tg* (°C)	$Td_{10\%}$[b] (°C)
0	31	0.14	157	428
0.013	35	0.34	164	432
0.05	42	0.25	156	433
0.07	70	0.78	178	432
0.07[c]	45	0.16	171	422
0.1	48	0.29	169	439

[a] In NMP at 27 °C
[b] Temperature for 10% weight loss under nitrogen
[c] Polymerization run at 200 °C in NMP

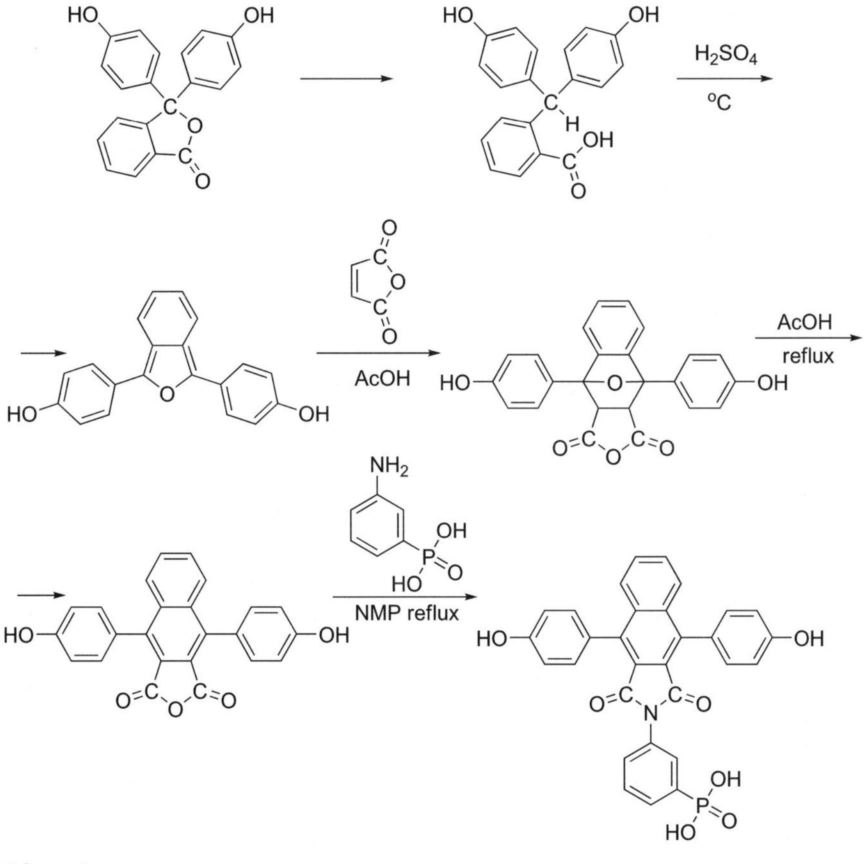

Scheme 4

dione – phenolphtaleine derivative [50, 51] – with *m*-aminophenylphosphonic acid (Scheme 4).

Poly(arylene ethers) containing phosphonic acid groups were prepared from the *bis*-phenol (mixed with *bis*-phenol A) and 4,4′-difluorodiphenyl sulfone [48, 49] in accordance with Scheme 5:

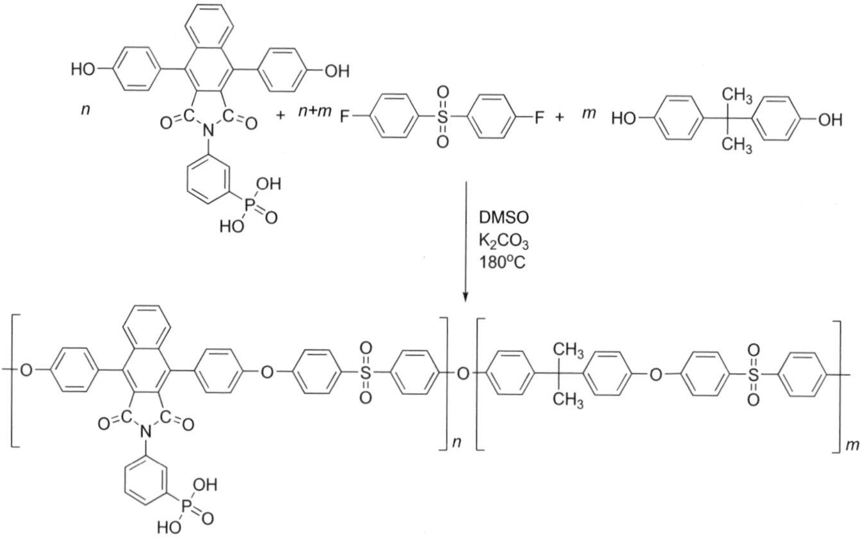

Scheme 5

The poly(arylene ethers) with various compositions were prepared.

Because of the very low nucleophilicity of phosphonic acid salts, high molecular weight polymers were obtained in a prolonged time of 5 hrs. The phosphorus-31 nuclear magnetic resonance (^{31}P-NMR) spectrum of polymers showed a single peak at 12.286 ppm indicating the presence of phosphonic acid group in the polymer chain.

Thus, general properties of the polymers were obtained, as given in Table 4.

As listed in Table 4, all the polymers have high molecular weights. Therefore, these polymers can easily be cast into tough clear and flexible films. Polymer based on 3-[4,9-*bis*(4-hydroxyphenyl)-1,3-dioxo-(1,3-dihydrobenzo[*f*]isoindolo-2-yl)-phenyl]-phosphonic acid only has the lowest molecular weight since the polymer precipitated during the polymerization reaction. Tg_s of the copolymers increase linearly with increasing phosphonic acid group content. The introduction of ionic moiety into the polymer generally leads to an increase of the interaction among the polymers chains; therefore, homopolymer ($n_c = 10.0$) has no detectable Tg. The thermogravimetric data indicates that the polymers listed in Table 4 are thermooxidatively stable, while the TGA_{onset} of these polymers decrease with the increase of phosphonic acid content. Nevertheless 5% weight loss temperature ($TGA_{5\%}$)

Table 4 Properties of polymers

n %	*m* %	3_{inh}	EM[a]	Water[b] Absorption %	*Tg* (°C)	TGA_{onset} (°C)	$TGA_{5\%}$ (°C)	Proton Conductivity ($S\,cm^{-1}$)
25	75	0.67	518	3.2	254	289.8	502.4	6.78×10^{-6}
50	50	0.60	596	4.7	277	245.7	498.0	2.96×10^{-5}
75	25	0.54	673	6.4	315	229.1	480.5	1.86×10^{-5}
100	0	0.43	750	7.5	ND	194.2	476.1	1.32×10^{-5}

[a] EM-equivalent molecular weights that are defined as the equivalent weight per phosphonic acid group
[b] Water absorption in hot water (12 hrs)

is much higher than TGA_{onset} for these polymers, demonstrating that these polymers are more stable than Nafion [52].

The proton conductivity of copolymer (50 : 50) was determined to be $2.96 \times 10^{-5}\,S\,cm^{-1}$. It is three orders of magnitude lower than that of Nafion ($10^{-2}\,S\,cm^{-1}$ under moisture). Surprisingly, there was no further increase in conductivity when the phosphonic acid moiety was larger than 50%. The conductivity decreased by a factor of two to $1.32 \times 10^{-5}\,S\,cm^{-1}$, because the increase in *Tg* values leads to the decrease in conductivity.

New diamine containing phosphonic acid groups – (4-[4,5-*bis*(3-aminophenyl)-1*H*-imidazol-2-yl]-phenyl)phosphonic acid diethyl ester – [53] was

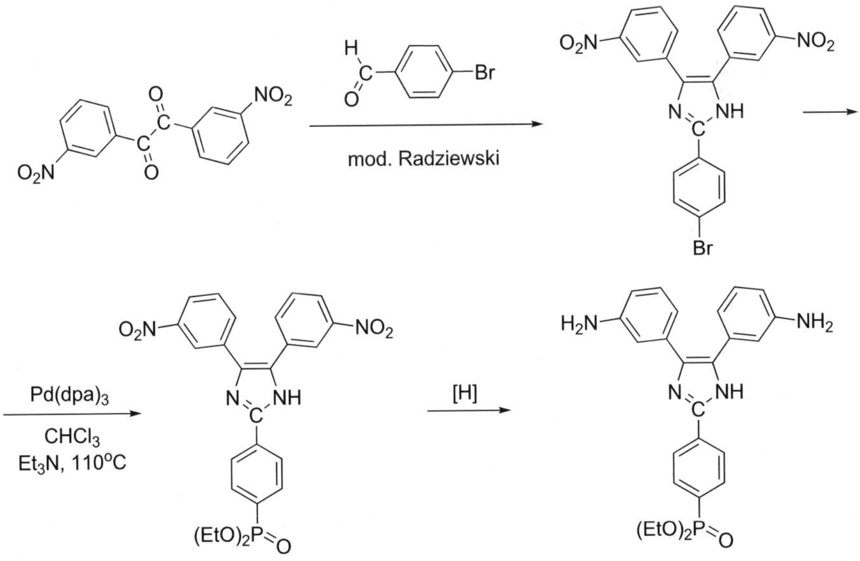

Scheme 6

prepared in accordance with the scheme, including the interaction of 3,3′-dinitrobenzyl with *p*-bromobenzaldehyde [54] P – C coupling and reduction of the diethyl phosphonate (Scheme 6).

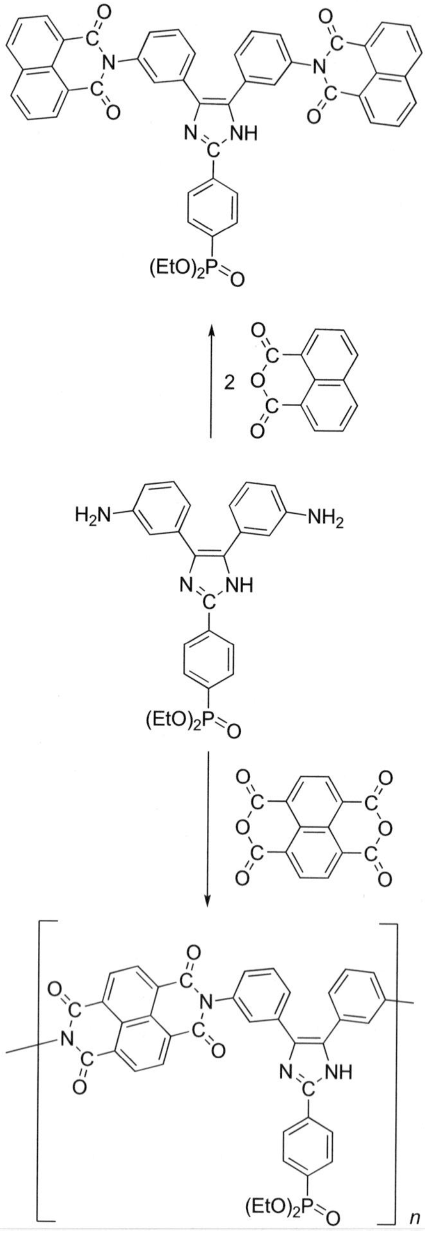

Scheme 7

Interaction of this diamine with two-fold molar amount of naphthalic anhydride and equimolar amount of naphthalene-1,4,5,8-tetracarboxylic acid dianhydride led to the formation of *bis* (model) compound and poly-naphthylimide-containing phosphonic acid groups, respectively (Scheme 7).

Phosphonated polybenzimidazoles were prepared by the interaction of 3,3′,4,4′-tetraaminodiphenylsulfone with 3,5-dicarboxylbenzenephosphonic acid (DCP) (Scheme 8) [55].

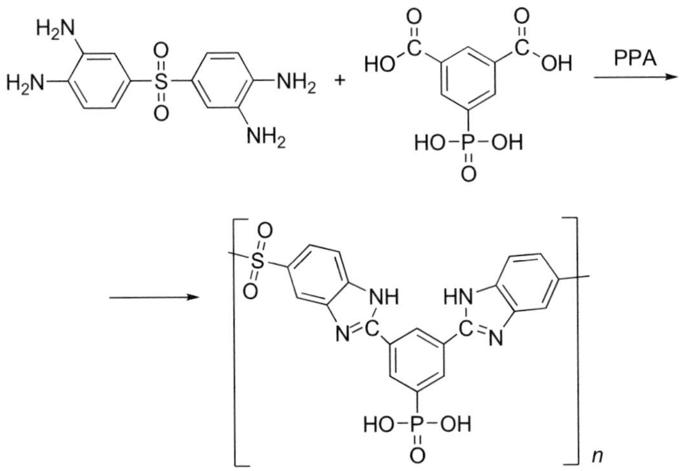

Scheme 8

The proton conductivity of such polymers was close to 10^{-2} S cm^{-1}, and it was higher than that of sulfonated polymers. Although the phosphonated structure has advantage of increasing the proton conductivity in ion-containing polybenzimidazole, incorporation of DCP in these polymers decreases the solubility in *N*-methyl-2-pyrrolidone (N-MP).

In order to obtain the membrane in which both conductivity and solubility coexist, copolymers containing sulfonic and phosphonic acid groups at the same time were prepared (Schemes 9, 10) [55]. As dicarboxylic acids containing sulfonic acid groups 5-sulfoisophthalic acid monosodium salt (SIA) and 2-sulfoterephthalic acid monosodium salt (STA) were used.

The copolymer with STA : DCP = 50 : 50 showed good solubility in N-MP, but only similar level of conductivity to the sulfonated polymer. When the content of DCP was increased to STA/DCP = 30 : 70, the conductivity jumped to similar level of phosphonated polymer, while maintaining the solubility in N-MP. Further increase of DCP content to STA/DCP = 15 : 85 induced the decrease in solubility. There is a narrow region where good conductivity and solubility coexist in these copolymers. Ion-containing polybenzimidazoles showed excellent stability against hydrolysis, radical attack and dimensional change in wet/dry cycles.

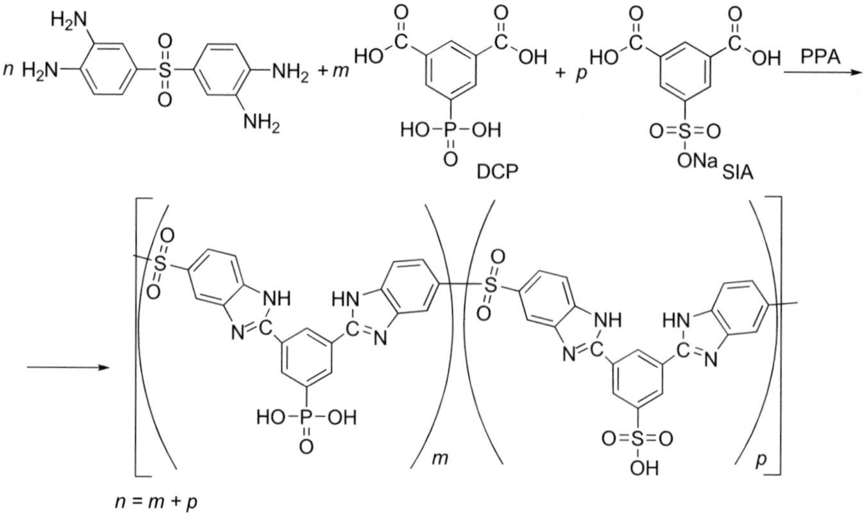

$n = m + p$

Scheme 9

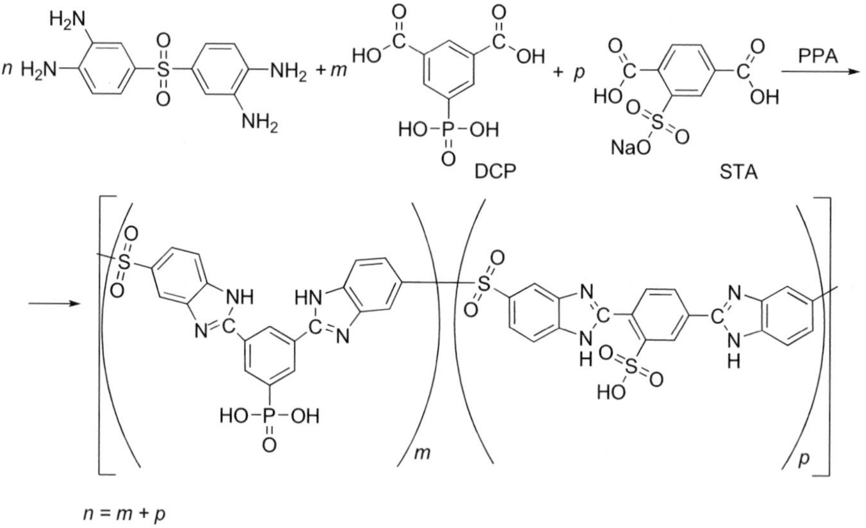

$n = m + p$

Scheme 10

In a recent preliminary report [56], Shanze et al. described poly(phenylene ethynylene) which feature phosphonate groups appended to the polymer backbone (Scheme 11).

The phosphonated polymer was prepared via a neutral precursor polymer, which was soluble in organic solvents, enabling the material to be characterized by nuclear magnetic resonance (NMR) and gel permeation chromatog-

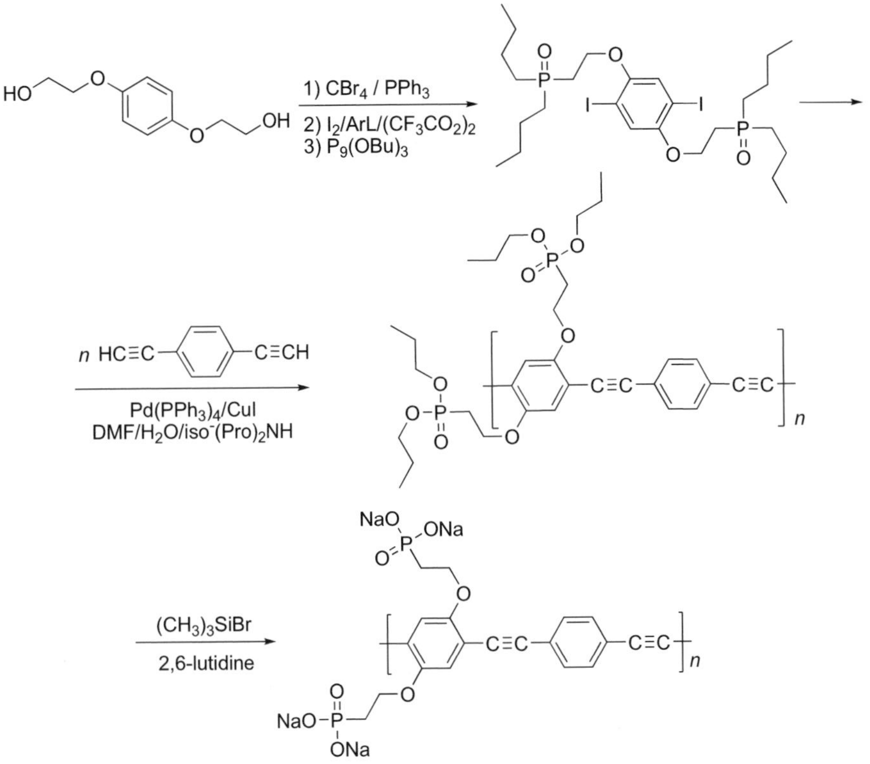

Scheme 11

raphy (GPC). Sonogashira polymerization of phosphonate monomer and 1,4-diethynylbenzene afforded neutral polymer in a 46% yield. Analysis of the neutral precursor polymer indicated $M_W = 18.3$ kD and polydispercity 2.9.

The target polymer was prepared by bromotrimethylsilane-induced cleavage of the *n*-butyl phosphonate ester group in neutral precursor polymer. After neutralization of the reaction mixture with aqueous sodium hydroxide, target polymer exhibited good solubility in water.

Generally, preparation of phosphonic acid functionalized monomers and their polymerization are problematic, because of the strong aggregation and condensation of phosphonic acid units which result in poor yields [7, 33, 56–58]. Consequently, a more attractive strategy is to phosphorylate existing polymers [59].

2.2
Phosphorylation of Existing Polymers

A number of procedures are well established for the preparation of aryl phosphonic acids by introducing phosphonic functional groups into aromatic compounds of low molecular weight [60], but some of them proved to be inappropriate for a polymeric starting material. The Friedel–Crafts reaction using phosphorus trichloride is, for example, a common efficient and low-cost procedure for direct phosphonation of aromatic rings. Systematic studies of this reaction on polystyrenes showed, however, that the trifunctionality of phosphorus trichloride leads to cross-linked products regardless of reaction conditions [61, 62].

Nevertheless, the possibility of modification of polysulfones main chains by Friedel–Crafts reaction with phosphorus trichloride and $SnCl_4$ as a catalyst was demonstrated by Ziaja et al. [63].

Recently, Jakoby et al. [64, 65] reported a synthesis of phosphonated polyphenylsulfone. This group has focused their efforts on two straightforward reactions for the synthesis of aryl phosphonic acids from brominated or iodinated aromatic starting materials.

The bromination of different polysulfones, including polyphenylsulfone, was described in details by Guiver et al. [66]. The characterization of the product by elemental analysis gave a degree of substitution of 2 Br atoms per repeating unit of the polymeric structure.

The electron-donating effect of the ether linkages of the main chain activates the phenylene rings of the *bis*-phenol part towards nuclephilic attack, and they may, thus, be used as positions for halogenation, such as bromination [66].

After bromination, two different procedures were tested for a bromine-phosphorus exchange. The first one is the Michaelis-Arbuzow reaction in the presence of transition metal salt catalysts [41, 42]; the second one is a P – C coupling reaction catalyzed by Pd (0) complexes [43–45]. Both reactions are usually carried out without any solvent. It was found that the reaction conditions of the P – C coupling reaction could be adapted to the application of polymeric substrates [65].

The Michaelis-Arbuzow reaction was adapted to polymeric substrates using high-boiling solvents, such as dimethylacetamide (DMA) and N-MP. The application of these polar solvents was necessary, because triethyl phosphite is a very strong precipitation medium for polymeric substrate. It turned out that the Michaelis–Arbuzow reaction was very slow, either due to the dilution of the components in the mixture or due to the sterical hindrance of reaction sites on the polymer chain. As a consequence, the reduction of the metal catalyst by triethyl phosphite, which is a common and well-known side reaction for this procedure [41], was favored and became predominant. The rapid reductive deactivation process of the catalyst could be observed by color changes of the reaction mixture. After adding the first drops of triethyl phosphite to the mixture at 150 °C, the green active catalyst species were converted into colorless Ni(0) complexes or black Pd metal particles within 1 hour.

The Pd(O)-catalyzed P – C bond formation is another convenient modern method for the preparation of aryl phosphonic acids from aryl bromides or iodides and diethyl phosphite (DEP). The reaction conditions had to be modified again for polymeric substrates, because the conversion is usually carried out with liquid aryl halides at 90 °C in the absence of an additional solvent. The reaction for dibrominated polyphenylsulfone is depicted in Scheme 12:

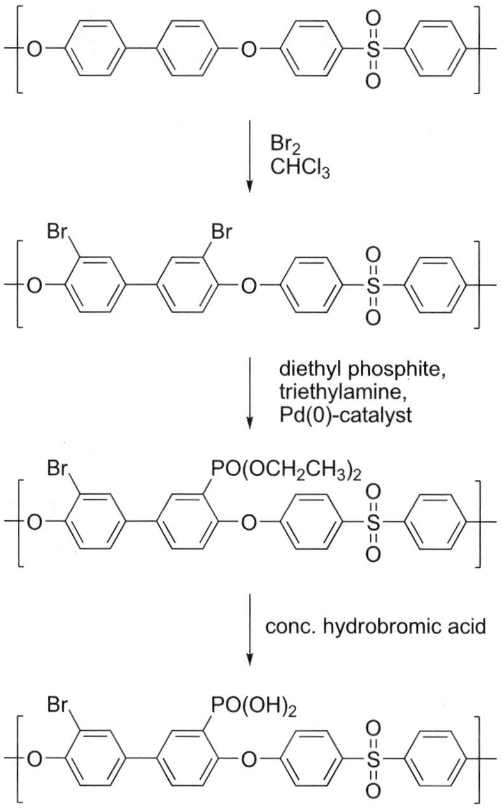

Br_2
$CHCl_3$

Br Br

diethyl phosphite,
triethylamine,
Pd(0)-catalyst

Br $PO(OCH_2CH_3)_2$

conc. hydrobromic acid

Br $PO(OH)_2$

Scheme 12

Diethyl phosphite turned out to be a less efficient precipitation medium for the dibrominated polyphenylsulfone than triethyl phosphite.

Systematic investigations showed that the polymer is soluble in mixtures of diethyl phosphite and *o*-dichlorobenzene (DEP : DCB = 1 : 2.5 v/v or 2 : 1 v/v), and insoluble in any of the pure components. Two different Pd(O) complexes with triphenyl phosphine or dibenzylideneacetone (dba) were used as ligands. Triethylamine (TEA) was required for the removal of the HBr side product. It is well known that the molar rations of the components, the choice of the catalyst, and the reaction temperature have a strong influence on the yield of the phosphonation product [67, 68]. Therefore, systematic variations of these parameters and their effect on the degree of phosphonation of the polymer material were investigated (Table 5).

Table 5 Influence of reaction conditions on the substitution (phosphonation) degree (DS – number of phosphoric groups per monomer unit)

Product no.	Temp. °C	Catalyst	Procedure	Solvent	DS [a]	Product weight [c] g
1	90	$Pd(PPh_3)_4$	2	DCB	0.40	8.5
2	120	$Pd(PPh_3)_4$	2	DCB	0.58	7.9
3	140	$Pd(PPh_3)_4$	2	DCB	decomposition	
4	120	$Pd_2(dba)_3 \cdot CHCl_3$	2	DCB	0.77	8.5
5	120	$Pd_2(dba)_3 \cdot CHCl_3$	3	DPE	0.88	9.2
6 [b]	120	$Pd_2(dba)_3 \cdot CHCl_3$	4	none	0.79	8.3
7	110	$Pd(PPh_3)_4$	2	pyridine	decomposition	

[a] Determined by ^{1}H NMR spectroscopy
[b] Post-phosphonation of product 4 in the absence of any additional solvent
[c] Starting from 10 g polymer

Due to solubility problems of the polymer, the reaction had to be started in presence of only 9–13 equivalents of diethyl phosphite and 1–2 equivalents of triethylamine, related to aryl bromide units. As the solubility of the polymer improved during phosphonation, the concentration of DEP and TEA could gradually be increased, until, finally, a molar ratio of ArBr : DEP : TEA = 1 : 22 : 4 resulted, which comes close to the reaction conditions described in [69].

An increase of the reaction temperature from 90 to 120 °C enhances the degree of phosphonation. If the temperature was further increased to 140 °C, the yellow color of the reaction mixture turned dark, and decomposition was observed. The Pd catalyst was obviously more active in the absence of triphenyl phosphine ligands. The highest degrees of phosphonation were achieved by using diphenyl ether as a solvent, instead of *o*-dichlorobenzene.

All phosphonated products could easily be converted into the free phosphonic acids by acid hydrolysis of the esters.

Concerning the polymer stability, the TGA curves show weight loss in the temperature range between 225 and 325 °C. According to the literature [34, 70], a weight loss between 250 and 350 °C is attributed to the fragmentation of the ethyl phosphonate units into the free acid and ethylene, and possibly a subsequent formation of anhydride groups from two phosphonic acids. This explains why the first weight loss in TGA curves was not found for free phosphonic acid samples.

The decomposition of the polymeric free phosphonic acids only started above 350 °C. Unreacted bromine atoms are also believed to split above 250 °C [70].

All the polymers with free phosphonic acid pendant groups are insoluble in methanol and water. They are soluble in DMA or DMSO up to a DS of

0.58, whereas the highly phosphonated sample with a DS above 0.75 did not completely dissolve any longer. In DMA, these samples only formed a highly swollen polymer gel. It turned out, however, that the addition of 2 vol % of concentrated hydrobromic acid to the mixtures led to rapid dissolution. The gelation is supposed to be caused by strong ionic clusters in the highly functionalized material. These clusters are destroyed by the mineral acid.

For film casting, a solution with 8 wt % of polymer in DMA was prepared in the presence of 2 vol % of concentrated HBr. The resulting polymer films were mechanically stable and clear. These films are under investigation as solid electrolyte membranes for fuel cell applications.

Transformation of the brominated polyethersulfones to the phosphonated polyethersulfones was carried also using Ni catalysts. Mulhaupt et al. [71] have prepared phosphonated polyethersulfones via the treatment of the brominated polyethersulfone with tris-(trimethylsilyl)phosphate (TMSP) catalyzed with $NiCl_2$ and subsequent methanolysis of the reaction product (Scheme 13).

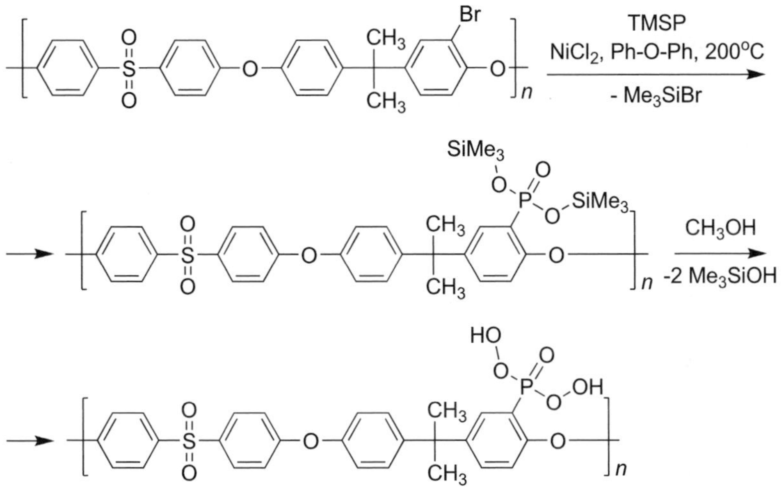

TMSP, $NiCl_2$, Ph-O-Ph, 200°C; - Me_3SiBr

CH_3OH; -2 Me_3SiOH

Scheme 13

Resulting polymers dissolve in N-MP or N-MP containing 2–5 vol % conc. HCl. In accordance with TGA data, degradation of polymers with arylphosphonic acids groups started at temperature >325 °C. Analog of this polymer containing sulfonic acid groups started to desulfonate at >200 °C. Depending on the conversion degree, protonic conductivity of the phosphonated polyethersulfone was ranged between 1.2–8.7 mSm cm^{-1}, which was lower than the conductivity of the polyethersulfone containing sulfonic acid groups (20 mSm cm^{-1}).

Another attractive way of functionalizing poly(arylene ether sulfone) is lithiation, followed by the reaction of the lithiated sites with an electrophile.

Savignac et al. [72] made an extensive study on the synthesis of phosphonates by nucleophilic substitution at phosphorus, a so-called $S_NP(V)$ reaction. They showed that organolithium compounds are the reagents of choice for the synthesis of phosphonates, because these compounds generally minimize the formation of side products.

In the frame of work [59],the feasibility of employing an $S_NP(V)$ reaction between lithiated polysulfones and either dialkylchlorophosphates or diarylchlorophosphates to phosphonate polysulfones without the use of a catalyst was investigated (Scheme 14):

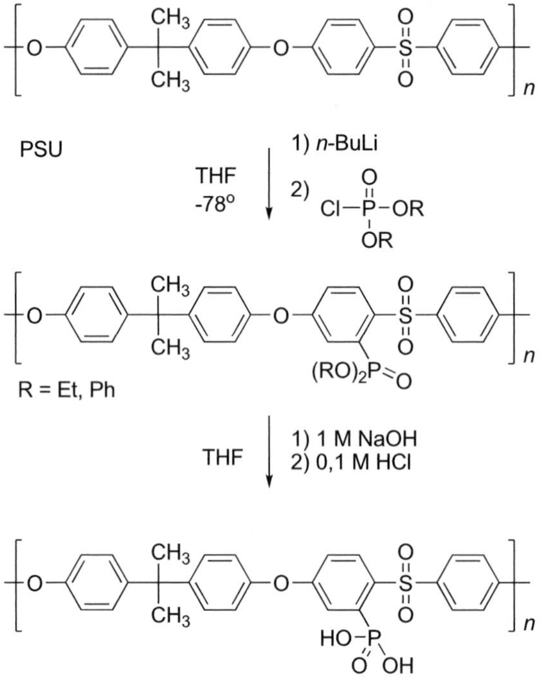

PSU

THF
-78°

1) *n*-BuLi
2) Cl–P(=O)(OR)–OR

R = Et, Ph

$(RO)_2P{=}O$

THF

1) 1 M NaOH
2) 0,1 M HCl

HO–P(=O)–OH

Scheme 14

Polysulfone, PSU, is an amorphous aromatic polymer that consists of alternating bisphenol A and biphenyl sulfone segments. The strong electron-withdrawing effect of the sulfone units gives a slight acidic character to the *ortho*-to-sulfone aromatic hydrogen. Thus, the carbon at these positions may be metalyzed by the use of a strong base, such as *n*-BuLi [73].

The use of lithium chemistry is a versatile and convenient way of modifying PSU. Moreover, lithiation and a subsequent reaction with an electrophile do not give rise to chain scission or any other degradation in the polysulfone main chain [74].

Polysulfones carrying benzoyl(difluoromethylenephosphonic) acid side groups [75] were prepared in accordance with (Scheme 15):

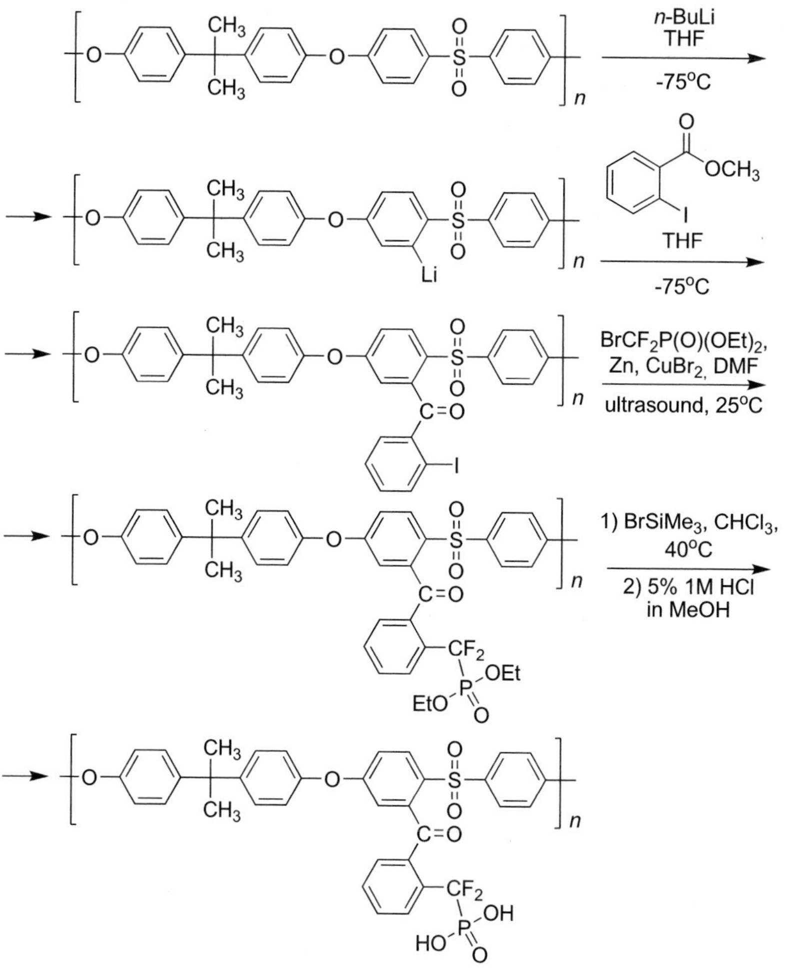

Scheme 15

In the first step, polysulfones were lithiated and reacted with methyl iodobenzoates to prepare *p*- and *o*-iodobenzoyl polysulfones. Next, the phosphonated polysulfones were prepared via CuBr – mediated cross-coupling reactions between the iodinated polymer and [(diethoxyphosphinyl)difluoromethyl]zinc bromide. Finally, dealkylation with bromotrimethylsilane afforded highly acidic $-CF_2-PO_3H_2$ derivatives. The replacement of the iodine atoms by $-CF_2-PO_3H_2$ units was almost quantitative in the case of *o*-iodobenzoul polysulfone. Membranes based on ionomers having 0.90 mmol of phosphonic acid units/g of dry polymer demonstrated protonic conductivities up to 5 mSm cm^{-1} at 100 °C. Thermogravimetry revealed that aryl-$CF_2-PO_3H_2$ arrangement decomposed at approximately 230 °C via cleavage of the C – P bond [75].

Alternative approach to the phosphonated PSU is based on lithiation of brominated PSU, because the halogen-lithium exchange is favored over the proton-lithium exchange at low temperatures [66] (Scheme 16):

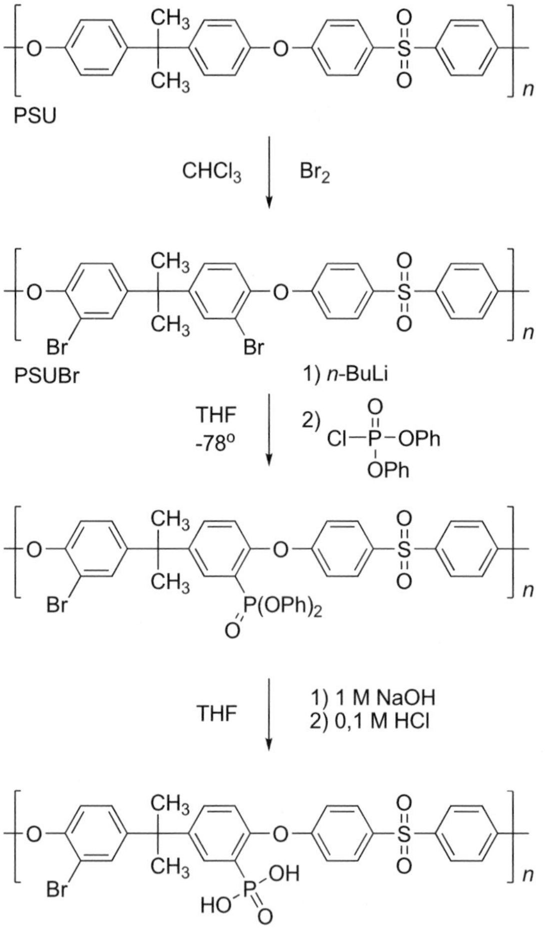

PSU

$CHCl_3$ Br_2

PSUBr

THF -78° 1) *n*-BuLi 2) Cl–P(O)(OPh)–OPh

THF 1) 1 M NaOH 2) 0,1 M HCl

Scheme 16

Along with poly(arylene ether sulfones), some other polymers were phosphonated using above-mentioned procedures.

Hay and coworkers [76] reported on a catalyzed P – C coupling reaction to synthesize phosphonated poly(arylene ethers) (Scheme 17).

Starting *bis*-propylcarbomate-masked derivatives of mono- and dibromo-tetraphenylhydroquinone were prepared in accordance with the Scheme 18.

Highly fluorinated poly(aryl ether) with a phosphonic acid group was prepared by Guiver et al. [77] in accordance with the Scheme 19.

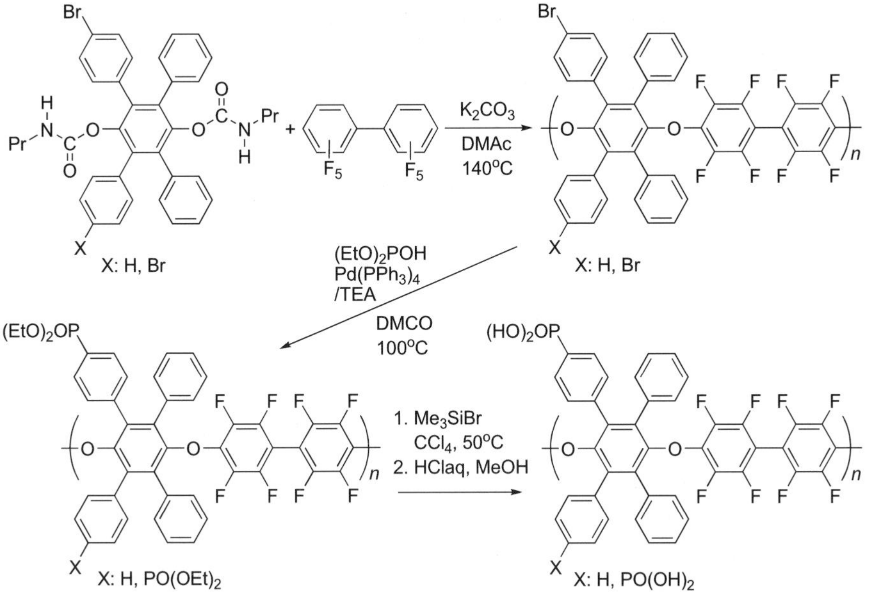

Scheme 17

The polymer obtained has high thermal stability – no obvious decomposition is observed until the main chain decomposition at ∼420 °C. Polymer is readily soluble in DMA, DMF, DMSO, and N-MP. The high solubility of the polymers enables thin films to readily be prepared.

Because of its higher temperature stability than its $-SO_3H$ analog and the possibility of conductivity at low humidity levels, polymer with phosphonic acid groups ($-PO_3H_2$) has been studied as potential candidate for use as PEM material. It showed promising conductivity up to 2.6×10^{-3} S cm^{-1} in water at room temperature, and 6.0×10^{-3} S cm^{-1} at 95% relative humidity at 80 °C. It also exhibits an extremely low methanol permeability value of 1.07×10^{-8} cm^2 S^{-1}.

An attempt to synthesize ethylphosphorylated polybenzimidazole was reported [2] using the Scheme 20.

The substitution reaction at the NH sites of benzimidazole rings was performed successfully, but the resulting polymer appeared to be insoluble in organic solvents. The reason for this could be aggregation of phosphoric acid groups during the substitution reaction. Ethylphosphorylated PBI exhibited high proton conductivity (10^{-3} S cm^{-1}) even in the pellet form. According to the results obtained, the presence of polar phosphoric acid groups enhances the proton conductivity of polymer electrolytes.

Polybenzimidazole, whose phosphoric acid sites are linked to aromatic rings by means of alkylene spacers, were prepared by deprotonation of start-

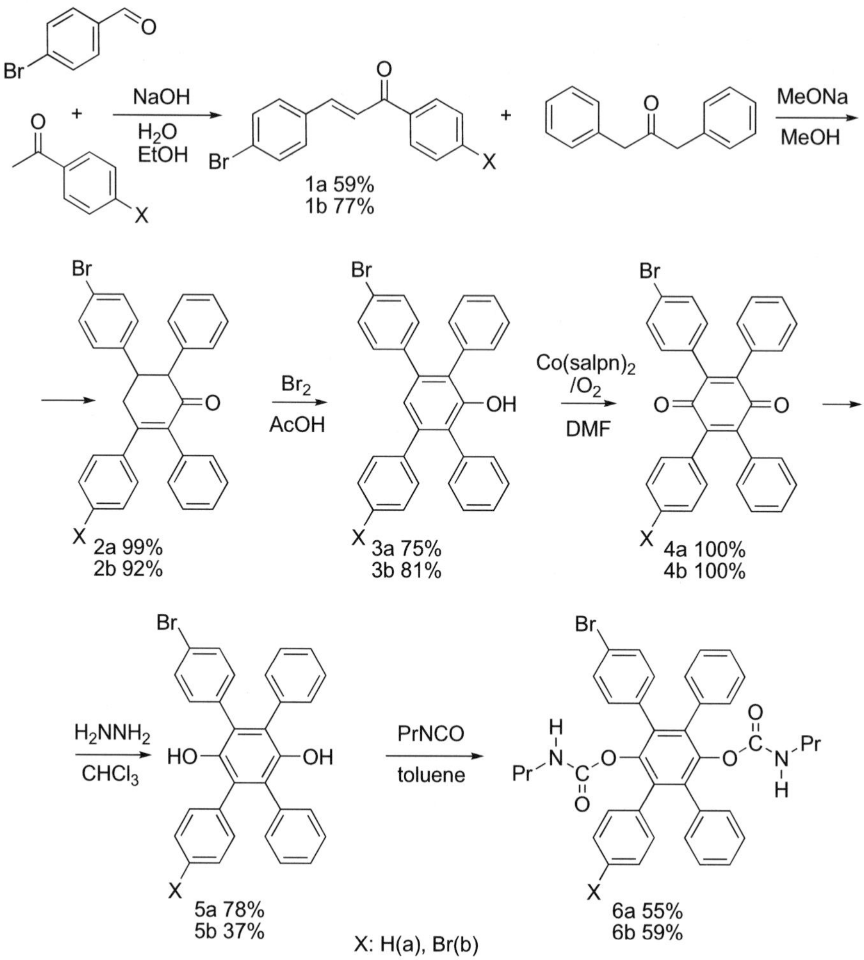

Scheme 18

ing polymers and subsequent conversion with chloro- or bromoalkylphosphonates [58].

The instability of benzylic methylene groups in an oxidizing environment is, however, a severe shortcoming of these materials for fuel cell applications.

N-substituted polybenzimidazoles, containing phosphonic acid groups which are linked to benzimidazole cycles by means of ethylenic groups, were prepared by treatment of different polybenzimidazoles with diethylvinylphosphonate [78] in accordance with Scheme 21.

Allcock and coworkers [79–81] reported on phosphonation of polyphosphazenes. Incorporation of pendent phosphate groups into aryloxyphosphazenes through phosphorylation (phosphorus-oxygen-carbon linkages) led to the formation of polymers containing phosphate ester bonds.

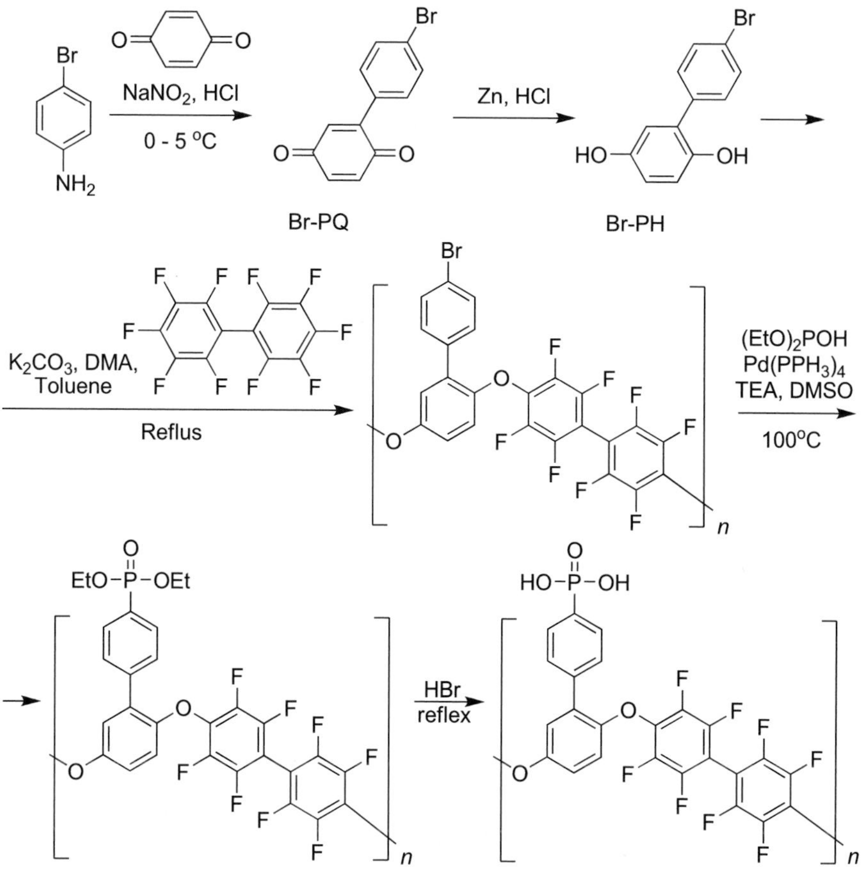

Br

NH_2

$NaNO_2$, HCl

0 - 5 °C

Br-PQ

Zn, HCl

HO OH

Br-PH

K_2CO_3, DMA,
Toluene

Reflus

F F F F F F F F

O

n

$(EtO)_2POH$
$Pd(PPH_3)_4$
TEA, DMSO

100°C

EtO–P–OEt

HBr
reflex

HO–P–OH

Scheme 19

DMA
LiH
85°

N H Li

$Cl(CH_2)_2PO(OH)_2$
NEt_3

DMAA, 21°C

$PO(OH)_2$ $PO(OH)_2$

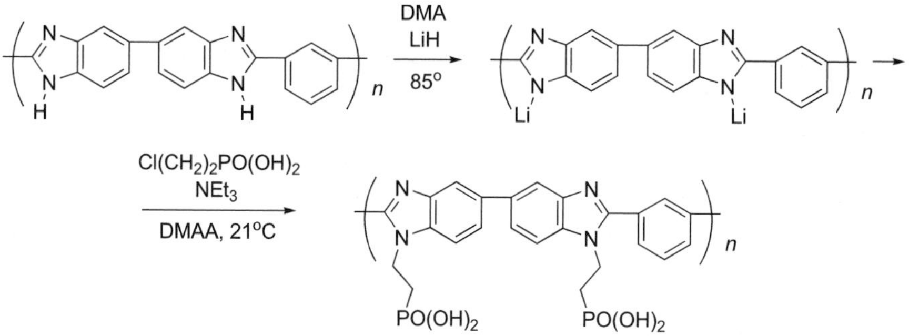

Scheme 20

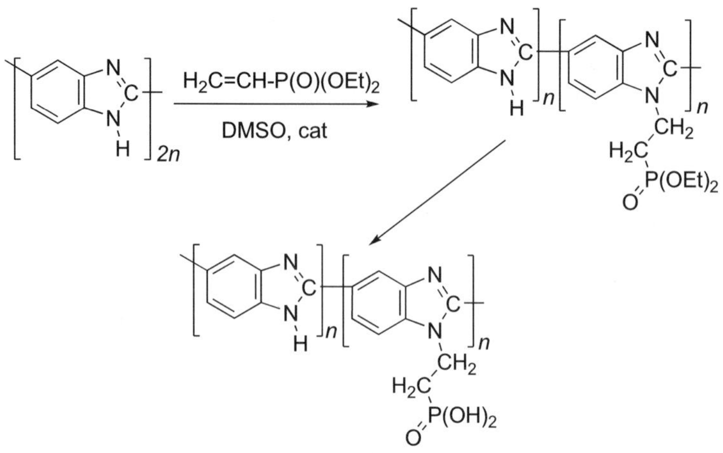

H_2C=CH-P(O)(OEt)$_2$

DMSO, cat

Scheme 21

However, phosphate ester bounds are known to have limited thermal stability and are subject to chemical attack by hydroxide ions, which may result in hydrolysis of the phosphorous-oxygen-carbon linkage and cleavage of pendent phosphate group [82].

As a result, the incorporation of pendent phosphate groups into aryloxyphosphazenes attached through phosphonate (phosphorus-carbon) linkages was examined. The use of direct phosphorus-carbon bond ensures greater thermal and chemical stability [82, 83], which should provide materials with more versatile properties.

The linkage of pendent phosphate groups to aryloxyphosphazenes is of interest, because the conversion of pendent dialkyl phosphate groups to phosphoric acid groups could lead to interesting and useful proton conducting membranes.

Two methods for the incorporation of dialkyl phosphonate units into the side groups of aryloxyphosphazenes were developed [80].

In the first approach, small molecule cyclic (Scheme 22) and high polymeric (Scheme 23) phosphazenes bearing bromomethylene – phenoxy side groups are treated with a sodium dialkyl phosphite, as shown in Scheme 22.

This method yields well-defined products at both the small-molecule and high polymeric level.

In the second route, a previously attempted procedure [64] was used, in which small-molecule cyclic (Scheme 24) and high polymeric (Scheme 25) phosphazenes bearing bromophenoxy side groups are treated with *n*-butyllithium, followed by a dialkyl chlorophosphonate, as shown in Scheme 24.

This method was improved for the synthesis of aryloxyphosphazenes with pendent dialkylphosphonate groups, and it led to well-defined products at both the small-molecule and polymeric level.

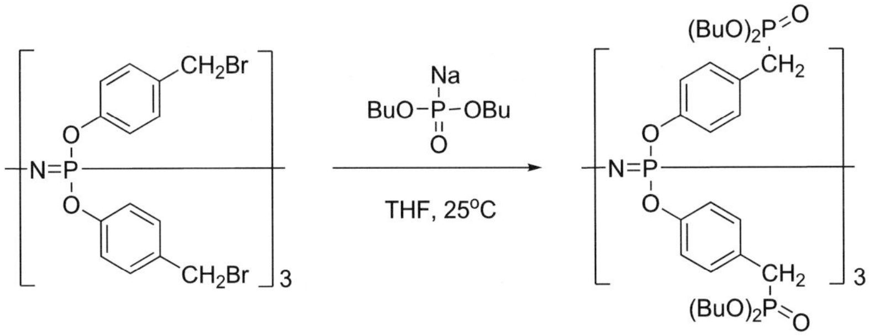

Scheme 22

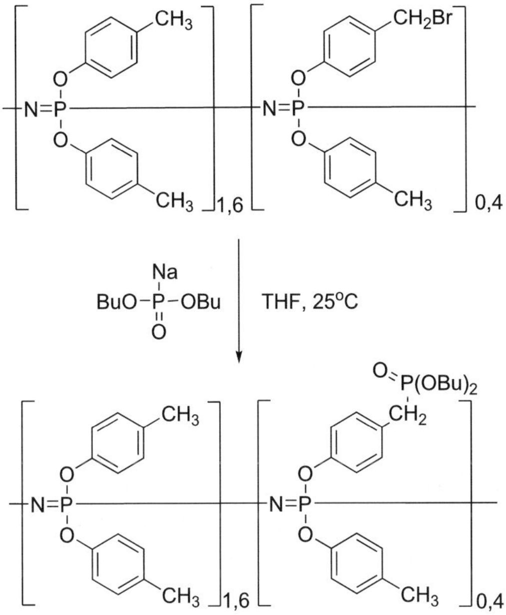

Scheme 23

Conversion of these pendent dialkyl phosphonate groups to phosphonic acid groups is a potential route to proton-conductive fuel cell membranes.

Phenylphosphonic acid functionalized poly-[aryloxyphosphazenes] were prepared [81] in accordance with the Scheme 26.

Phosphonation of poly(aryloxyphosphazenes) via lithiophenoxy intermediates was realized via addition of *tert*-butyllithium to a solution of the polymer in THF at –75 °C. The lithiated polymers were then treated with diphenyl chlorophosphate, and subsequent basic hydrolysis and acidification yielded phenylphosphonic groups, as shown in Scheme 26.

Initial attempts to phosphonate the lithiated polymers via the dropwise addition of diphenyl chlorophosphate resulted in precipitation of the polymer

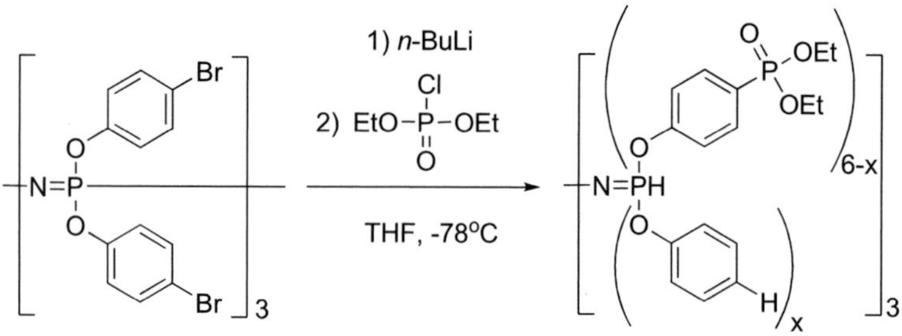

Br
O
N=P
O
Br 3
1) n-BuLi
Cl
2) EtO–P–OEt
O
THF, -78°C
O
P
OEt
OEt
O
N=PH
O
H
x
6-x
3

Scheme 24

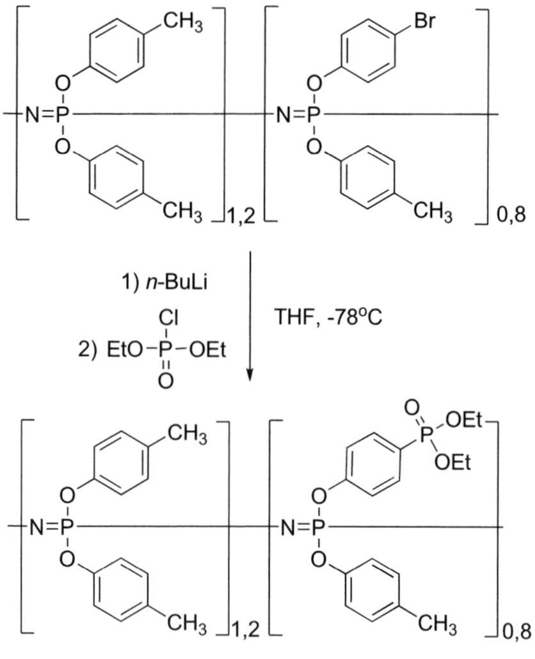

CH_3
O
N=P
O
CH_3 1,2
Br
O
N=P
O
CH_3 0,8
1) n-BuLi
Cl
2) EtO–P–OEt
O
THF, -78°C
CH_3
O
N=P
O
CH_3 1,2
O
P
OEt
OEt
O
N=P
O
CH_3 0,8

Scheme 25

and formation of insoluble product. However, soluble product was obtained by the rapid addition of diphenyl chlorophosphate. On the basis of ^{31}P-NMR spectroscopy data, it was concluded that intramolecular coupling reaction takes place; in this reaction diphenyl phosphonate ester sites on the polymer react with a second lithio-phenoxy side groups on the same polymer backbone, resulting in a phosphorus bridge between two aryloxy side groups, as shown in Scheme 27.

This was attributed to an intramolecular reaction rather than the analogous intermolecular cross-linking reaction, due to the high solubility of the polymers.

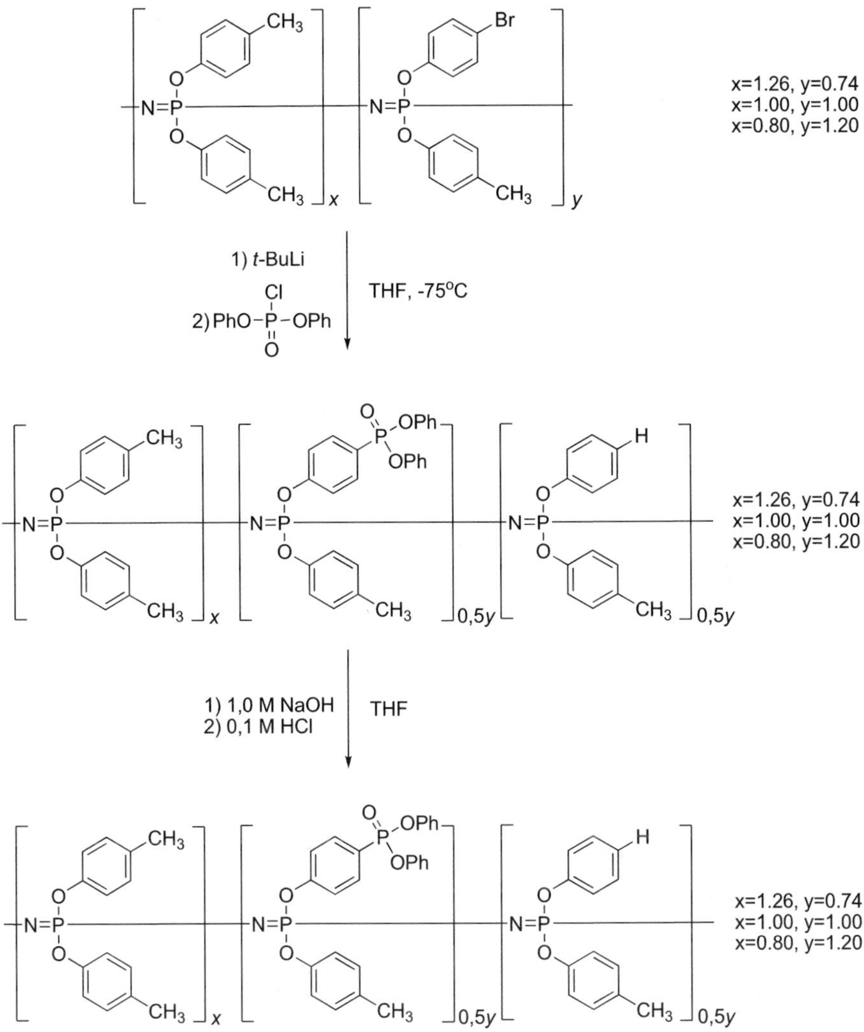

x=1.26, y=0.74
x=1.00, y=1.00
x=0.80, y=1.20

1) t-BuLi
2) PhO–P(Cl)(=O)–OPh
THF, -75°C

x=1.26, y=0.74
x=1.00, y=1.00
x=0.80, y=1.20

1) 1,0 M NaOH
2) 0,1 M HCl
THF

x=1.26, y=0.74
x=1.00, y=1.00
x=0.80, y=1.20

Scheme 26

The overall efficiency of the phosphonation method, with respect to percent conversion of the initial bromine atoms present, was determined by ^{31}P-NMR spectroscopy.

Phosphonation of polymers via lithiophenoxy intermediates proceeded with ~50% conversion of the bromophenoxy side groups to diphenyl phosphonate ester side groups. The total percentage of lithiophenoxy groups that had reacted with the diphenyl chlorophosphate was 70%. However, only ~50% of the lithiophenoxy groups resulted in formation of the diphenyl phosphonate ester.

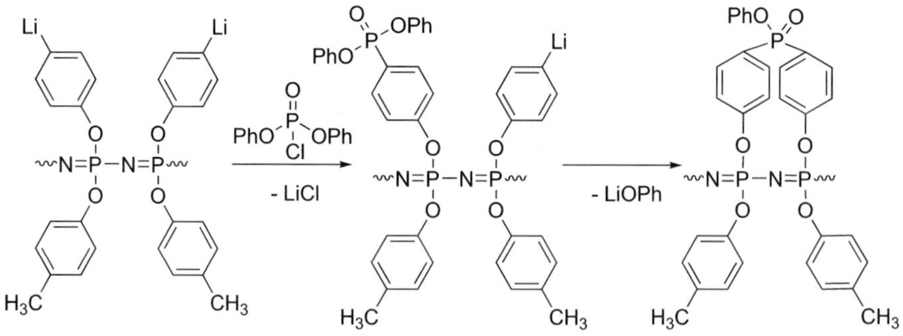

Scheme 27

Elemental analysis revealed that >90% of the initial bromine atoms had been lithiated. Thus, rather than incomplete lithiation, the limiting factor in the overall reaction appears to be the side reaction of the highly reactive lithiophenoxy intermediate, which may occur during the addition of diphenyl chlorophosphate. This is also consistent with observations made using diethyl chlorophosphate [81].

Hydrolysis of the diphenyl phosphonate ester groups was accomplished by treatment of a THF solution of the phosphonated polymers with 1.0 M NaOH over 24 h. During this reaction, the polymer precipitates from solution, at which point little further hydrolysis take place. No change in the molecular weight of the polymers was detected, confirming the stability of the phosphazenes backbone to the basic hydrolysis conditions.

The advantage of this synthesis protocol is the freedom to tune the polymer composition and properties by varying the initial aryloxy side group ration, the degree of lithiation, and the amount of ester hydrolysis.

These polymers have been evaluated as membrane materials for fuel cell applications [84, 85]. They have low methanol diffusion characteristics; methanol diffusion coefficients for these membranes were found to be at least 12 times lower than that for Nafion 117 and 6 times lower than that for a cross-linked sulfonated polyphosphazene membrane [83, 84]. Membranes were found to have IEC values between 1.17 and 1.47 mequiv/g and proton conductivities between 10^{-2} and 10^{-1} S/cm. They appear to be superior to sulfonated polyphosphazene analogues [84, 85].

3 Conclusions

To summarize, the aforesaid shows that heterochain and heterocyclic polymers carrying phosphonic acid groups can be thought of as candidates for fuel cell applications. Of especially great interest are phosphorylated polyethers, polysulfones, polybenzimidazoles, and polyphosphazenes prepared on the basis of phosphonic acid functionalized monomers or by phosphorylation of existing polymers. Further investigations are required to design proton-conducting materials exhibiting long-term thermal stability and mechanical strength, capable of operating at high temperatures.

Acknowledgements This work was supported by the North Atlantic Treaty Organization (NATO) (Grant 981 672).

References

1. Watkins S (1993) Fuel Cell Systems. Plenum Press, New York
2. Zawodzinski TA, Deraum C, Rodzinski S, Sherman RT, Smith UT, Springer TE, Gottesfeld S (1993) J Elecrochem Soc 140:1041
3. Gottesfeld S, Zawodzinski TA (1997) In: Akire RC, Gerischer H, Kobb DM, Tobias CW (eds) Advances in electrochemical science and engineering, (5). Wiley, New York
4. Savadogo OJ (1998) J New Mater Electrochem Syst 1:47
5. Kordesch KV, Simader GR (1995) Chem Rev 95:191
6. Kreuer KD (1996) Chem Mater 8:610
7. Rikukawa M, Sanui K (2000) Prog Polym Sci 25:1463
8. Kerres JA (2001) J Membr Sci 185(1):3
9. Costamanga P, Srinivasan S (2001) J Power Sources 102:242
10. Steele BCH, Heinzel A (2001) Nature 414:345
11. Roziere J, Jones DJ (2003) Annu Rev Mater Res 33:503
12. Yang HC (2003) Nengyuan Jikan 33(2):127
13. Jannasch P (2003) Curr Opin Colloid Interface Sci 8:96
14. Hickner MA, Ghassemi H, Kim YS, Einsla BR, McGrath JE (2004) Chem Rev 104:4587
15. Rusanov AL, Likhachev DY, Mullen K (2002) Russ Chem Rev 71:761
16. Rusanov AL, Likhachev D, Kostoglodov PV, Mullen K, Klapper M (2005) Adv Polym Sci 179:83
17. Dobrovol'skiy YA, Volkov EV, Pisareva AV, Fedotov YA, Likhachev DY, Rusanov AL (2007) Russ J Gen Chem 77:766
18. Dobrovolskiy YA, Jannasch P, Lafitte B, Belomoina NM, Rusanov AL, Likhachev DY (2007) Russ J Electrochem 43(5):489
19. Javaid Zaidi SM (2003) Arab J Sci Eng 28(2B):183
20. Lafitte B, Karlsson LE, Jannasch P (2002) Macromol Rapid Commun 23:896
21. Karlsson LE, Jannasch P (2004) J Membr Sci 230:61
22. Wang F, Hickner M, Kim YS, Zawodzinski TA, McGrath JE (2002) J Membr Sci 197:231
23. Harrison WL, Wang F, Mecham JB, Bhanu VA, Hill M, Kim YS, McGrath JE (2003) J Polym Sci, Part A: Polym Chem 41:2264
24. Gao Y, Robertson GP, Guiver MD, Jian XG (2003) J Polym Sci, Part A: Polym Chem 41:497
25. Gao Y, Robertson GP, Guiver MD, Jian XG, Mikhailenko SD, Wang KP, Kaliaguine S (2003) J Polym Sci, Part A: Polym Chem 41:2731
26. Genies C, Mercier R, Sillion B, Cornet M, Gebel G, Pineri M (2001) Polymer 42:359
27. Wang F, Li J, Chen T, Xu J (1999) Polymer 40:795
28. Li Q-F, He R-H, Jensen JO, Bjerrum NJ (2003) Chem Mater 15:4896
29. Cerfontain H (1968) Mechanistic Aspects in Aromatic Sulfonation and Desulfonation; Interscience Monographs on Organic Chemistry. Interscience Publ New York

30. Lafitte B, Jannasch P (2007) Adv Fuel Cells 1(3):119
31. Rusanov AL, Solodova EA, Bulycheva EG, Abadie MJM, Voytekunas VY (2007) Russ Chem Rev 76:1073
32. Nolte R, Ledjeff K, Bauer M, Mulhaupt R (1993) BHR Group Conf Ser Publ 3:381
33. Xu X, Cabasso I (1993) Polym Mater Sci Eng 68:120
34. Kotov SV, Pedersen SD, Qui WM, Qui ZM, Burton DJ (1997) J Fluorine Chem 82(1):13
35. Jaffe HH, Fredman LD, Doak GO (1953) J Am Chem Soc 75:2209
36. Lassegues JC, Grondin J, Hernandes M, Maree B (2001) Solid State Ionics 145:37
37. Yamabe M, Akiyama K, Akatsuka Y, Kato M (2000) Eur Polym J 36:1035
38. Kosolapoff GM (1942) J Am Chem Soc 64:2982
39. Balthazor TM, Miles JA, Stults BR (1978) J Org Chem 43:4538
40. Foa M, Cassar L (1975) J Chem Soc, Dalton Trans, p 2753
41. Tavs P (1970) Chem Ber 103:2428
42. Balthazor TM, Grabiak RC (1980) J Org Chem 45:5425
43. Hirao T, Masunagawa T, Ohshiro Y, Agawa T (1981) Synthesis 1:56
44. Hirao T, Masunagawa T, Yamada N, Ohshira Y, Agawa T (1982) Bull Chem Soc Jpn 55:909
45. Yan YY, RajanBabu TV (2000) J Org Chem 65:900
46. Miyatake K, Hay AS (2001) J Polym Sci, Part A: Polym Chem 39:1854
47. Wang ZY, Carvalho NH, Hay AS (1991) J Chem Soc Chem Commun 17:1221
48. Meng YZ, Tjong SC, Hay AS, Wang SJ (2001) J Polym Sci, Part A: Polym Chem 39:3218
49. Meng YZ, Tjong SC, Hay AS, Wang SJ (2003) Eur Polym J 39(3):627
50. Meng YZ, Hill AR, Hay AS (2001) Polym Adv Technol 12:206
51. Meng YZ, Abu-Yousef IA, Hill AR, Hay AS (2000) Macromolecules 33(25):9185
52. Gierke TD, Munn GE, Wilson FC (1981) J Polym Sci, Part B: Polym Phys 19:1687
53. Alvares-Gallego Y, Lozano AE, Ferreiro JJ, Nunes SP, Abajo Jde (2002) In: Kricheldorf HR (ed) Polycondensation 2002. 25th Hamburger Macromolekulares Symposium, Hamburg 1:113
54. Akutsu F, Inoki M, Sawano M, Kasashima Y, Naruchi K, Miura M (1998) Polymer 39:6093
55. Sakaguchi Y, Kitamura K, Takase S (2005) In: Abstracts of Pacific Polymer. Conference IX, Mani Hawaii E-16
56. Pinto MR, Reynolds JR, Schanze KS (2002) Am Chem Soc Polym Prepr 43(1):139
57. Stone C, Daynard TS, Hu LQ, Mah C, Steck AE (2000) J New Mater Electrochem Syst 3(1):43
58. Samsone MJ, Onorato FJ, Ogata N, Hoechst Celanese Corp (1997) US Patent 5 599 639
59. Lafitte B, Jannasch P (2005) J Polym Sci, Part A: Polym Chem 43:273
60. Yasuda M, Yamashita T, Shima K (1990) Bull Chem Soc Jpn 63:938
61. Alexandratos SD, Strand MA, Quillen DR, Walder AJ (1985) Macromolecules 18:829
62. Popescu F (1969) Rev Roum Chim 14:1525
63. Ziaja J, Balogh L, Trochimczuk WW (1998) In: Prace Naukowe Inst Podstaw Elecktrotech Electrotechnol Politech Wroclawskiej 34:153
64. Lehmann D, Meier-Haack J, Vogel C, Taeger A, Pereira Nunes S, Paul D, Peinemann K-V, Jakoby K (2003) Patent WO/2003/030289
65. Jakoby K, Peinemann KV, Nunes SP (2003) Macromol Chem Phys 204:61
66. Guiver MD, Kutowy O, ApSimon JW (1989) Polymer 30:1137
67. Park J, Lee JPH (1997) Bull Korean Chem Soc 18:1130
68. Skoda-Foldes R, Kollar L, Horvath J, Tuba Z (1995) Steroids 60(12):791
69. Hirao T, Kohno S, Ohchira Y, Agawa T (1983) Bull Chem Soc Jpn 56:1881
70. Cabasso I, Jagur-Grodzinski J, Vofsi D (1974) J Appl Polym Sci 18:1969

71. Bock T, Mülhaupt R, Möhwald H (2006) Macromol Rapid Commun 27:2065
72. Eymery F, Iorda B, Savignac P (1999) Tetrahedron 55:13109
73. Guiver MD, ApSimon JW, Kutowy O (1988) J Polym Sci, Part C: Polym Lett 26:123
74. Guiver MD, Robertson GP, Yoshikawa M, Tam CM (2000) ACS Symp Ser 744:137
75. Lafitte B, Jannasch P (2007) J Polym Sci, Part A: Polym Chem 45(2):269
76. Miyatake K, Hay AS (2001) J Polym Sci, Part A: Polym Chem 39:3770
77. Liu B, Robertson GP, Guiver MD, Shi Z, Navessin T, Holdcroft S (2006) Macromol Rapid Commun 27:1411
78. Rybkin YY (2005) In: PhD Thesis INEOS RAS Moscow
79. Allcock HR, Taylor JP (2000) Polym Eng Sci 40:1177
80. Allcock HR, Hofmann MA, Wood RM (2001) Macromolecules 34:6915
81. Allcock HR, Hofmann MA, Ambler CM, Morford RV (2002) Macromolecules 35:3484
82. Goldwhite H (1981) In: Introduction to Phosphorus Chemistry. Cambridge University Press Cambridge England
83. Hutchinson DW (1991) In: Townsen LB (ed) Chemistry of Nucleosides and Nucleotides. Plenum Press, New York
84. Allcock HR, Hofman MA, Ambler CM, Lvov SN, Zhou XY, Chalkova E, Weston J (2002) J Membr Sci 201:47
85. Fedkin MV, Zhou XY, Hofmann MA, Chalkova E, Weston JA, Allcock HR, Lvov SN (2002) Mater Lett 52:192

Adv Polym Sci (2008) 216: 157–183
DOI 10.1007/12_2007_130

Published online: 1 February 2008

Polyphosphazene Membranes for Fuel Cells

Ryszard Wycisk (✉) · Peter N. Pintauro

Department of Chemical Engineering, Case Western Reserve University,
10900 Euclid Avenue, Cleveland, OH 44106-7217, USA
ryszard.wycisk@case.edu

Abstract Polyphosphazenes possess numerous properties that are attractive for PEM fuel-cell applications, including thermal and chemical stability and the unlimited possibility for side-group functionalization. There are some impressive results in the literature, in particular where sulfonated polyphosphazenes are used in a direct methanol fuel cell. There is much less data on polyphosphazene fuel-cell membranes with phosphonic acid or sulfonimide side groups, but preliminary physical and transport property data on these materials is encouraging. Clearly, more work is needed to truly exploit the full potential of this important class of inorganic polymers, including the development of new polymer synthesis schemes and the use of combinatorial methods to screen the unlimited number of side group-substituted polymers. Herein an overview of prior research on the synthesis and use of acid-functionalized polyphosphazenes in proton exchange membrane fuel cells is presented.

Keywords Fuel cells · Polyphosphazenes · Proton-conducting polymers · Sulfonation

Abbreviations

MEA	Membrane-electrode assembly
PAN	Polyacrylonitrile
PB3MPP	Poly[bis(3-methylphenoxy)phosphazene]
P3MP4EPP	Poly[(3-methylphenoxy)(4-ethylphenoxy)phosphazene]
P4EPPP	Poly[(4-ethylphenoxy)(phenoxy)phosphazene]
PBPP	Poly[bis(phenoxy)phosphazene]
PDCP	Poly(dichlorophosphazene)
PBI	Polybenzimidazole
SPBPP	Sulfonated poly[bis(phenoxy)phosphazene]
PVDF	Poly(vinylidene fluoride)
POP	Polyorganophosphazene
SPOP	Sulfonated polyorganophosphazene

1 Introduction

The proton-conducting membrane in a hydrogen/air or direct methanol fuel cell (DMFC) performs a number of critically important functions. It physically separates the anode and cathode to prevent an electrical short circuit, it separates the fuel and oxidant to eliminate a chemical short circuit, and it provides pathways for the flow of protons from the anode (where they are generated) to the cathode (where they are consumed in the oxygen reduction reaction). The general requirements of the ion exchange membrane in a proton-exchange membrane (PEM) fuel cell include [1]: high ionic conductivity (with zero electronic conductivity) under cell operating conditions, long-term chemical and mechanical stability, low fuel (H_2 or methanol) and O_2 gas crossover, interfacial chemical/mechanical compatibility with catalyst layers, and low cost.

Recent fuel-cell membrane research efforts have been focused on three areas [2]: (1) membranes for hydrogen/air fuel cells that operate above 100 °C at low humidity conditions, (2) high proton conductivity and low methanol permeability membranes for direct methanol fuel cells and (3) low-cost alternatives to perfluorosulfonic acid membranes for both DMFCs and hydrogen/air fuel cells.

The most widely studied fuel-cell membrane is DuPont's Nafion®, a copolymer of tetrafluoroethylene and perfluoro(4-methyl-3,6-dioxa-7-octene-1-sulfonic acid). Nafion is the membrane material of choice for most proton-exchange membrane fuel cells that operate at a temperature $\leq$80 °C. While Nafion offers high conductivity combined with exceptional chemical and mechanical stability [3], it suffers from several critical drawbacks. When used in a direct methanol fuel cell, Nafion shows significant methanol leakage (crossover from the anode to the cathode) with the resultant reduction in fuel-cell performance. To overcome this shortcoming the methanol concentration in the anode feed is usually reduced to 0.5–2.0 M, which necessitates either a highly diluted methanol fuel cartridge or the addition of a complex on-board methanol dilution subsystem. For automotive applications, a high-temperature (100–140 °C) fuel cell that operates at ambient pressure and less than 100% relative humidity would be highly desirable (such a fuel cell would be compatible with present-day automotive radiator cooling systems and with the use of reformate fuels with trace amounts of CO). Unfortunately, Nafion dehydrates with a resultant loss in proton conductivity at such temperature and humidity conditions. Additionally, the cost of Nafion (about $600/m^2) is too high for any serious (large-scale) commercialization for automotive or stationary power. DuPont's recent R&D efforts have been directed toward decreasing the thickness of Nafion (to lower ohmic losses in a fuel cell) and improving Nafion durability. Cost projections indicate a Nafion price drop to $5–20/m^2 for an annual fuel-cell production rate of 10 million m^2/year [4]. Nevertheless, there is a strong drive in the U.S., Europe, and Japan to develop alternative membrane polymers that would offer cost and performance benefits as compared to Nafion.

A variety of membrane materials have been investigated for possible use in hydrogen/air fuel cells and DMFCs, including sulfonated polyimides, sulfonated poly(arylene ethers), sulfonated poly(ether ether ketones), and phosphoric acid-doped polybenzimidazole [5–12]. Many of these membrane materials suffer from poor oxidative stability, excessive swelling when wet and brittleness when dry, and only moderately high proton conductivity. Thus, no one polymer has yet emerged as the preferred proton exchange membrane material for fuel-cell applications. Polyphosphazene-based membranes also are potential candidates for PEM fuel cells [13], although they have not been studied to the same extent as those polymeric materials listed above. Herein, a summary is presented of polyphosphazene polymer synthesis and functionalization, membrane fabrication and characterization techniques, and fuel-cell test data.

2 Polyphosphazenes

Polyphosphazenes [14–17] are inorganic polymers, with a backbone consisting of alternating phosphorus and nitrogen atoms, with two side groups attached to each phosphorus (Fig. 1). The side groups can be organic, inorganic or organometallic. A very broad range of groups can be easily incorporated into the polyphosphazene chain, which creates unlimited possibilities for derivatization and fine tuning the properties of the resultant material. The initial interest in polyphosphazenes as high-performance elastomers has expanded to solid polymer electrolytes for batteries, membranes for gas and liquid separations, optically active polymers, biomaterials and proton-exchange membranes for fuel cells.

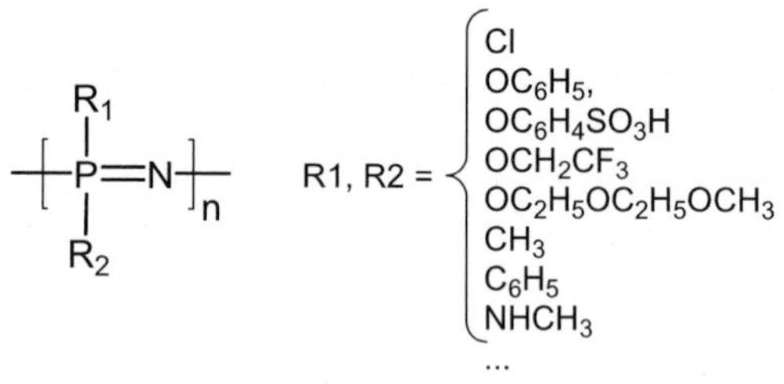

$$\left[P(R_1)(R_2)=N \right]_n \quad R1, R2 = \begin{cases} Cl \\ OC_6H_5, \\ OC_6H_4SO_3H \\ OCH_2CF_3 \\ OC_2H_5OC_2H_5OCH_3 \\ CH_3 \\ C_6H_5 \\ NHCH_3 \\ \ldots \end{cases}$$

Fig. 1 General structure of polyphosphazenes. In most cases R1 and R2 are aryloxy groups

There are several synthetic schemes for polyphosphazene synthesis but the most widely used method is the thermal ring-opening polymerization of chlorophosphazene cyclic trimer, developed by Allcock [18–20] and shown schematically in Fig. 2. The scheme begins with the reaction of phosphorus pentachloride and ammonium chloride resulting in hexachlorocyclotriphosphazene (step a in Fig. 2). After a multi-stage purification involving sublimation, this compound is polymerized (step b) by heating at 250 °C in an evacuated and sealed glass tube for 24–72 h. Next, the resultant linear poly(dichlorophosphazene) (henceforth abbreviated as PDCP) is isolated, by either sublimation or extraction of low molecular weight species, or by dissolution and precipitation. The molecular weight of the polymer is often very high, $M_w \sim 10^6$, but with a broad molecular weight distribution. In the final step (c), poly(dichlorophosphazene) is dissolved in an inert solvent, usually tetrahydrofurane (THF), and mixed with the appropriate nucleophilic reagent(s). The product polyphosphazene is isolated by precipitation.

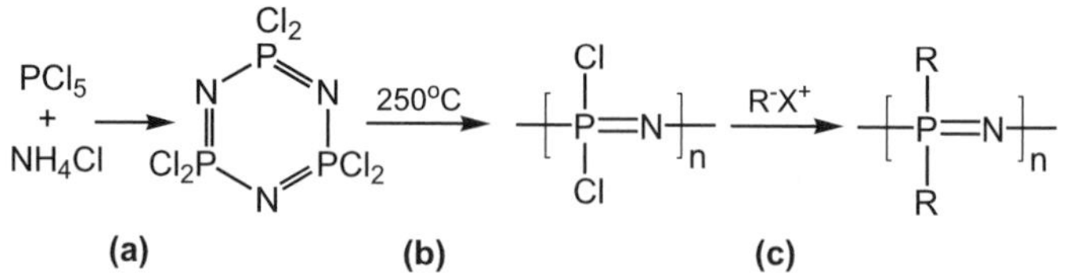

$$\begin{matrix} PCl_5 \\ + \\ NH_4Cl \end{matrix} \xrightarrow{(a)} (NPCl_2)_3 \xrightarrow[(b)]{250^{\circ}C} \left[P(Cl)_2=N \right]_n \xrightarrow[(c)]{R^-X^+} \left[P(R)_2=N \right]_n$$

Fig. 2 Polyphosphazene synthesis scheme developed by Allcock: **a** trimer synthesis, **b** ring-opening polymerization, and **c** macromolecular nucleophilic substitution

Other methods of polyphosphazene synthesis have also been reported [17]. The most interesting is the "living" cationic condensation polymerization of phosphoranimines [21, 22] as it offers access to polyphosphazenes with controlled molecular weight, low polydispersity, and more complex architectures, e.g. block or graft copolymers. At the present time, only moderate molecular weights ($M_w = 10^4$–10^5) are available via the condensation route.

2.1
Problems with the Synthesis and Stabilization of Poly(dichlorophosphazene)

Although there are numerous papers in the literature on the synthesis and characterization of various polyphosphazene derivatives, the preparation of a high-quality polyorganophosphazene with reproducible physicochemical characteristics requires experience and attention to experimental details. The key challenges are: (1) minimization of branching and crosslinking during the later stages of the ring-opening polymerization of the trimer, (2) stabilization of the PDCP material, and (3) complete replacement of chlorine during the nucleophilic substitution step.

Careful control of the reaction conditions is required to achieve high conversion of the ring-opening reaction while minimizing branching and crosslinking at the final stages of PDCP polymerization. Andrianov et al. [23] concluded that out of two possible crosslinking mechanisms, one involving cationic side-chain growth and the other via intermolecular condensation between P – Cl and P – OH groups (Fig. 3), experimental results supported the former mechanism.

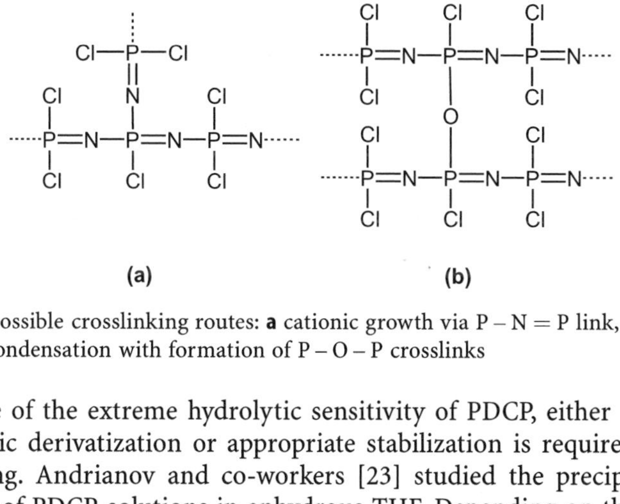

(a)

(b)

Fig. 3 Two possible crosslinking routes: **a** cationic growth via P – N = P link, and **b** intermolecular condensation with formation of P – O – P crosslinks

Because of the extreme hydrolytic sensitivity of PDCP, either immediate nucleophilic derivatization or appropriate stabilization is required to avoid crosslinking. Andrianov and co-workers [23] studied the precipitation/gel formation of PDCP solutions in anhydrous THF. Depending on the polymer concentration, the presence of minute amounts of water (as little as 0.0009% H_2O) resulted in polymer crosslinking within several days. The addition of diglyme, however, resulted in a dramatic increase in the stability of PDCP solutions. In [20], the authors reported no sign of crosslinking in pure diglyme, even after 4 years of storage.

To ensure proper hydrolytic stability of the final polyphosphazene material, substitution of all the chlorine atoms of PDCP is necessary. Stewart et al. [24] found that ^{31}P NMR spectroscopy was well suited for tracking the progress of nucleophilic substitution. For example, during the synthesis of poly[bis(phenoxy)phosphazene] (PBPP), the initial NMR spectrum consisted of three signals: the most intense one at –16 ppm originating from PN(OPh)Cl species, another at –17.5 ppm due to presence of $PN(OPh)_2$, and the third signal at –17.8 ppm, reflecting resonance of $PNCl_2$ units. After refluxing at 115 °C for 37 h in toluene/diglyme solution, the intensity of the PN(OPh)Cl and $PNCl_2$ signals decreased significantly and after 55 h, the two peaks disappeared completely.

2.2 Sulfonated Polyphosphazenes

For a polyphosphazene to be used as the membrane material in a fuel cell, it must possess some kind of protogenic functionality, to allow the polymer to conduct protons. The most widely used protogenic group is the sulfonic acid moiety which can be incorporated into the macromolecular chain either at the polymer synthesis step (Fig. 4a) or later, during postsulfonation (Fig. 4b). While the postsulfonation reaction is generally easier to carry out, it is the direct synthesis of the functionalized polyphosphazene that should give better control over the final material properties and minimize unwanted side

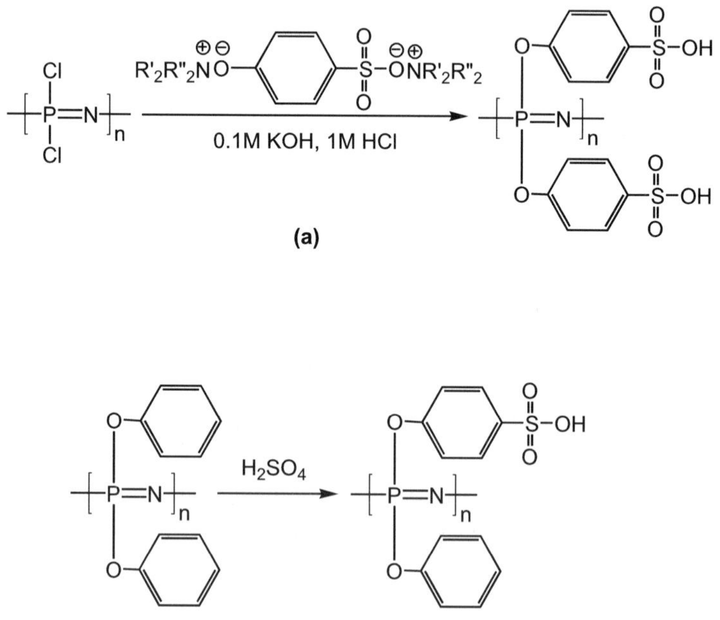

Fig. 4 Synthesis of sulfonated polyphosphazenes by: **a** direct substitution of PDCP with a protected sulfonic acid nucleophile, and **b** postsulfonation using concentrated sulfuric acid

reactions, including degradation of molecular weight. Unfortunately, there are only two reported attempts to use a direct sulfonation route through the replacement of chlorine atoms of the PDCP precursor with a sulfonic-acid-containing nucleophile [25, 26]. Problems arose from the fact that either alkoxy or aryloxy sulfonate nucleophiles acted as difunctional reagents towards PDCP leading to an unstable product that later underwent degradation.

2.2.1 Direct Synthesis of Sulfonated Polyphosphazenes

A partial solution to the problem of difunctional reactivity of alkoxy sulfonate nucleophiles during the macromolecular substitution of PDCP was proposed by Ganapathiappan et al. [25] and is presented in Fig. 5. Here the disodium salt of 2-hydroxyethanesulfonic acid is reacted with an excess of linear PDCP in the presence of a phase-transfer catalyst. A partially substituted, crosslinked semi-product is obtained, which is then treated with another nucleophile, the monofunctional sodium salt of 2-(2-methoxyethoxy)ethanol, to displace the sulfonate groups and chlorines attached to phosphorous atoms. It was found that the amount of sulfonate groups incorporated into the polyphosphazene was generally 50% of that initially used in the reaction mixture.

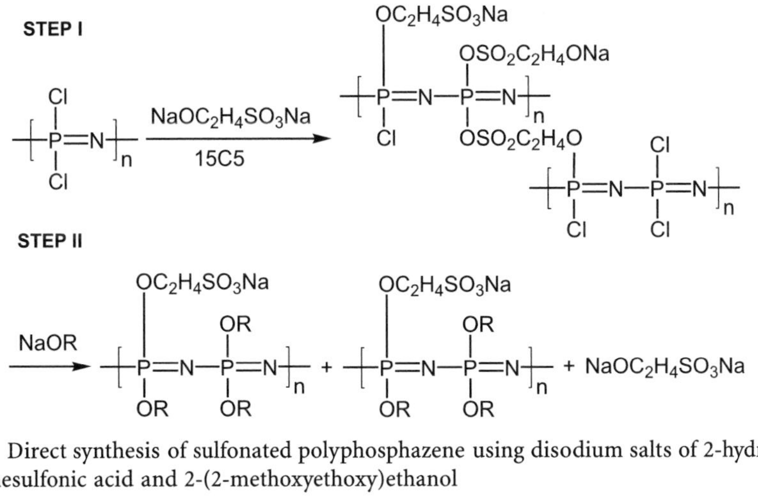

Fig. 5 Direct synthesis of sulfonated polyphosphazene using disodium salts of 2-hydroxyethanesulfonic acid and 2-(2-methoxyethoxy)ethanol

2.2.2 Non-Covalent Protection of Sulfonic Groups

Recently, Andrianov et al. [26] published on an elegant protection scheme (Fig. 4a) that eliminates unstable intermediates and makes direct sulfonation a viable option. When the dimethyldipalmitylammonium salt of hydroxybenzenesulfonic acid was used in the nucleophilic substitution rather than the disodium salt, the reaction yielded high molecular weight polyphosphazene sulfonate. The protective groups were removed through reaction with potassium hydroxide. Polyphosphazene copolymers containing sulfophenoxy and ethylphenoxy substituents were also prepared using the new protection method.

2.2.3 Post-Sulfonation of Polyphosphazenes

Although the single-step, direct synthesis is promising, postsulfonation of various poly[(aryloxy)phosphazenes] is the most practical way to prepare sulfonated derivatives. On the basis of the work of Montoneri and co-workers [27, 28], Wycisk et al. [29] synthesized and characterized for the first time proton-conducting membranes from sulfonated poly[(aryloxy)phosphazenes]. The general scheme is shown in Fig. 6. Dichloroethane was employed as the polymer solvent and sulfur trioxide (SO_3) as the sulfonating agent. Poly[(3-methylphenoxy)(phenoxy)phosphazene] was found to be the best starting material in terms of the ease of controlling the sulfonation degree and in attaining the highest ion-exchange capacity for a water insoluble polymer (Fig. 7a). Depending on the molar ratio of SO_3 to the polymer mer, water insoluble membranes were prepared with an ion-exchange capacity ranging from near 0 to 2.3 mmol/g, an AC impedance in 0.1 M NaCl between 48 kohm m and 0.04 ohm m, and water swelling from 0.1 to 0.9 g/g.

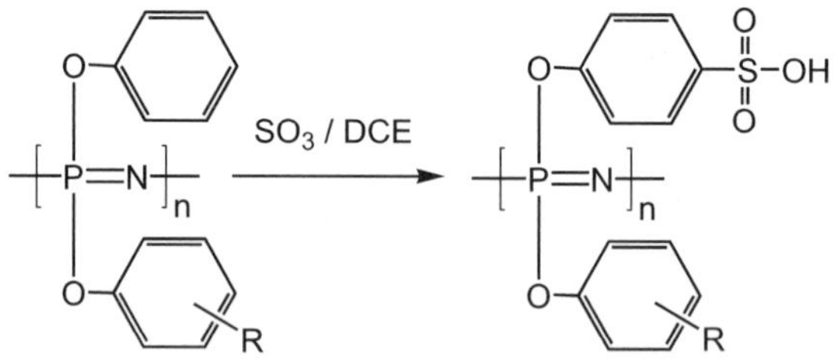

Fig. 6 Sulfonation scheme used by Pintauro and co-workers for sulfonation of various polyaryloxyphosphazenes (see [28–32])

Tang et al. [30] and Guo et al. [31] found that sulfonated poly[bis(3-methylphenoxy)phosphazene] (denoted as SPB3MPP) (Fig. 7b) offered an attractive combination of good proton conductivity, crosslinkability and low methanol permeability; features especially important for direct methanol fuel-cell membranes. UV-crosslinked membranes prepared from SPB3MPP of IEC 1.4 mmol/g exhibited a methanol diffusion coefficient on the order of 1.0×10^{-7} cm^2/s, which was significantly smaller than that in a Nafion

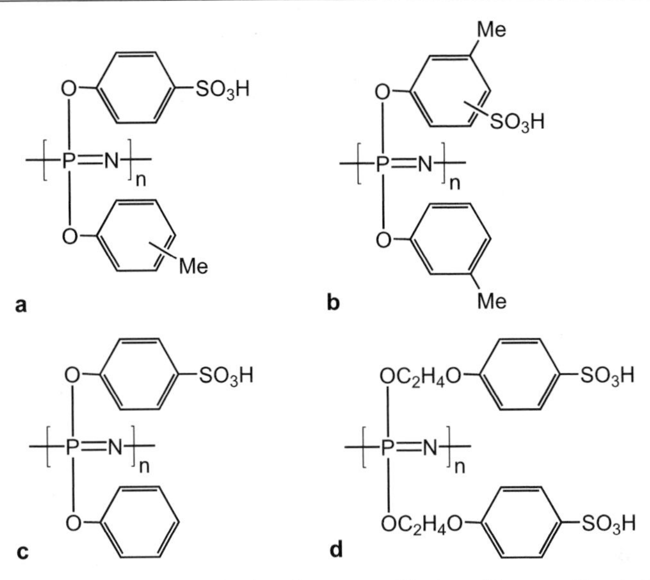

Fig. 7 Example structures of polyphosphazene sulfonic acids: **a** sulfonated poly[(methylphenoxy)(phenoxy)phosphazene], **b** sulfonated poly[bis(3-methylphenoxy)(phenoxy)-phosphazene], **c** sulfonated poly[bis(phenoxy)phosphazene] and **d** sulfonated poly[bis(2-phenoxyethoxy)phosphazene]

117 membrane. Additionally, the oxidative stability of the membranes (tested in Fenton's reagent) surpassed that of commercial styrenic cation-exchange membranes and was comparable to the stability of Nafion perfluorosulfonic acid (less then 5% weight loss over 24 h in a 3% H_2O_2/4 ppm Fe^{+2} solution at 68 °C). These materials were studied intensively (e.g., NMR and IR spectroscopy [30], proton conductivity and water diffusivity measurements [31, 32] and small-angle X-ray analyses [32]) and direct methanol fuel-cell results were later published [33] (DMFC results are presented below in Sects. 4.1 and 4.2).

More recently, another polyphosphazene membrane was reported [34]. It was fabricated from a blend of sulfonated poly[bis(phenoxy)phosphazene] (SPBPP) (Fig. 7c) and polybenzimidazole. SPBPP was prepared by sulfonation of PBPP using concentrated sulfuric acid, which served as both the solvent and sulfonating agent. Depending on the sulfonation time, polymers with an IEC from 0.2 to 3.1 mmol/g were obtained. SPBPP with an IEC of up to 1.4 mmol/g were insoluble in cold and hot water, while polymers with an ion exchange capacity greater than 2.0 mmol/g dissolved in water at room temperature. A discussion of the fuel-cell performance of sulfonated poly[bis(phenoxy)phosphazene] is presented below, in Sect. 4.3.

An interesting polyphosphazene derivative, sulfonated poly[bis(2-phenoxyethoxy)phosphazene], was reported by Paulsdorf et al. [35] (Fig. 7d). Living cationic polymerization of $Cl_3P=NSi(CH_3)_3$ monomer was used to obtain PDCP, which was reacted with 2-phenoxyethanol. The aryl substituents were then sulfonated with concentrated sulfuric acid. The ionic conductivity of the polymer in the sodium salt form was tested in a water-saturated nitrogen atmosphere at temperatures in the range of 20–60 °C. The highest conductivity (7.1×10^{-2} S/cm) was found at 40 °C. This polymer has not been tested in a fuel cell.

2.3 Phenylphosphonic Acid Functionalized Polyphosphazenes

The synthesis of polyphosphazenes with phosphonic acid functionalities was studied by Allcock and co-workers [36–38]. In initial experiments [36], dialkyl phosphonate units were incorporated into poly[(aryloxy)phosphazenes] containing either bromomethanephenoxy or lithiophenoxy (via lithiation of bromophenoxy) side groups. Later, the approach was extended to poly[(aryl-

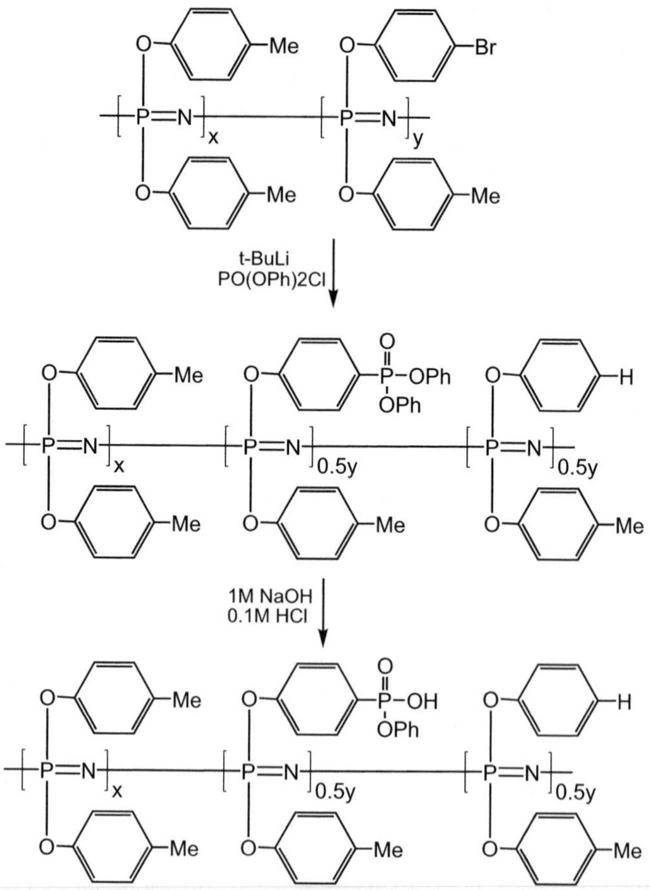

Fig. 8 Synthetic route to a phosphonated polyphosphazene

oxy)phosphazenes] that were functionalized with diphenyl phosphonate groups and subsequently hydrolyzed to phenylphosphonic acid [37, 38].

The general scheme is shown in Fig. 8. The polyphosphazene-containing bromophenoxy side-groups were lithiated and then reacted with diphenyl chlorophosphate. The resultant diphenyl phosphonate ester groups were hydrolyzed with 1 M NaOH to give primarily a monoester/monosalt which was acidified with 0.1 M HCl to give phenylphosphonic acid moieties. It was surprising that full hydrolysis to the diacid was impossible, but the authors claimed this was due to limited solubility and precipitation of the sodium salt during the conversion. It was found that approximately 50% of the bromophenoxy side groups were converted to phenoxy diphenylphosphonate functions. The freedom to modify the polymer side group composition and the extent of phosphonation and hydrolysis were the advantages of the phosphonation scheme, according to the authors.

Membranes were fabricated from phosphonated poly[(4-methylphenoxy)(4-bromophenoxy)phosphazenes] with an IEC between 1.17 and 1.43 mmol/g by solution casting using DMF as the solvent [39, 40]. Some of the membranes were crosslinked by exposure to ^{60}Co gamma radiation. The proton conductivity of the cast films varied from 0.01 to 0.1 S/cm and the equilibrium water swelling was between 11 and 32% (a higher conductivity and greater swelling was observed for a higher IEC membrane). Methanol diffusion coefficients were measured in 3.0 M and 12.0 M aqueous methanol solutions at 80 °C and 2.8 bar. With 3.0 M methanol, the measured diffusivity was between 0.14×10^{-6} and 0.27×10^{-6} cm^2/s, which is over 20-times smaller than that in Nafion 117. With 12.0 M methanol, the diffusion coefficients were between 0.77×10^{-6} and 1.51×10^{-6} cm^2/s (2–4-times less than that in Nafion 117). Unfortunately, no fuel-cell test data were reported with these phosphonated polyphosphazene membranes.

2.4 Polyphosphazenes with Sulfonimide Side Groups

As an alternative to sulfonic or phosphonic acid functionalization, the sulfonimide group has been shown to possess high acid strength and thus to provide good proton-conducting capability [41]. Fluoropolymers containing sulfonimide functions were originally developed by DesMarteau and co-workers [41, 42]. Feiring et al. [43] reported on novel aromatic polymers with pendant sulfonimide groups as potential electrolytes for lithium batteries and fuel cells. Hofmann et al. [44] published data on the synthesis and characterization of a polyphosphazene that contained sulfonimide groups. The synthetic reaction scheme for adding sulfonimide sidechains to poly(dichlorophosphazene) is shown in Fig. 9. First, a suitable sulfonimide-group-containing phenoxide nucleophile was prepared. At the same time PDCP is reacted with sodium 4-methylphenoxide to displace ca. 50% of chlor-

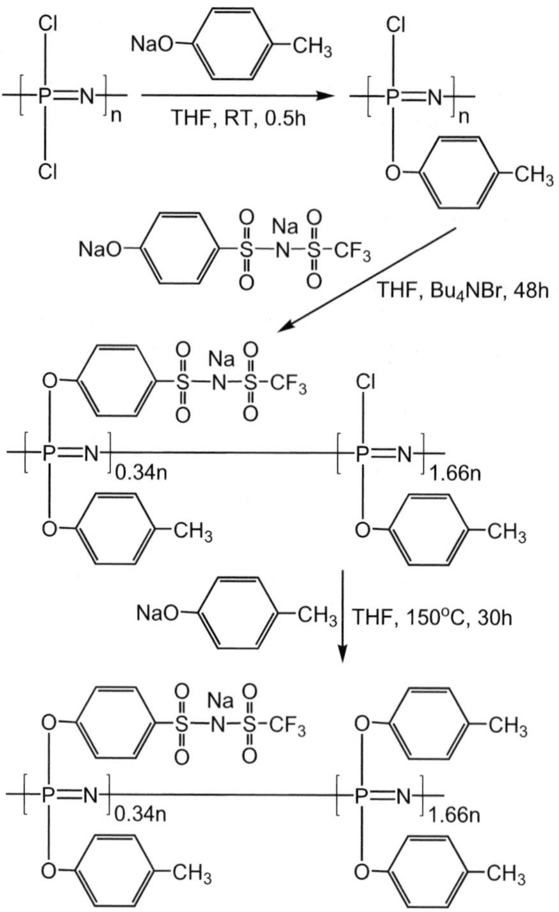

Fig. 9 Synthesis of polyphosphazene with sulfonimide side groups

ine atoms. The sulfonimide-bearing sodium phenolate was added to the partially substituted polychlorophosphazene precursor. Finally, the remaining chlorine atoms were displaced by treatment with sodium 4-methylphenoxide. Membranes were cast from dioxane and were crosslinked using γ-radiation. The polymer IEC was 0.99 mmol/g and the water swelling decreased from 119% through 73% to 42% when the irradiation dose was increased from 0 to 20 and further to 40 MRads. Accordingly, the proton conductivity varied from 0.049 S/cm (0 MRad) to 0.071 S/cm (20 MRad) and to 0.058 S/cm (40 MRad). The membranes were later tested in hydrogen fuel cells [45]; this data is presented below, in Sect. 4.1.

In a follow-up study, Lvov et al. [46] prepared membranes of good physicochemical characteristics from blends of sulfonimide functionalized polyphosphazene and PVDF. For example, a membrane prepared from polyphosphazene of IEC = 1.2 mmol/g containing 20% PVDF and crosslinked using 40 MRads of γ irradiation showed a proton conductivity (in water at 25 °C) of 0.052 S/cm and low room-temperature water swelling (30%). There is no fuel-cell performance data with this blended membrane.

2.5 Crosslinking and Blending of Sulfonated Polyphosphazenes

Pintauro and co-workers [47] investigated three routes to stabilize and control the water and methanol swelling properties of sulfonated polyphosphazene membranes: (1) crosslinking with UV light or e-beam radiation, (2) blending with an inert polymer, e.g., PVDF or PAN, and (3) a combination of the above through blending/complexation with PBI. Covalent crosslinking was accomplished via hydrogen abstraction from the benzylic carbon of alkylphenoxyside groups and the subsequent recombination of the macroradicals (Fig. 10a). Blending of sulfonated poly[bis(3-methylphenoxy)phosphazene] (SB3MPP) with inert polymers like PVDF or PAN resulted in phase-separated structures with a domain size in the micrometer range (Fig. 11a,b). Blends of sulfonated poly[bis(phenoxy)phosphazene] (SPBPP) and polybenzimidazole (PBI) exhibited a nanodomain morphology (Fig. 11c); good compatibility

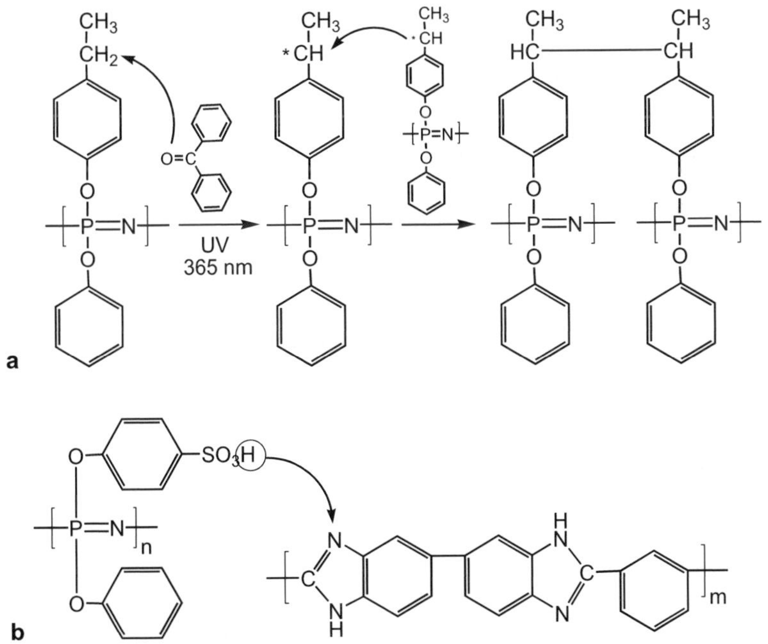

Fig. 10 Two methods of polyphosphazene crosslinking: **a** using benzophenone and UV irradiation, and **b** via acid–base complexation with PBI

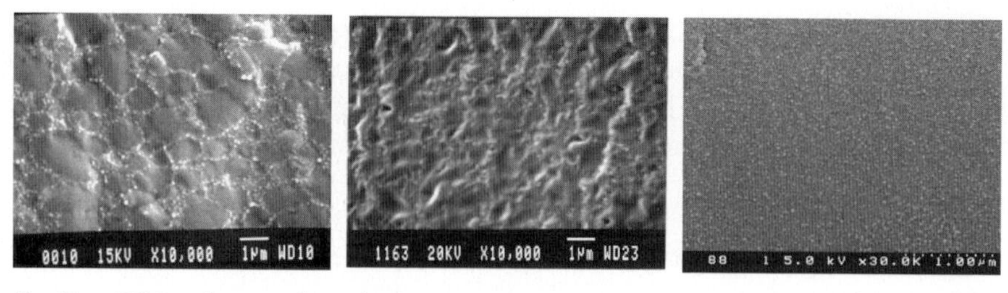

Fig. 11 SEM micrographs of phase-separated sulfonated polyphosphazene blends: **a** SPB3MPP-PVDF, **b** SPB3MPP-PAN, and **c** SPBPP-PBI

being the results of energetically favorable interactions between the acidic protons of the sulfonic groups and the basic nitrogens of the imidazole units of PBI (Fig. 10b).

While promising results were reported with all of the above systems, especially encouraging direct methanol fuel-cell performance data were obtained with membrane prepared from blends of sulfonated poly[bis(phenoxy)phosphazene] (SPBPP) and PBI [34]. The effect of PBI content on room-temperature membrane proton conductivity, room-temperature water uptake, and methanol permeability (at 60°) is shown in Fig. 12.

It is seen in Fig. 12a that the conductivity of neat 1.4 mmol/g SPBPP (0.08 S/cm) decreased with addition of PBI to a value of 0.005 S/cm when the membrane contained 12% PBI. The drop in conductivity was associated with three factors: (1) a simple dilution effect, where an increase in weak protonic conductor (PBI) content in a mixture with a good conductor (SPBPP) leads to a decrease in the effective conductivity of the blend, (2) complexation of PBI with the polyphosphazene, which resulted in reduced water uptake

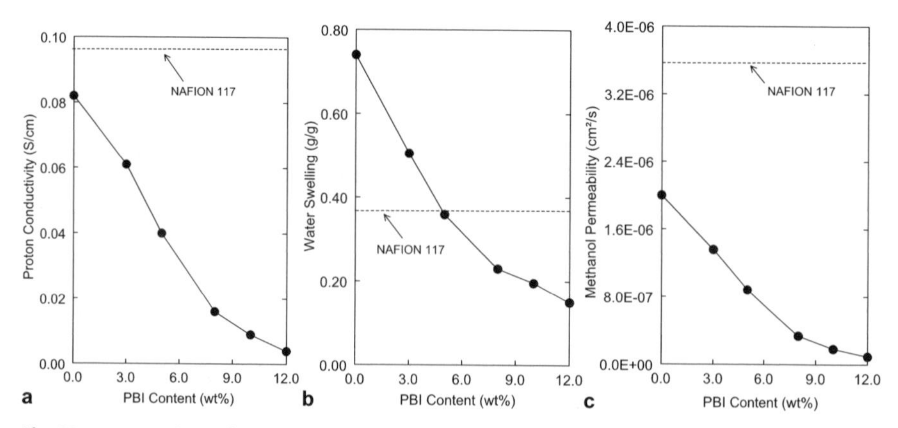

Fig. 12 Properties of SPBPP-PBI blended membranes as a function of PBI content: **a** proton conductivity in water at room temperature, **b** equilibrium water swelling at room temperature, and **c** methanol permeability at 60 °C, for 1.0 M methanol

(lower equilibrium swelling), and (3) sulfonic acid proton immobilization by interaction with the basic sites of PBI, which lowered the effective IEC of the blend and reduced the number of available carriers for charge transport. The effect of PBI content on membrane water swelling is presented in Fig. 12b. The strong decrease in swelling with increasing PBI concentration at relatively low PBI contents can be regarded as proof that the SPBPP component of the blend is being immobilized via complexation with the resultant characteristics equivalent to those of a crosslinked polymer. At room temperature, a pure 1.4 mmol/g SPBPP membrane absorbed 0.75 g of water per 1 g of dry polymer. When the PBI concentration was increased to 12%, the water content dropped to 15 wt %, which is probably the lower swelling limit for proton conduction (it is important to note that water is necessary to dissociate sulfonic acids and to create proton transporting channels; a reduction in membrane water content below the percolation limit will result in negligible proton conductivity). Methanol permeabilities, measured in a standard two-compartment cell at 60 °C with 1.0 M methanol, are plotted in Fig. 12c as a function of the membrane PBI content. The permeability decreased from 2.0×10^{-6} cm^2/s for a pure 1.4 mmol/g SPBPP film to 1.0×10^{-7} cm^2/s for a 1.4 IEC blend containing 12 wt % PBI (these permeabilities are 3–20 times lower than the methanol permeability in Nafion 117 at 60 °C).

Although electron microscopy testing revealed a submicrometer size domain morphology which proved good compatibility between the sulfonated polyphosphazene and PBI, differential scanning calorimetry was employed to investigate the glass-transition behavior of the blend. For miscible polymer blends, it is known that the glass transitions of the individual components disappear and only one glass transition is observed [48]. This is generally regarded as proof of polymer miscibility. Samples of SPBPP-PBI blends of different PBI content were analyzed by DSC using a Mettler Toledo DSC 822 Calorimeter at a heating rate of 10 deg/min under a nitrogen atmosphere. The sample pans were punctured to allow water vapor removal. Each measurement involved two heating cycles. In the first cycle, the sample was heated from room temperature to 200 °C, in order to remove any traces of water. In the second cycle, spanning -40 °C to 200 °C, the actual thermogram was recorded. Because of the fact that the glass transition of PBI occurs at 399 °C, which is above the decomposition temperature of SPBPP (ca. 320°), only the thermal transitions of the polyphosphazene were investigated.

The resultant curves are shown in Fig. 13. It can be seen that upon sulfonation, the glass-transition temperature (T_g) of the polyphosphazene increases from -3 °C (base polymer) to 59 °C (1.4 mmol/g IEC), and the melting endotherm disappears which indicates loss of crystallinity. Upon blending with PBI (in the 5–10 wt % range), the T_g decreases, but at 20 and 50 wt % PBI, the glass-transition temperature of the blend increased substantially to 67 and 90 °C, respectively. This apparently confusing behavior can be explained by realizing that the glass-transition temperature change was a superposition of

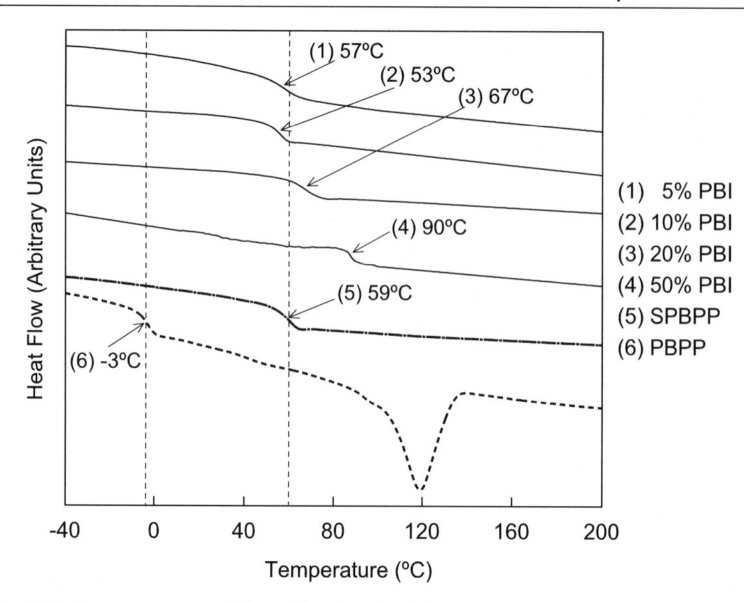

Fig. 13 DSC thermograms of four blends of sulfonated poly[bis(phenoxy)phosphazene] with PBI (SPBPP-PBI), unsulfonated poly[bis(phenoxy)phosphazene] (PBPP) and pure sulfonated poly[bis(phenoxy)phosphazene] (SPBPP) with an IEC = 1.4 mmol/g. The *two vertical broken lines* depict the glass transitions of the PBPP (–3 °C) and SPBPP (59°)

two effects: the depression in T_g due to a reduction in the IEC of the blend and an elevation in T_g due to mixing sulfonated poly[bis(phenoxy)phosphazene] with the high glass-transition temperature PBI material. If protons of the sulfonic groups of the SPBPP are immobilized, then they will no longer participate in the hydrogen-bonded network of SPBPP and thus there will be an increase in the polyphosphazene's backbone flexibility. On the other hand, favorable interactions with PBI will lead to miscibility, and will drive the T_g of the blend towards 399 °C (the T_g of neat PBI). So the net effect can be either a depression or an increase of the blended polymer's T_g, depending on the PBI content.

Although the three polyphosphazene-based blended systems described above, i.e. SB3MPP-PAN, SB3MPP-PVDF and SPBPP-PBI, differed to some extent regarding morphological details and macroscopic appearance, there was one important similarity, which is shown in Fig. 14. As can be seen, there is a single, linear correlation between water swelling and proton conductivity of the different membrane blends. The following general conclusion can be drawn from these results: that proton conductivity is primarily controlled by the membrane's water content (which is related to the film's effective IEC) and not by the phase structure or specifics regarding the chemical composition of the blend's components. While this may be expected when comparing phase-separated SPB3MPP-PAN and SPB3MPP-PVDF systems, the fact that

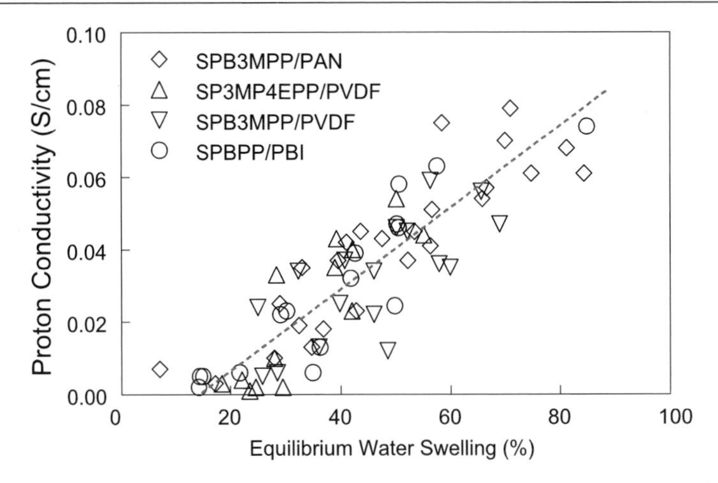

Fig. 14 Proton conductivity versus water swelling correlation for three types of blended membranes with sulfonated polyphosphazene

the SPBPP-PBI blends also obey this "universal" conductivity-swelling correlation is somewhat surprising.

3 Polyphosphazene Membranes for Hydrogen Fuel Cells

There is only one example in the literature of polyphosphazene performance in a proton-exchange membrane (PEM) hydrogen fuel cell. Allcock and Lvov [45] tested a sulfonimide polyphosphazene membrane in a hydrogen/oxygen fuel cell at room temperature and at 80 °C. The membrane-electrode-assembly (MEA) was fabricated from a 100 μm thick sulfonimide polyphosphazene membrane that was crosslinked with γ-radiation (40 MRad). The polymer IEC was 0.99 mmol/g, with an equilibrium water swelling of 42%, and a proton conductivity of 0.058 S/cm. The anode and cathode were prepared from carbon-supported platinum (20% Pt on Vulcan XC-72R) at a Pt loading of 0.33 mg/cm^2. The electrodes were hot pressed onto the membrane at 65 °C and 400 psi for 30 s. As a reference, a Nafion 117 MEA was also prepared with the same electrode catalyst at a loading of 0.26 mg/cm^2 for the anode and 0.48 mg/cm^2 for the cathode. For Nafion, the electrodes were hot pressed at 125 °C and 1400 psi for 2 min.

Fuel-cell tests were performed at 22 °C without humidification of either the hydrogen or oxygen streams, and at 80 °C with full humidification of both gases (humidifiers were kept at 95 °C). The H_2 and O_2 gasses were pressurized to 3 and 5 bar, respectively, during testing at 22 °C and to 3 and 3.3 bars for tests at 80 °C. The fuel-cell performance results are shown in Fig. 15. At both the high and low temperatures, the maximum power density with the

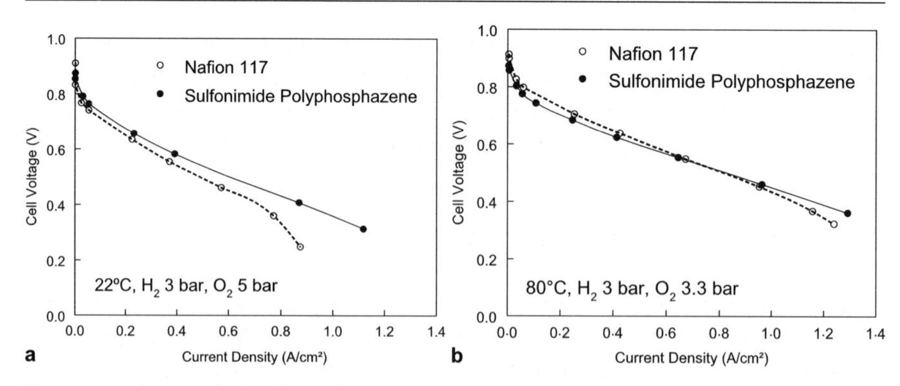

Fig. 15 Hydrogen fuel-cell performance curves with a sulfonimide polyphosphazene proton-exchange membrane at 22 °C (**a**) and 80 °C (**b**)

polyphosphazene MEA exceeded that with Nafion 117. The difference was more pronounced at the lower temperature (Fig. 15a). At 22 °C the limiting (maximum) current density was 1.12 A/cm^2 and the maximum power density was 0.36 W/cm^2. There was only a modest increase in the limiting current density (1.29 A/cm^2) and maximum power density (0.47 W/cm^2) when the temperature was increased to 80 °C.

The improved performance of the sulfonimide polyphosphazene MEA, as compared to that with Nafion 117, was probably the result of the lower areal resistance of the former (the areal resistance is defined as the ratio of membrane thickness to proton conductivity). While the conductivity of the polyphosphazene membrane was approximately 60% that of Nafion 117 (at room temperature), its thickness was half that of Nafion 117. If temperature does not alter significantly the thickness to conductivity ratio, then the areal resistance of the sulfonimide MEA is estimated to be 15–20% less than that of Nafion 117. It should also be noted that there was no long-term stability analysis of the sulfonimide membranes. Such tests are critical in evaluating new membranes for fuel cells.

4 Polyphosphazene Membranes for Direct Methanol Fuel Cells (DMFCs)

The direct methanol fuel-cell performance of three polyphosphazene-based membrane systems is presented below. The first system deals with blends of PVDF with either sulfonated poly[(3-methylphenoxy)(4-ethylphenoxy)phosphazene] (SP3MP4EPP) or sulfonated poly[(4-ethylphenoxy)(phenoxy)phosphazene] (SP4EPPP), where the membranes were prepared by solution casting mixtures with subsequent crosslinking using electron-beam irradiation (60 MRad). The second membrane system was based on UV-photocrosslinked blends of sulfonated poly[bis(3-methylphenoxy)phosphazene] (SPB3MPP) and PAN, where benzophenone initiator was added to the polymer solution prior to film casting. The third membrane group was prepared from blends of sulfonated poly[bis(phenoxy)phosphazene] (SPBPP) and PBI.

4.1
Membranes from SPB3MPP-PVDF Blends

DMFC tests were performed using a 5 cm^2 MEA (geometric electrode area) at 60 °C with 1.0 M methanol as the anode feed and ambient pressure air as the oxidant. To avoid problems (i.e., poor fuel-cell performance) related to the fabrication of the electrodes (binder content) and electrode attachment to the membranes (hot pressing conditions), these tests were performed with a sandwiched MEA where the polyphosphazene membrane was positioned between two Nafion 112 half-MEAs (an anode hot pressed to one Nafion 112 membrane and the cathode hot pressed to the second Nafion film). Methanol crossover was determined, at open circuit potential, by measuring the concentration of CO_2 in the air exhaust using a Vaisala GMP222 sensor (any methanol that passed through the membrane, from the anode to the cathode, would be oxidized chemically at the cathode, producing CO_2).

The fuel-cell performance results of the sandwiched MEAs, along with a reference *i*–*V* curve recorded with a Nafion 117 MEA, are shown in Fig. 16.

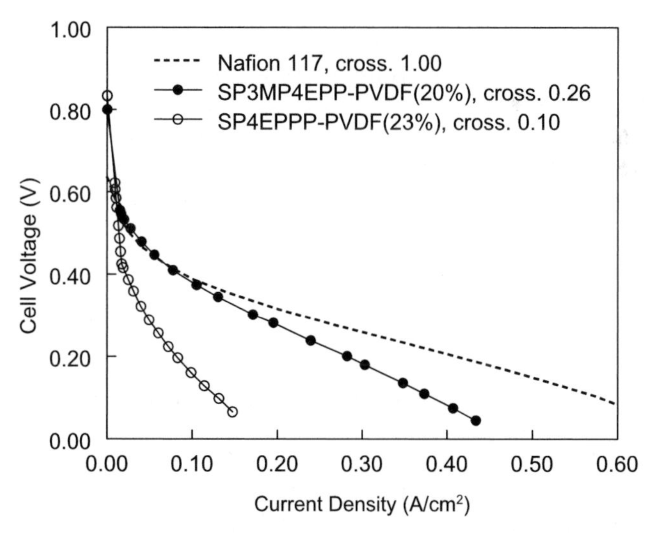

Fig. 16 Direct methanol fuel-cell performance curves with blended membranes composed of PVDF and either sulfonated poly[(3-methylphenoxy)(4-ethylphenoxy)phosphazene] (SP3MP4EPP) or sulfonated poly[(4-ethylphenoxy)(phenoxy)phosphazene] (SP4EPPP). 1.0 M methanol feed, 60 °C, air at ambient pressure and 500 sccm. *Cross* denotes the methanol crossover flux (mol/cm^2 min) at *open circuit*, relative to that in Nafion 117

As compared to Nafion 117, the methanol crossover was lower by a factor of four, but the maximum power density was reduced by 31% for the sandwich MEA with a SP3M4EPP-PVDF membrane. In the case of the SP4EPPP-PVDF membrane, the methanol crossover was approximately 10-times less than that for Nafion 117, but the maximum power density was very small, about 20% that of Nafion. Long-term stability tests revealed a slow decrease of proton conductivity of the blended membranes, which was associated with leaching-out of the high-IEC polyphosphazene fraction due to inefficient crosslinking.

4.2 Membranes from SPB3MPP-PAN Blends [33]

Membranes were prepared from sulfonated poly[bis(3-methylphenoxy)phosphazene] (PB3MPP, IEC = 2.1 mmol/g) that was blended with polyacrylonitrile (PAN, at either 40 wt% or 48 wt%) and then UV crosslinked using benzophenone (5%) as the photoinitiator (benzophenone was dissolved in the SPB3MPP-PAN membrane casting solution and remained dissolved in the polymeric film after solvent evaporation). The effective IEC of the membranes was 1.15 (40% PAN) and 1.00 (48% PAN). MEAs with a geometric area of 5 cm^2 were prepared by painting a catalyst ink onto carbon cloth (ELAT, E-TEK) and then hot pressing the catalyzed electrodes onto a SPB3MPP-PAN membrane at 120 °C and 125 psi for 5 min. The catalyst loading for both the anode (Pt – Ru, 1 : 1) and the cathode (Pt) was 4 mg/cm^2.

Steady-state current density-voltage data were collected using a single-cell DMFC test station (Scribner Series 890B) with mass flow and temperature control. The cell was operated at 60 °C, with 1.0 M methanol (20 ml/min) and humidified air (150 sccm at 30 psi back pressure). The methanol crossover flux was determined by measuring the carbon dioxide concentration in the cathode air exhaust at open circuit with a Vaisala GMM12B CO_2 detector. The initial (6–8 h) performance curves for the two SPB3MPP-PAN-based MEAs and for a Nafion 117 MEA (as a reference) are shown in Fig. 17. The open circuit potential for both polyphosphazene MEAs was higher than that with Nafion 117, indicating lower methanol crossover. While the performance of the polyphosphazene MEAs at low current densities (<50 mA/cm^2) was better than that with Nafion, the maximum power output was highest with Nafion. The MEA prepared with the polyphosphazene blend membrane of higher effective IEC performed well and delivered a maximum power density of 51.4 mW/cm^2 (70% that of a Nafion 117 MEA), with a methanol crossover flux that was smaller than Nafion by a factor of three. Better methanol barrier properties were observed with the lower IEC polyphosphazene membrane, with a crossover flux six-times smaller than that in Nafion 117. Unfortunately, the electrochemical performance of this membrane was

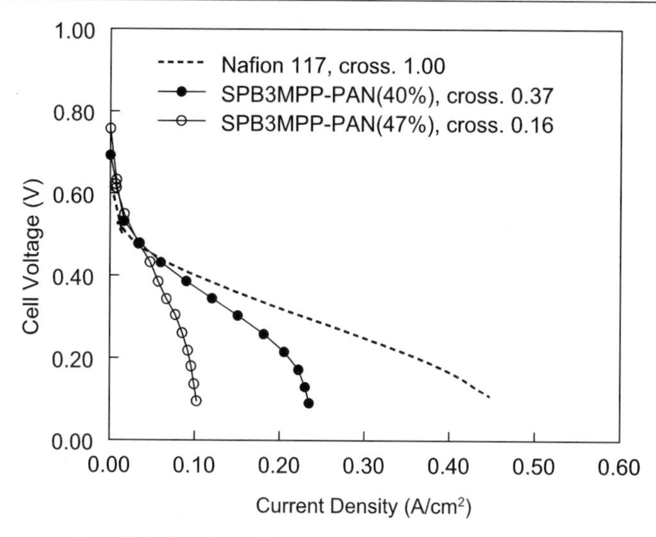

Fig. 17 Direct methanol fuel-cell performance curves with blended membranes containing sulfonated poly[bis(3-methylphenoxy)phosphazene] and PAN. 1.0 M methanol feed, 60 °C, air at 50 sccm and 30 psi back pressure. *Cross* denotes the methanol crossover flux (mol/cm^2 min) at *open circuit*, relative to that in Nafion 117

poor, with a maximum power density of only 23.4 mW/cm^2 (30% that of a Nafion MEA).

4.3 Membranes from PBI-Doped SPBPP [34]

A series of proton-conducting fuel-cell membranes were prepared from blends of sulfonated poly[bis(phenoxy)phosphazene] (SPBPP) and polybenzimidazole (PBI), where the latter, being a polymer base, was used as a complexing agent that effectively crosslinked the membrane. Depending on the SPOP ion-exchange capacity (1.2 or 1.4 mmol/g) and the amount of added PBI (5–12 wt %), the resulting membranes had a room-temperature proton conductivity in the range 0.01–0.08 S/cm, an equilibrium water swelling from 18–75%, and a methanol permeability (at 60 °C and 1.0 M methanol) that ranged from 1.2×10^{-6}–1.2×10^{-7} cm^2/s (all parameters decreasing with increasing PBI content). The membranes (82–120 μm in thickness) were tested in a direct methanol fuel cell, operating at 60 °C with a 1.0 M aqueous methanol feed solution and ambient pressure air at 500 sccm. A Scribner Series 890B fuel-cell test station with mass flow and temperature control was employed for these experiments. Methanol crossover flux was determined by measuring the carbon dioxide concentration in the cathode air exhaust at open circuit with a Vaisala GMM12B or GMM220A CO_2 detector. MEAs (5 cm^2 electrode area) were prepared by painting catalyst ink onto carbon cloth (ELAT, E-TEK) and then hot pressing the catalyzed electrodes to a membrane at 80 °C and 125 psi for 3 min. The catalyst loading for both the anode (Pt – Ru, 1 : 1) and the cathode (Pt) was 4 mg/cm^2. Two types of membrane-electrode assembly (MEA) were examined: (i) a sandwich design, where a SPOP-PBI film was placed between two half-MEAs prepared from Nafion 112 and (ii) a standard design, with direct hot-pressing of catalyst electrodes onto a polyphosphazene-based proton-exchange membrane.

Comparable current-voltage DMFC characteristics were obtained with the two MEA configurations (sandwich and direct electrode hot pressing), as shown in Fig. 18. The sandwich-type MEA is attractive for preliminary fuel-cell screening of new membrane materials; the use of such a configuration circumvents the need to identify the proper electrode catalyst/binder composition and hot-pressing conditions. The fuel-cell performance curves for three sandwiched SPBPP-PBI membranes and a sandwiched Nafion 112 reference are presented in Fig. 18a. The polyphosphazene curves are ordered according to the PBI content of the inner SPBPP-PBI film, indicating that membrane resistance was the primary factor in determining the shape of the plots (i.e., the slope of the plots in the IR regime). The DMFC with membranes containing 3% PBI generated nearly as much power as that with a Nafion membrane MEA. The most visible drop in performance occurred for membranes containing 8 and 12% PBI.

Representative *i–V* curves for MEAs with four different polyphosphazene membranes are shown in Fig. 18b, along with a reference plot of a Nafion 117-based MEA. For these tests, the electrodes were hot pressed directly to the SPBPP/PBI membrane. DMFC performance with a 1.2 mmol/g IEC SPBPP membrane with 3 wt % PBI was essentially identical to that with Nafion 117,

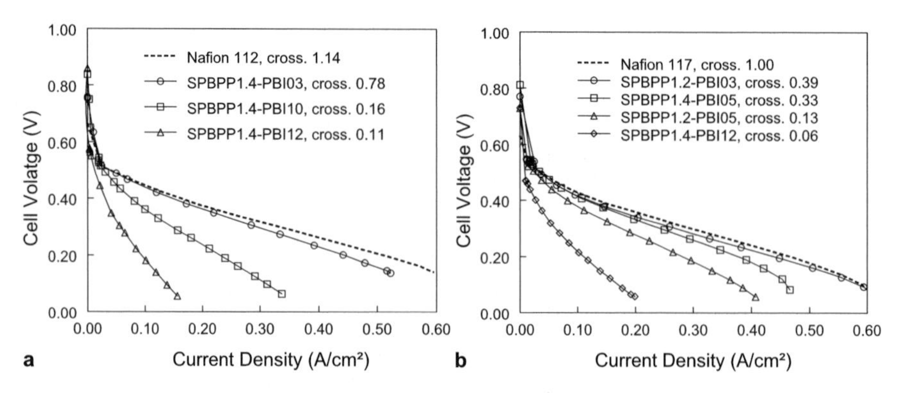

Fig. 18 Direct methanol fuel-cell performance curves with membranes composed of PBI-doped sulfonated poly[bis(phenoxy)phosphazene] using: **a** sandwich MEA design, and **b** direct hotpressing of electrodes. 1.0 M methanol feed, 60 °C, ambient pressure air at 500 sccm. *Cross* denotes the methanol crossover flux (mol/cm^2 min) at *open circuit*, relative to that in Nafion 117

but the methanol crossover was 2.5-times lower than that of Nafion. As the effective IEC of the doped polyphosphazene decreased (higher PBI content at a given SPBPP IEC or lower SPBPP IEC at a given PBI content), the methanol permeability decreased. The best methanol barrier material (17-times lower methanol permeability, as compared to Nafion) was a film containing 1.4 mmol/g SPBPP with 12 wt % PBI. Unfortunately, this highly blocking film had a high areal resistance which resulted in low power output during fuel cell operation. Thus, depending on the ion-exchange capacity of the SPOP (either 1.2 or 1.4 mmol/g) and the PBI content of the membrane (which ranged from 5% to 12%), the maximum DMFC power density ranged from 23 to 89 mW/cm^2 (as compared to 96 mW/cm^2 with Nafion 117), while the methanol crossover was between 4.2×10^{-6} mol/cm^2 min and 5.9×10^{-6} mol/cm^2 min (versus 1.17×10^{-5} mol/cm^2 min with Nafion 117).

For many practical applications, especially those dealing with portable power, a high methanol feed concentration is preferred in a DMFC. The effect of methanol feed concentration on the DMFC performance of MEAs containing PBI-doped sulfonated poly[bis(phenoxy)phosphazene] (SPBPP) membranes was determined under the following fuel-cell operating conditions: 60 °C temperature, 2 ml/min methanol flow rate, and ambient pressure air at 500 sccm (humidified at 70 °C). The results are summarized in Fig. 19, where the maximum power density is compared for Nafion 117 and two SPBPP-PBI membranes (1.2 mmol/g IEC + 3% PBI with a thickness of 82 μm and 1.4 mmol/g IEC + 4% PBI with a thickness of 141 μm) at three methanol feed concentrations (1.0, 5.0, and 10.0 M). For Nafion, there is a steady and significant decline in the maximum power with concentration, from 96 mW/cm^2 at 1.0 M methanol to 9 mW/cm^2 at 10.0 M. The loss in power is due to the increase in methanol crossover flux with increasing methanol feed concen-

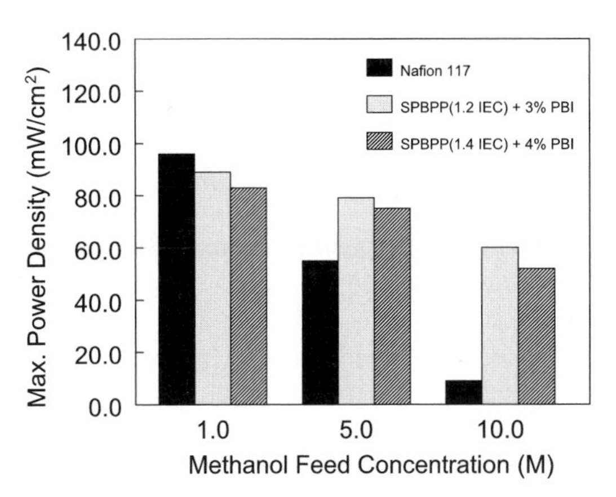

Fig. 19 The effect of methanol feed concentration on the maximum power density in a direct methanol fuel cell, for Nafion 117 and two PBI-doped sulfonated poly[bis(phenoxy)phosphazene] membranes

tration. The PBI-doped polyphosphazene membranes, on the other hand, are better methanol blockers and the loss in power with increasing methanol feed concentration is much less severe. Thus, the maximum power density in a polyphosphazene-based DMFC was more than 6-times greater than that with Nafion 117 for a 10.0 M methanol feed.

In lifetime experiments, an MEA was tested in a DMFC for 100 h, at 60 °C with 1.0 M methanol and ambient pressure air. To extend the longevity of the electrodes, the fuel cell was operated in a load cycling mode [49], i.e., repeated cycling of 59 min with the current on (at 0.10 A/cm^2) and 1 min at open circuit. During the cycling, the voltage was continuously recorded. The MEA was prepared by hot-pressing electrodes directly onto a SPBPP14-PBI08 membrane (1.4 mmol/g polyphosphazene with 8 wt% PBI). It was observed that the cell voltage decreased slowly, at a rate of approximately 0.8 mV/h, from an initial value of 0.39 V to 0.31 V (at 0.1 A/cm^2) after 100 h. At the same time, the cell resistance, as measured by a current interrupt method, slowly increased. Methanol crossover remained unchanged for the entire 100–h experiment. The initial cell voltage could be fully recovered by lowering the temperature to 25 °C and washing the membrane with deionized water for 2 h (either by removing the MEA from the fuel-cell test fixture or by an in-situ procedure where the methanol feed solution was replaced by room-temperature water). After this regeneration step and upon further constant current operation, the cell voltage once again began to decrease. At the present time, the observed voltage loss is associated with an interfacial resistance between the membrane and the Nafion binder in the electrodes, which increases over time due to a growing difference in swelling between Nafion and the SPBPP-PBI film. It may be possible to correct this problem by adjusting the membrane IEC and PBI content (e.g., adding more PBI to the blend should stiffen the membrane and minimize/eliminate the slow, upward drift in swelling by methanol).

5 Patents on Proton-Conducting Polyphosphazenes

There are three important patents and patent applications related to the use of polyphosphazene-based proton-exchange membranes in fuel cells. The first patent, authored by Pintauro and Tang and issued in April 2002 [50], deals with the use of a polyphosphazene membrane with acidic functional groups in a proton-exchange membrane fuel cell (either a hydrogen/air or direct methanol fuel cell). The patent describes membranes composed of sulfonated PB3MPP with/without crosslinking and blending, where the IEC of the polyphosphazene is between 0.8 and 1.9 mmol/g.

The second polyphosphazene fuel-cell patent was authored by Allcock and co-workers at the Pennsylvania State University [51] and was issued on July 6,

2004. This patent deals with proton-conducting membranes having improved resistance to methanol crossover. The membranes are obtained by solution casting followed by solvent evaporation from a solution containing an organic solvent, a polymer (preferably a polyphosphazene), and an oxoacid. It is claimed that a particularly useful application for these polymeric membranes is in methanol fuel cells.

One of the recent attempts to patent polyphosphazenes for fuel cells is an application filed by Honda Motor Co., Ltd., Tokyo on January 20, 2005 [52]. The invention deals with the fabrication of membranes composed of highly sulfonated polyalkylphenoxyphosphazenes, for possible use in a hydrogen/air or direct methanol fuel cell. Methods of synthesizing polyphosphazenes with an IEC as high as 4.9 mmol/g are described and the proton conductivity of the resulting films is presented.

The latest patent application comes from researchers representing Toyota and Case Western Reserve University [53]. The invention describes the use of multiaromatic ring-sulfonated polyphosphazenes as fuel-cell electrode binders.

6 Future Prospects for Polyphosphazenes in Fuel Cells

As shown by the research groups of Allcock and Pintauro over the past decade, interesting functionalizations of polyphosphazenes and good fuel-cell performance with the resultant materials can be realized, but reproducibility and polymer availability are lingering issues. The synthesis of the poly(dichlorophosphazene) precursor by thermal ring-opening polymerization is a rather capricious process, with inconsistent final molecular weights and variable branching and/or crosslinking. The use of "living" ionic polymerization (as reported by Allcock et al. [21]) might be a promising way to design appropriately substituted polyphosphazenes, including block copolymers. Such a synthetic route would avoid the "randomness" and dispersion of the thermal ring-opening polymerization process and, eventually, the macromolecular substitution route for adding side groups to the P – N backbone.

Having a methodology for the reproducible synthesis of poly(dichlorophosphazene) or, better yet, developing an ionic polymerization scheme would shorten the time to produce new phosphazene polymers and establish structure/property correlations associated with such new materials. While the unlimited functionalization capability of polyphosphazenes can be regarded as a great advantage, it can also be viewed as an impediment to the further development of these polymers for PEM fuel cells. With so many possible side group choices, which should be investigated first and what are the anticipated properties of such new materials? Combinatorial chemistry and high-throughput syntheses might be the best options for identifying new polyphosphazene polymers in the future, especially for fuel-cell applications where there are stringent property requirements.

References

1. Gasteiger HA, Mathias MF (2005) In: Proceedings of the Electrochemical Society, vol 2002-3 (Proton-Conducting Membrane Fuel Cells III Symposium), The Electrochemical Society, Pennington NJ, p 1
2. Pintauro PN, Wycisk R (2005) Fuel Cell Membranes. In: Li N, Fane T, Matsuura T, Ho W (eds) Membranes: Manufacturing Utilizing Six Sigma and Applications. Wiley, New York
3. Banerjee S, Curtin DE (2004) J Fluorine Chem 125:1211
4. Mathias MF, Makharia R, Gasteiger HA, Conley JJ, Fuller TJ, Gittleman CJ, Kocha SS, Miller DP, Mittelsteadt CK, Xie T, Yan SG, Yu PT (2005) The Electrochem Soc Interf 14:24
5. Rikukawa M, Sanui K (2000) Prog Polym Sci 25:1463
6. Li Q, He R, Junsen JO, Bjerrum NJ (2003) Chem Mater 15:4896
7. Haile SM (2003) Acta Mater 51:5981
8. Jannasch P (2003) Curr Opin Colloid Interf Sci 8:96
9. Savadogo O (2004) J Power Source 127:135
10. Hickner MA, Ghassemi H, Kim YS, Einsla BR, McGrath JE (2004) Chem Rev 104:4587
11. Hogarth WHJ, Diniz da Costa JC, Lu GQ (2005) J Power Source 142:223
12. Smitha B, Sridhar S, Khan AA (2005) J Membr Sci 259:10
13. Allcock HR, Wood RM(2006) J Polym Sci Part B 44:2358
14. Singler RE, Schneider NS, Hagnauer GL (1975) Polym Eng Sci 15:321
15. Gleria M, De Jaeger R (2001) J Inorg Organomet Polym 11:1
16. Allcock HR (2003) Chemistry and Applications of Polyphosphazenes, 1st edn. Wiley, Hoboken, NJ
17. Wycisk R, Pintauro PN (2003) Polyphosphazenes. In: Mark HF (ed) Encyclopedia of Polymer Science and Technology, vol. 7. Wiley, Hoboken, NJ, p 603
18. Allcock HR, Kugel RL (1965) J Am Chem Soc 87:4216
19. Allcock HR, Kugel RL, Valan KJ (1966) J Inorg Chem 5:1709
20. Allcock HR, Kugel RL (1966) Inorg Chem 5:1716
21. Allcock HR, Crane CA, Morrissey CT, Nelson JM, Reeves SD, Honeyman CH, Manners I (1996) Macromolecules 29:7740
22. Allcock HR, Crane CA, Morrissey CT, Olshavsky MA (1999) Inorg Chem 38:280
23. Andrianov AK, Chen J, LeGolvan MP (2004) Macromolecules 37:414
24. Stewart FF, Peterson ES, Stone ML, Sinler RE (1997) 213 National Meeting of the American Chemical Society, San Francisco, USA
25. Ganapathiappan S, Chen K, Shriver DF (1988) Macromolecules 21:2299
26. Andrianov AK, Martin A, Chen J, Sargent J, Corbett N (2004) Macromolecules 37:4075
27. Montoneri E, Gleria M, Ricca G, Pappalardo GC (1989) Makromol Chem 190:191
28. Montoneri E, Gleria M, Ricca G, Pappalardo GC (1989) J Macromol Sci Chem A26:645
29. Wycisk R, Pintauro PN (1996) J Membr Sci 119:155
30. Tang H, Pintauro PN, Guo Q, O'Connor S (1999) J Appl Polym Sci 71:387
31. Guo Q, Pintauro PN, Tang H, O'Connor S (1999) J Membr Sci 154:175

32. Tang H, Pintauro PN (2001) J Appl Polym Sci 79:49
33. Carter R, Wycisk R, Yoo H, Pintauro PN (2002) Electrochem Solid-State Lett 5:A195
34. Wycisk R, Lee JK, Pintauro PN (2005) J Electrochem Soc 152:A892
35. Paulsdorf J, Burjanadze M, Hagelschur K, Wiemhofer HD (2004) Solid State Ionics 169:25
36. Allcock HR, Hofmann MA, Wood RM (2001) Macromolecules 34:6915
37. Allcock HR, Hofmann MA, Ambler CM, Morford RV (2002) Macromolecules 35:3484
38. Allcock HR, Hofmann MA, Ambler CM, Lvov SN, Zhou XY, Chalkova E, Weston J (2002) J Membr Sci 201:47
39. Fedkin MV, Zhou X, Hofmann MA, Chalkova E, Weston JA, Allcock HR, Lvov SN (2002) Mater Lett 52:192
40. Zhou X, Weston J, Chalkova E, Hofmann MA, Ambler CM, Allcock HR, Lvov SN (2003) Electrochim Acta 48:2173
41. Koppel IA, Taft RW, Anvia F, Zhu SZ, Hu LQ, Sung KS, DesMarteau DD, Yagupolskii LM, Yagupolskii YL, Ingatev NV, Kondratenko NV, Volkonskii AY, Vlasov VM, Notario R,Maria PC (1994) J Am Chem Soc 116:3047
42. DesMarteau DD (1995) J Fluorine Chem 72:203
43. Feiring AE, Choi SK, Doyle M, Wonchoba ER (2000) Macromolecules 33:9262
44. Hofmann MA, Ambler CM, Maher AE, Chalkova E, Zhou XY, Lvov SN, Allcock HR (2002) Macromolecules 35:6490
45. Chalkova E, Zhou X, Ambler C, Hofmann MA, Weston JA, Allcock HR, Lvov SN (2002) Electrochem Solid-State Lett 5:A221
46. Lvov S, Chalkova E, Pague M, Allcock H, Ambler C, Maher A, Wood R (2003) 204th Meeting of The Electrochemical Society, Orlando, USA
47. Pintauro PN, Wycisk R (2004) Sulfonated polyphosphazene membranes for direct methanol fuel cells. In: Gleria M, De Jaeger R (eds) Phosphazenes: a worldwide Insight. Nova Science Publishers, New York, 25:591
48. Fox TG (1956) Bull Am Phys Soc 2:123
49. Knights SD, Colbow KM, St-Pierre J, Wilkinson DP (2004) J Power Source 127:127
50. Pintauro PN, Tang H (2002) US Patent 6 365 294
51. Allcock HR, Hofmann MA, Lvov SN, Zhou XY, McDonald D (2004) US Patent 6 759 157
52. Akita H (2005) US Patent Application 2005/0 014 927
53. Li W, Muldoon J, Hamaguchi H, Tsujiko A, Wycisk RJ, Lin J, Pintauro PN (2007) US Patent Application 2007/0 015 040

Adv Polym Sci (2008) 216: 185–258
DOI 10.1007/12_2008_155

Published online: 10 July 2008

Sulfonated Polyimides

Catherine Marestin[1] · Gérard Gebel[2] (✉) · Olivier Diat[2] · Régis Mercier[1]

[1]Laboratoire des Matériaux Organiques à Propriétés Spécifiques, UMR 5041, Chemin du Canal, Echangeur de Solaize, 69360 Solaize, France

[2]Structure et Propriétés des Architectures Moléculaires, Groupe Polymères Conducteurs Ioniques UMR 5819 (CEA-CNRS-UJF), INAC, CEA-Grenoble, 17 rue des Martyrs, cedex 9, 38054 Grenoble, France
gerard.gebel@cea.fr

Abstract Sulfonated polyimides have been designed to be used as proton conducting membranes in fuel cells. These materials present most of the required properties for this application, including a high level of ionic conductivity, a low gas and methanol permeability, and good mechanical properties. However, they exhibit a low stability when immersed in liquid water and in hydrogen peroxide solutions at elevated temperature due to a high sensitivity of the imide functions to hydrolysis. The aim of this article is to review the different routes of synthesis, the membrane-specific properties, the structural and transport property characteristics, and finally their behavior in fuel cells in terms of performance and stability.

Keywords Degradation · Fuel cells · Polymer synthesis · Proton exchange membranes · Structure · Sulfonated polyimides · Transport properties

1 Introduction

Proton exchange membrane fuel cells (PEMFCs) are considered as the most promising power source for automotive transportation in order to preserve oil resources and reduce significantly greenhouse gas emissions. During the last 10 years, this technology was significantly improved and numerous prototypes were unveiled by most of the important car constructors. The large-scale commercialization is, however, postponed primarily because of the difficulties encountered in the production, storage, and distribution of hydrogen and because PEMFCs still suffer from high production costs. Moreover, the operation of prototypes has revealed major reliability problems of the main components and especially of the membrane–electrode assembly (MEA). A fuel cell converts directly the chemical energy of hydrogen and oxygen into electricity through two electrochemical reactions on electrodes deposited on both sides of a proton exchange membrane (PEM). The main roles of this membrane are to convey the protons from the anode to the cathode with a minimum of resistance and to separate the reactants as efficiently as possible. Therefore, the membranes should exhibit very good mechanical and conductive properties whatever the condition of operation and the external conditions. The benchmark materials for PEMFCs are perfluorosulfonated ionomers, of which the most famous representative is Nafion from E.I. du Pont de Nemours. The Nafion membrane was originally designed to be used as a selective separator in electrolyzers for chlor-alkali production [1, 2]. Therefore, in addition to a tremendous ionic conductivity, these membranes exhibit an exceptionally high stability in corrosive media. The chemical stability issues from the perfluorinated nature of the polymer, and the conducting properties are attributed to a specific morphology of the ionic domains related to the highly hydrophobic properties of the perfluorinated backbone. Nowadays, many chemical companies commercialize perfluorosulfonated membranes and MEAs, which can be an alternative to Nafion for fuel cell application: Asahi Glass (Flemion™), Asahi Kasei (Aciplex™), Solvay Solexis (Hyflon™), Gore (Gore select™), 3M, ...

One of the major drawbacks of Nafion is its dehydration at elevated temperature ($T > 80\,^{\circ}\mathrm{C}$) when the atmosphere is not fully humidified [3]. Ionomer membranes are constituted of polymer chains bearing sulfonic groups. They require a minimum water content to become sufficiently proton conductive for application in fuel cells. The ionic groups cluster together to form ionic domains embedded in the hydrophobic polymer matrix (especially when perfluorinated [4, 5]). In the presence of few water molecules, the ionic groups become ionized and solvated [6]. The additional water molecules can swell the ionic domains and protons can move freely through the membrane when these domains are percolated. As a consequence, the ionic conductivity is directly related to the membrane microstructure. In the case of Nafion, the structure has been recently shown to be formed of an isotropic distribution of bundles of elongated polymer particles [7, 8]. The ionic groups are located at the solvent/polymer interface of the elongated highly hydrophobic particles which are likely to be ribbonlike. In such a model, the ion pathways are already present in the dry state, favoring a fast solvent penetration and a good ionic conductivity even at low levels of hydration. This is one of the main advantages of this model compared to the commonly accepted Gierke model [9], which assumes a percolation of spherical clusters by small ionic channels, whereas this percolation threshold is not experimentally observed in the expected range of water contents. Whatever the structural model, the swelling of the ionic domain structure induces a macroscopic swelling with 15% of dimensional changes from the dry state to fully hydrated state and consequently a mechanical fatigue upon swelling–contraction cycles.

During the last 10 years, the main development in the field of perfluorinated materials was the use of thinner membranes in order to decrease the ohmic drops within the membrane. The membrane thickness was reduced by a factor of 10 (from 200 to 20 μm) and this is a practical limit in terms of the gas barrier properties and mechanical strength. The thickness reduction was not associated with an actual cost reduction since it is more difficult to produce large areas of defect-free very thin membranes. Efficient prototypes were produced based on these new materials since the thickness reduction permitted the car constructor specification to be attained in terms of global power and power density. However, it has been shown that the lifetime of these membranes in fuel cell operation was significantly reduced, especially under cycling conditions. Typically, a PEMFC with a relatively thick membrane, such as Nafion 117 or 115 (175 and 125 μm, respectively), can operate for up to 50 000 h under stationary conditions [10], while the lifetime with a thinner membrane under severe cycling electric loading conditions is limited to a few hundreds of hours [11]. The thinner membranes exhibit a lower resistance to the mechanical fatigue induced by the swelling–contraction and to heterogeneous mechanical loads leading to anticipated membrane rupture. Moreover, these undesirable effects are enhanced by the fact that the membrane operates close to its glass transition temperature (T_g). A promising development was the introduction of chemical modifications on the fluorine backbone in order to increase the T_g and thus the thermal and mechanical stability. Recently, the chemical analysis of the water produced by the cell revealed the presence of a nonnegligible amount of fluorine atoms and sulfate groups, which shows that Nafion is also subject to chemical degradation depending on the operating conditions [12, 13].

The development of alternative low cost and stable membranes operating at high temperatures is thus a priority. Low production costs involve obviously the use of conventional hydrocarbon polymers or eventually lightly fluorinated ones. Since aliphatic polymer chains are not stable in the highly oxidative media encountered in fuel cells, most of the alternative membranes are now aromatic polymers [14–18]. Aromatic polymers are not intrinsically proton conductors and they should be either doped or functionalized. Some polymers, such as polybenzimidazoles, can be doped by strong acids [19, 20], but elution is always possible in the presence of liquid water and the low ionic conductivity at low temperature only permits fuel cell operation at very high temperatures. Therefore, functionalization by grafting of sulfonic groups on the polymer chain is now preferred. The sulfonic group was chosen because of its strong acidity, which is expected to lead to a very good dissociation and consequently a good conductivity. Phosphonic polymers can also be used for maintaining proton conductivity at high temperature [21]. Aromatic polymers are thermostable polymers with very high glass transition temperatures, the value of which is enhanced by the introduction of sulfonic acid groups along the polymer chain. Indeed, the T_g values of sulfonated aromatic polymers are often larger than the degradation temperature. The glassy nature of the polymer matrix whatever the operating temperature is expected to limit the dimensional changes associated with water uptake and thus the sensitivity to physical ageing. A first idea can be the use of substituted poly(*para*-phenylene) backbones but the extreme rigidity of the polymer chain led to brittle membranes, especially in the dry state [22]. More flexible polymers introducing ketone, ether, or sulfone bridges between the aromatic rings were developed [15, 16] and the number of possible families was also increased using polyaromatic heterocyclic materials, such as polybenzimidazoles [23, 24], polybenzoxazoles [25], phthalazinones [26–28] and polyimides [29–36].

Two routes can be used to introduce sulfonate groups onto a polymer chain. The first and easiest procedure is the post-sulfonation of commercial polymers by dissolution in concentrated sulfuric acid or using strong sulfonating agents such as fuming sulfuric acid, chlorosulfonic acid, or sulfur trioxide. The degree of sulfonation is then controlled by the duration of the sulfonation reaction. However, it is difficult to extrapolate such a procedure on an industrial scale because of the large production of chemical wastes and the duration of the sulfonation reaction, which can exceed 100 h to reach the required sulfonation level [15]. Moreover, it has been shown that this reaction produces some defects on the polymer backbone such as polymer chain scissions, which limit the lifetime in PEMFCs and cannot be completely avoided even using mild sulfonation procedures [37]. One of the major drawbacks of the post-sulfonation method is the absence of control of the sulfonic group distribution along the polymer chains. The required number of sulfonic groups is a balance between obtaining a sufficient ionic conductivity and preserving good mechanical properties of the swollen state at elevated temperatures. On the contrary to post-sulfonation, the use of a combination of sulfonated and non-sulfonated monomers allows the synthesis of controlled block copolymers by a two-step direct polymerization. A nonrandom distribution of the ionic groups along the polymer chain usually leads to membranes with improved properties. Long nonionic sequences generate hydrophobic nodes enhancing the mechanical properties which restrict water swelling, and long ionic sequences favor the formation of well-defined ionic domains. Moreover, multifunctional monomers can be used to increase the phase separation, while one ionic group per monomer is a practical limit when the membranes are prepared by post-sulfonation.

Among the numerous families of polymers which have been proposed as possible alternative membranes to Nafion during the last 10 years [15, 16], the sulfonated polyimides (SPIs) are probably the material for which the most complete studies have been performed in terms of morphology, properties determination, and stability studies depending on the polymer chemical structure. SPIs were first developed 35 years ago for application as a cation-exchange material in electrodialysis [38], but these materials were not studied afterwards except for the work of Solomin et al. [39]. At the end of the 1990s, the interest in PEMFCs was growing and SPI membranes have engendered new interest as an alternative material to Nafion [29, 30, 40]. SPIs are mostly prepared by polycondensation of a dianhydride and a combination of disulfonated and non-sulfonated diamines. The first SPIs were based on five-membered imide rings (called phthalic SPIs) but these polymers exhibited a very low stability even in mild conditions. Six-membered-ring polyimides (naphthalenic type) were then designed leading to reasonable performances in fuel cell conditions [29]. Different series of polymers were then synthesized, characterized, and tested in fuel cells. Despite it soon appearing that their stability in fuel cells was limited, the research activity on SPI membranes is continuously growing, with a special emphasis on chemical modifications based on new monomer synthesis to increase the stability. The aim of this paper is to review the research performed on SPIs from both the chemistry and characterization points of view, and to try to extract from these data some general trends for the development of new membranes with high performance and stability.

2 Synthesis

2.1 Linear Sulfonated Polyimides

SPIs can be synthesized by either post-sulfonation of a polyimide or by direct polymerization of sulfonated monomers. While the first method is largely used for the sulfonation of aromatic polymers such as poly(aryl ether ether ketones) [41–44] or polysulfones [37], it is scarcely used to synthesize SPIs, probably because of the very low solubility of these polymers in common organic solvents and in sulfuric acid solutions.

2.1.1
Sulfonated Polyimides by Post-Sulfonation

No research work about the post-sulfonation of a naphthalenic structure has yet been reported and only few examples of the post-sulfonation of a phthalic polyimide are mentioned in the literature [45, 46]. Both mentioned articles are concerned with the sulfonation of the polyimide synthesized from bis[4-(3-aminophenoxy)phenyl]sulfone and PMDA, either in the presence of chlorosulfonic acid ($ClSO_3H$) or sulfur trioxide-triethyl phosphate ($2SO_3 \cdot$ TEP) (Fig. 1). Because of the very low solubility of the polyimide precursor, the post-sulfonation is performed in heterogeneous conditions.

According to the pioneering work of Robeson et al. [47], when the mild sulfonating agent $2SO_3 \cdot TEP$ was used at room temperature, only very low ion-exchange capacities (IECs) (< 0.41) are obtained. With the stronger chlorosulfonic acid, higher IECs can be obtained (0.41–1.29), but the authors mention some decrease of the polymer mechanical properties which they attribute to some chain scissions.

Another major problem related to the heterogeneous nature of the reaction is the fairly difficult control of the extent of the sulfonic groups grafted on the polymer backbone. More interesting is the direct polymerization of a sulfonated monomer. Indeed, in that case, the final IEC is fixed by the ratio of the sulfonated and non-sulfonated comonomers. Moreover, the introduction of SO_3H groups is much better controlled.

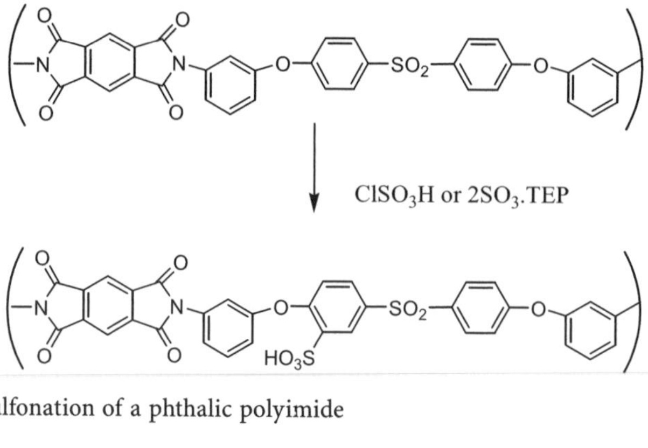

Fig. 1 Post-sulfonation of a phthalic polyimide

2.1.2 Sulfonated Polyimides by Direct Polymerization of a Sulfonated Monomer

As reported by Geniès et al. [48] from a study on model compounds, sulfonated naphthalenic polyimides prove to be more resistant to hydrolysis than their phthalic analogues. Few articles are devoted to the synthesis and characterization of sulfonated phthalic polyimides, whereas sulfonated naphthalenic polyimides have attracted much attention [53]. Rather few different naphthalenic dianhydride monomers are reported for the synthesis of SPIs (Fig. 2).

Indeed, besides the perylene tetracarboxylic dianhydride (PTDA) used by Lee et al. [54] and the bisimide dianhydride synthesized by Kim [55], only two new naphthalenic dianhydrides have been described. In this respect, 4,4′-ketone dinaphthalene 1,1′,8,8′-tetracarboxylic dianhydride (KDNTDA, Fig. 3) is obtained through a multistep synthesis, with a 19% overall yield [56].

BNTDA [57, 58] is synthesized by a biaryl coupling reaction (Fig. 4). The authors attribute the high hydrolytic stability of the resulting SPI to the specific structure of the dianhydride monomer. They argue that the noncoplanar binaphthyl rings disrupt the electronic conjugation of the dianhydride, therefore resulting in a higher electronic density of the imide linkage, which becomes less prone to a water nucleophilic attack.

However, it should be noted that most works related to SPI synthesis involve 1,4,5,8-naphthalenetetracarboxylic dianhydride (NTDA) as naphthalenic dianhydride. The sulfonic acid groups are incorporated into the

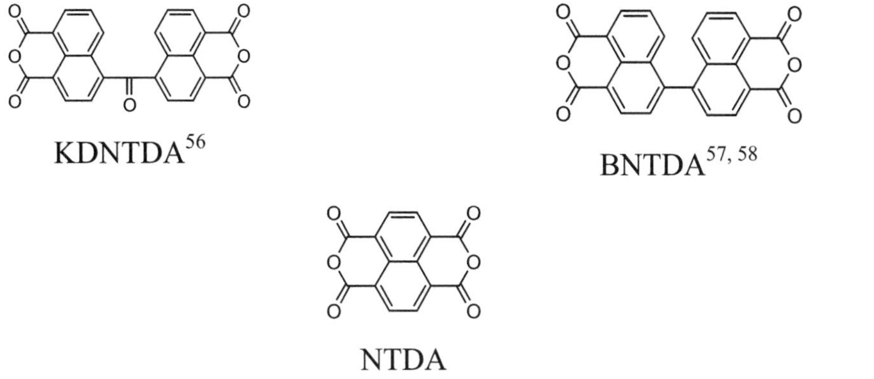

PTDA[54]

mBbNIA[55]

KDNTDA[56]

BNTDA[57, 58]

NTDA

Fig. 2 Naphthalenic dianhydrides for the synthesis of SPIs

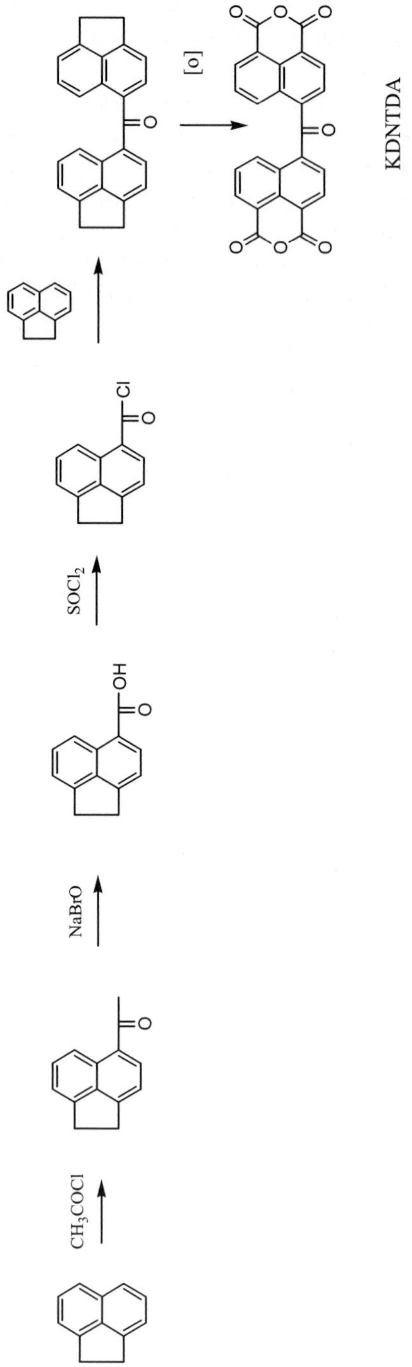

Fig. 3 Synthesis of KDNTDA

polymer backbone by the direct copolymerization of sulfonated monomers, and more precisely of sulfonated diamines. Very few of them are commercially available (Fig. 5).

If 5,5′-dimethylbenzidine-2,2′-disulfonic acid (DMBDSA) [59], 2,5-diaminobenzenesulfonic acid (DABSA) [60], and 4,4′-diaminostilbene-2,2′-disulfonic acid (DSDSA) [64] are scarcely used, 4,4′-diaminobiphenyl-2,2′-disulfonic acid (BDSA) has been the most employed in recent years.

From NTDA and BDSA monomers, a wide range of SPIs have been synthesized by changing the non-sulfonated diamine comonomer [30, 31, 36, 61–63], including aliphatic diamines [33, 65, 66]. Different properties (solubility, water uptake, conductivity, hydrolytic stability) can be fine-tuned by choosing carefully the structure of the non-sulfonated diamines. As developed in Sect. 3, it has been established that:

- Flexible non-sulfonated diamines increase the solubility of the final polymer as well as its hydrolytic stability.
- Bulky diamines increase the interchain space which induces a higher conductivity at high humidity rates as well as a higher hydrolytic stability.

Based on the research work reported by Faure [40], either random or block copolymers can be obtained depending on the synthesis pathway. When all monomers are reacted in one step, random copolymers are obtained. To tailor a blocklike structure, first the sulfonated diamines with a fraction of NTDA

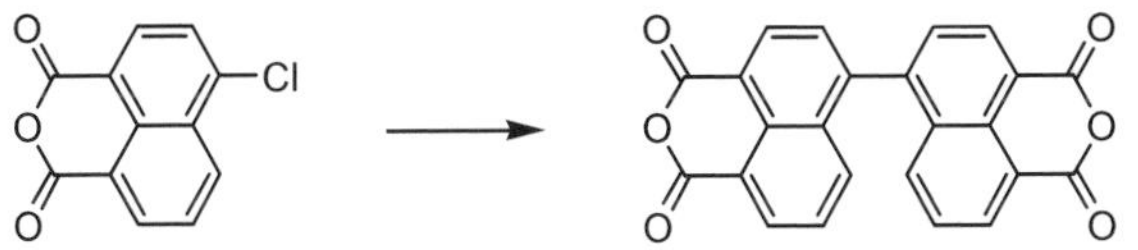

Fig. 4 Synthesis of BNTDA

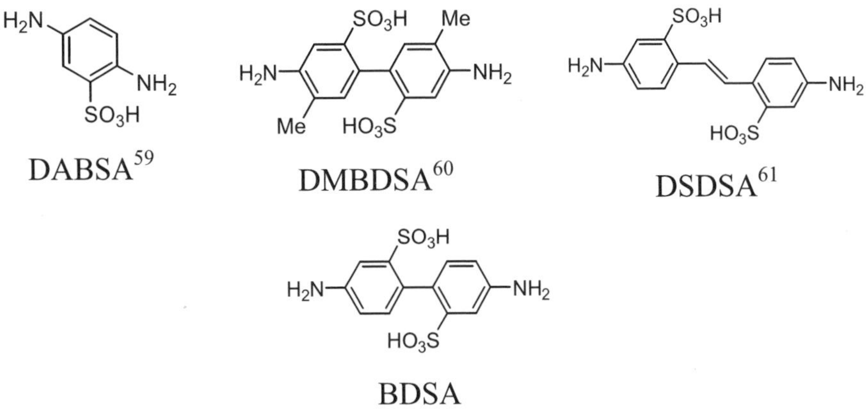

DABSA[59] DMBDSA[60] DSDSA[61] BDSA

Fig. 5 Commercially available sulfonated diamines

are polymerized. The ratio of the diamine and NTDA is calculated to reach a desired length of oligomers end-capped with amine groups. The final polymers are obtained by adding precise amounts of NTDA and non-sulfonated diamine to reach the desired ion content.

As could be expected, the membrane final properties (conductivity, swelling, hydrolytic stability) depend on the degree of sulfonation which is usually expressed as the IEC (see Sect. 3.1 for determination). More interestingly, it has been shown that for a given IEC, the properties also depend on the distribution of the sulfonic groups along the polymer backbone. This observation suggests that the structure of the polymer greatly influences the microstructure of the membrane formed and therefore its macroscopic properties, as detailed in Sect. 3 [67].

Despite the good performances obtained in a long-term fuel cell test at 60 °C (3000 h) [29], BDSA-based SPIs have been reported to display limited hydrolytic stability, especially at high temperature (> 80 °C) [68]. In order to improve the resistance of SPIs, recent interest has been devoted to the design of new sulfonated diamines. From this perspective, different groups have contributed to the synthesis of a wide range of new sulfonated monomers. As the sulfonic acid groups can either be directly linked to the polymer backbone or grafted on pendant alkyl or aryl groups, "main-chain" or "side-chain" monomers can be distinguished.

Most main-chain sulfonated diamines are synthesized by direct sulfonation of a non-sulfonated precursor (Fig. 6) or its dinitro analogue [69]. 4,4′-Diaminodiphenyl ether-2,2′-disulfonic acid (ODADS), 4,4′-bis(4-aminophenoxy)biphenyl-3,3′-disulfonic acid (BAPBDS) [71, 72], 2,2′-bis[4-(4-aminophenoxy)hexafluoroisopropane disulfonic acid (BAHFDS) [73, 74], 9,9-bis (4-aminophenyl)fluorene-2,7-disulfonic acid (BAPFDS) [35], and 2,2′-bis(4-aminophenoxy)biphenyl-5,5′-disulfonic acid (*o*-BAPBDS) [75] were synthesized accordingly. Because of the electrophilic nature of the sulfonation reaction, the sulfonic acid groups are usually introduced in the *ortho*/*para* position to the electron-donating substituents. Polymer-grade monomers are obtained with excellent yields (83, 85, and 90%, respectively, for BAPFDS, ODADS, and *p*-BAPBDS), except in some cases (*o*-BAPBDS, 48%).

A slightly different synthetic procedure consists of an aromatic nucleophilic substitution involving a sulfonated precursor. In this respect, McGrath and coworkers [25, 76] reported the synthesis of bis[4-(3-aminophenoxy) phenyl]sulfone-3,3′-disulfonic acid (43% overall yield) by directly reacting sulfonated dichlorodiphenylsulfone with 3-aminophenol (Fig. 7).

A similar method was used by Zhai [77] for the synthesis of BAPBDS and 4,4′-bis(4-aminophenylthio)benzophenone-3,3′-disulfonic acid (BAPTBPDS), as well as by Shobha et al. [78] for the synthesis of sulfonated bis(3-aminophenyl)phenyl phosphine oxide (SBAPPO). Because of the aromatic nucleophilic substitution reaction, this procedure leads to the incorporation of two ether bridges in the corresponding sulfonated diamines. This structure

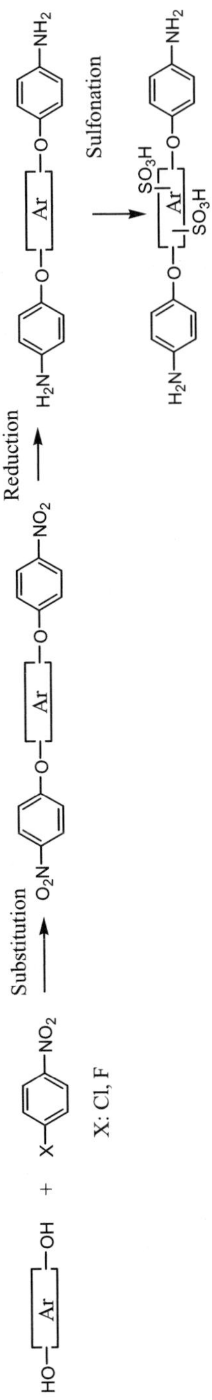
HO–[Ar]–OH
+
X–C6H4–NO2
X: Cl, F
Substitution
O2N
NO2
Ar
Reduction
H2N
NH2
Sulfonation
SO3H
SO3H

Fig. 6 Sulfonated diamine synthesis by sulfonation

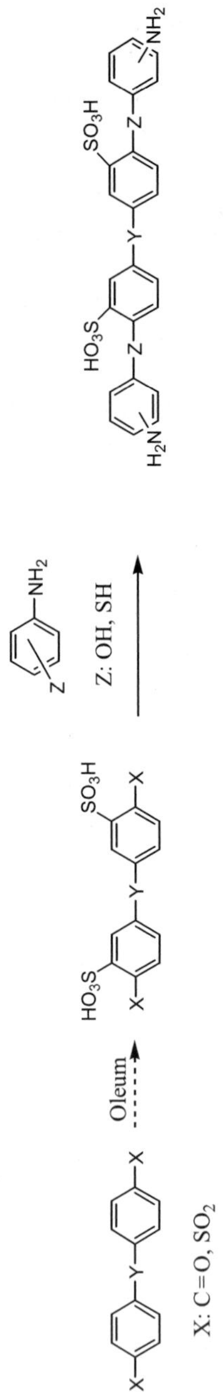

Fig. 7 Sulfonated diamine synthesis by aromatic nucleophilic substitution

both confers a good flexibility to the monomer and is expected to enhance the nucleophilic character of the amine groups.

By condensation of a biphenol with 2-fluoro-5-nitrobenzenesulfonic acid sodium salt [79, 80] or 5-amino-2-chlorobenzenesulfonic acid [81], some sulfonated diamines bearing the sulfonic group on the terminal amino phenyl ring can be synthesized (Fig. 8).

In order to enhance the sulfonic acidity and therefore improve the polymer proton conductivity, a different synthetic pathway has been elaborated to introduce the sulfonic acids in the *ortho* position to electron-withdrawing groups (sulfones, ketones). Based on a procedure widely developed by Jannasch et al. for the synthesis of proton exchange polyarylethers [82, 83], Chen et al. [84] designed an original and interesting monomer (Fig. 9). However, the harsh experimental conditions required (BuLi, gaseous SO_2) and relatively low overall yields (24.7%) should preclude the wide development of such monomers.

Okamoto et al. classified the different above-mentioned sulfonated diamines into two groups (Fig. 10): those having a sulfonic group directly bonded to the phenyl ring on which the amino group is present ("Type 1"), and those having the sulfonic group on a bridged phenyl ring ("Type 2"). According to the authors, such a classification relies on basicity considerations. As the sulfonic acid group is electron-withdrawing, the electron density of the benzene ring bearing such a group is lowered compared to the electron density of the unsubstituted aromatic amine, resulting in a decrease of the corresponding amine basicity. Considering that the reactivity of the condensation reaction of a diamine with a dianhydride increases with the basicity of the diamine, the authors argue that the reverse reaction (hydrolysis) should be less important for imide groups based on Type 2 diamines.

As reported for different aromatic and heterocyclic polymers, the introduction of side-chain sulfonic groups seems particularly interesting [85].

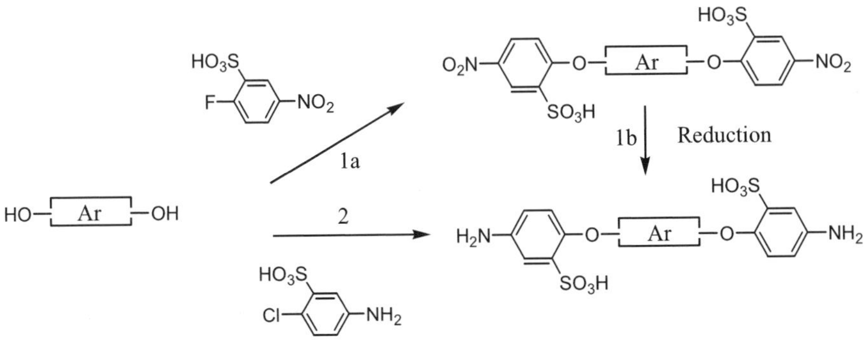

Fig. 8 Introduction of the sulfonic group on the amino phenyl ring by SNAr

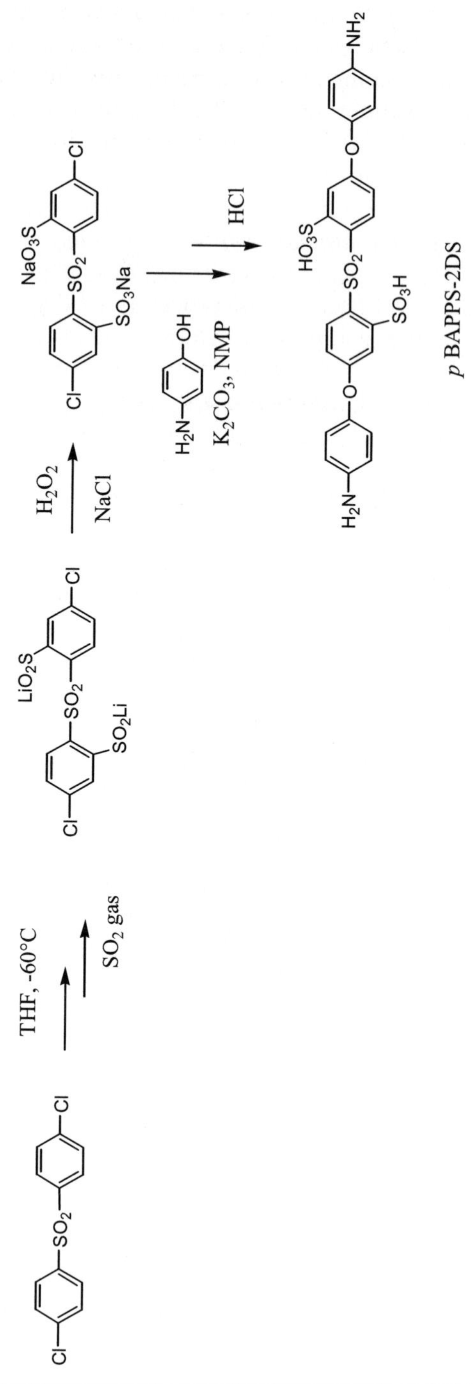

THF, -60°C

SO_2 gas

H_2O_2

NaCl

K_2CO_3, NMP

HCl

p BAPPS-2DS

Fig. 9 Sulfonation by lithiation/reaction with gaseous SO_2

Indeed, the side chains containing sulfonic acid groups are expected to give microphase separation structures, which could be favorable for proton transport, and undesired desulfonation reaction might be avoided if sulfonic groups are grafted onto alkyl spacers. Therefore, particular interest has been devoted to the synthesis of SPIs from diamine-containing sulfonic acid side

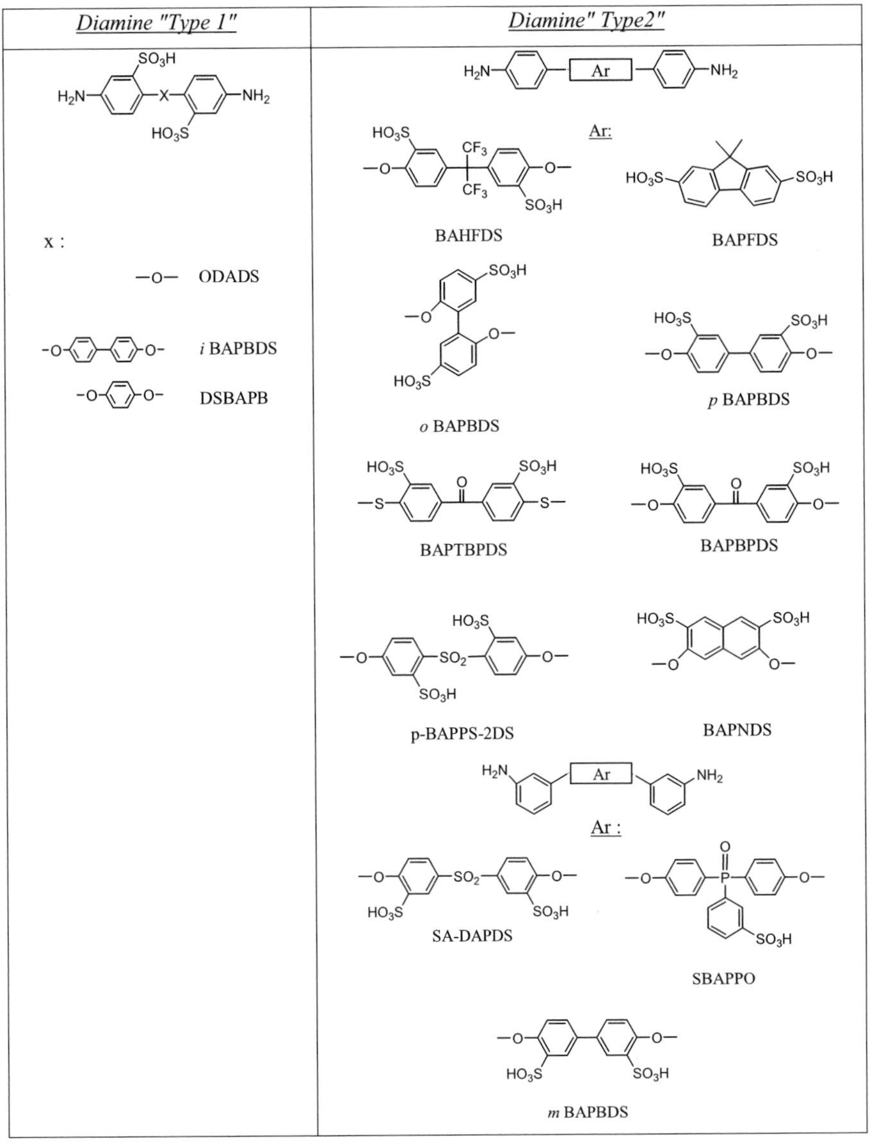

Fig. 10 Main-chain sulfonated diamines

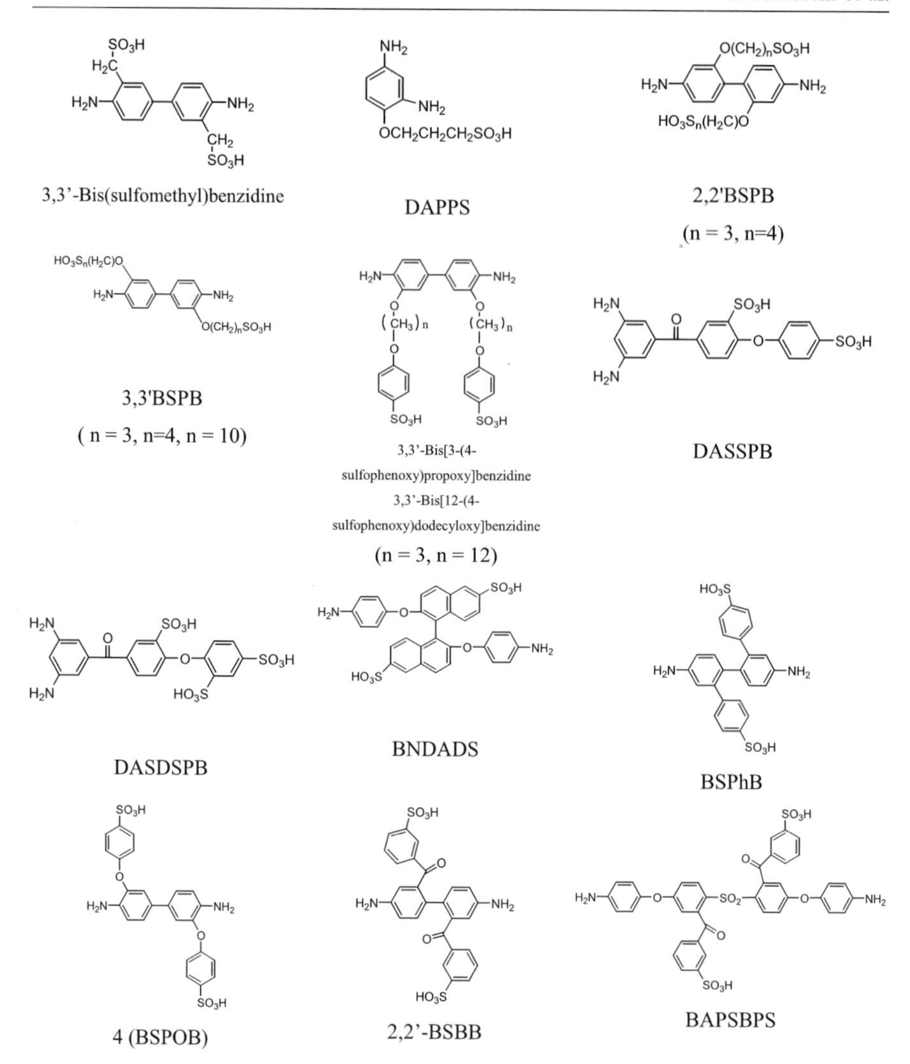

Fig. 11 Side-chain sulfonated diamines

chains (Fig. 11). Such monomers are usually obtained through multistep reactions. Detailed syntheses are reported hereafter.

Sulfonic acid groups can be grafted on alkyl side chains directly linked to a phenyl ring [86]. Accordingly, a new sulfonated diamine bearing a sulfomethylene group (3,3′-bis(sulfomethyl)benzidine) was synthesized in three steps (Fig. 12), with a 66% overall yield. Synthetic difficulties unfortunately preclude the synthesis of longer alkyl chain analogues.

The formation of alkoxy linkages enables the synthesis of longer sulfonic acid side chains. In this spirit, Yin et al. [87] reported the synthesis of DAPPS

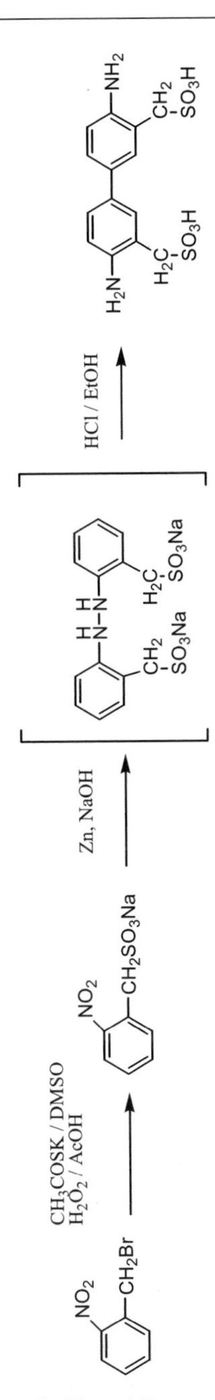

Fig. 12 Synthesis of 3,3′-bis(sulfomethyl)benzidine

in two steps. 2,4-Dinitrophenol is first reacted with 3-bromopropane sulfonic acid sodium salt. The resulting compound is then reduced (Fig. 13), giving the expected sulfonated diamine with a 44% overall yield. The limiting step seems to be the unfavored nucleophilic substitution, due to the electron-withdrawing effect of the nitro group in the *para* position to the phenol.

The same authors describe the synthesis of a BDSA-like monomer with a sulfonated side chain [88–90]. Depending on the reactant (2- or 3-nitrophenol), they succeeded in synthesizing two isomers, 2,2′-bis(3-sulfopropoxy)benzidine and 3,3′-bis(3-sulfopropoxy)benzidine (Fig. 14).

The four-step synthesis involves an aromatic nucleophilic substitution, the formation of an azo compound by a reduction reaction in the presence of zinc powder under basic conditions, and the reduction of this compound into a hydrazo analogue. The diamine monomer is then obtained by a rearrangement in acidic conditions. However, the overall yields of these syntheses remain rather low (respectively 39.5 and 30% for 2,2′- and 3,3′-isomers). Watanabe et al. [86] report a much longer aliphatic chain analogue ($n = 10$) by a similar procedure.

According to another synthetic pathway, initially described by Jönsson et al. [91, 92], Yasuda et al. [93] synthesized 2,2′- and 3,3′-BSPB (Fig. 15) diamines, as well as their analogues having butoxy side chains.

The position and the alkyl chain length do not seem to affect the final properties, except the polymer solubility. The introduction of alkoxy groups is reported to improve the hydrolytic stability of the polymers. However, the presence of these alkoxy groups seems to be at the expense of the oxidative stability, as the membranes are rapidly dissolved in Fenton's reagent [87, 93].

Watanabe and coworkers [86, 94] describe the synthesis of monomers having pendant sulfophenoxypropoxy groups (Fig. 16). This synthesis proceeds via a four-step reaction: first, the amino functions of the aminophenol reactant are protected by acetylation. The resulting compound is then bromoalkylated. Sulfonic acid groups (in the sodium salt form) are therefore grafted on the side chains by nucleophilic substitution, and a final acidic treatment leads to the desired monomer (12 and 46% yield, respectively, for chain lengths $n = 3$ and 12).

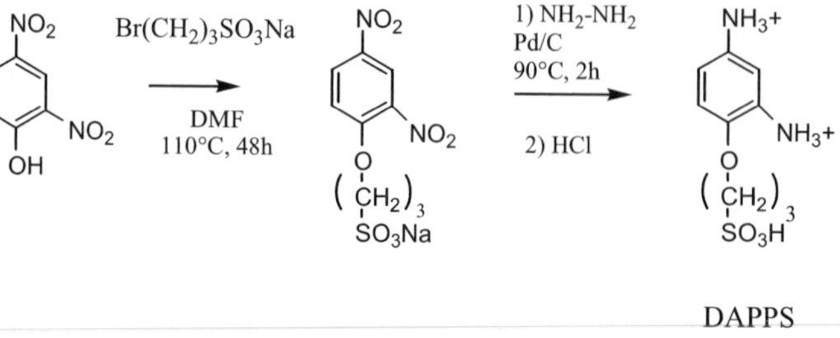

Fig. 13 Synthesis of DAPPS

Whereas the SPIs bearing the longest side chains have better properties than common main-chain or short side-chain SPIs (a higher oxidative and better hydrolytic stability, and a comparable proton conductivity) [86], they showed less water uptake and lower proton conductivity compared to SPIs

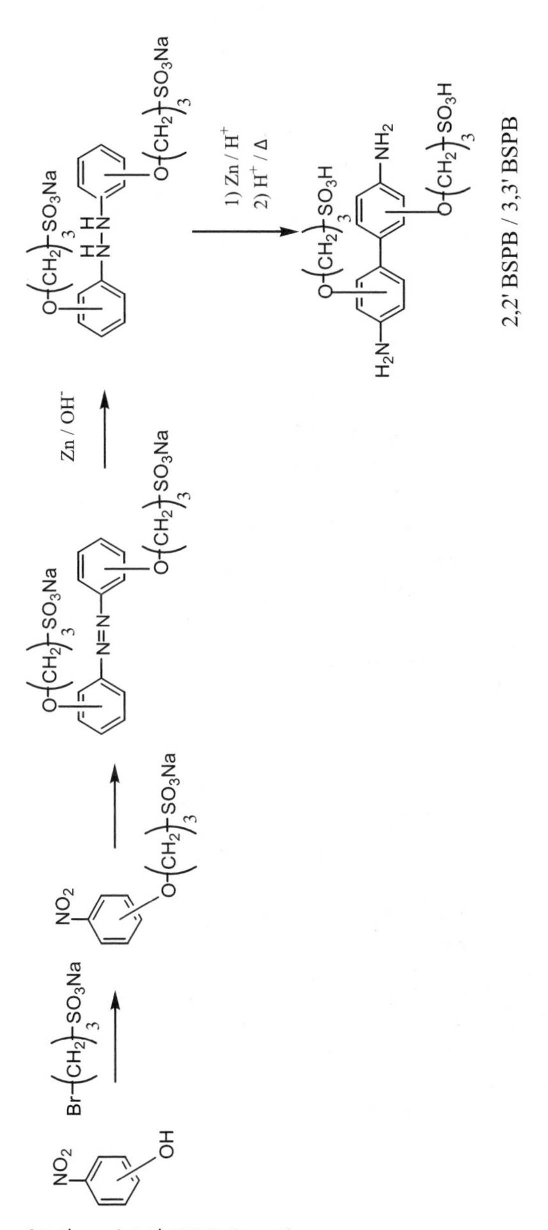

Fig. 14 Synthesis of 2,2′- and 3,3′-BSPB ($n = 3$)

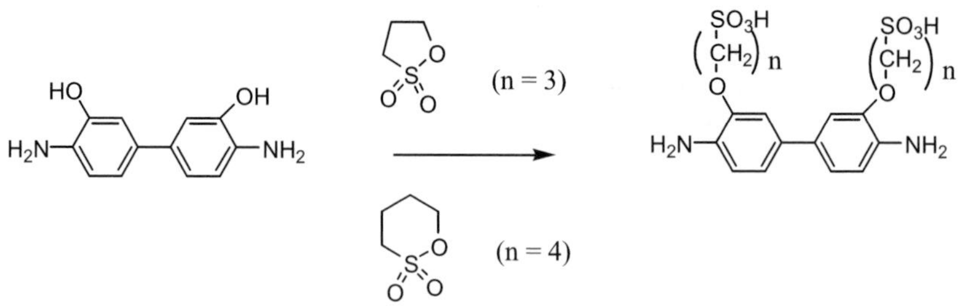

Fig. 15 Synthesis of 3,3′-BSPB ($n = 3$ and $n = 4$)

based on BSPB. Based on these results, the authors consider that concerning proton-conducting properties, there is an optimum side-chain length (three or four methylene units) [94].

Sulfophenoxy groups can be introduced either on a diamine monomer by aromatic nucleophilic substitution [95] involving 4-fluorobenzenesulfonic acid sodium salt or by direct sulfonation [96] of a compound having phenyl ether linkages (Fig. 17). Both methods were mentioned for the synthesis of 3,3′-BSPOB. The SPI obtained from 3,3′-BSPOB displays an outstanding water stability (> 2000 h at 100 °C). The argument of the authors to account for this hydrolytic stability is based on membrane morphological aspects.

A slightly different monomer in which sulfophenyl groups are present instead of sulfophenoxy or sulfopropoxy groups was obtained by direct sulfonation of 2,2′-diphenylbenzidine [34] (Fig. 18). Whereas the water stability of the membranes based on this monomer seems better than that of the membranes bearing sulfopropoxy groups, they remain less stable than those having BSPOB diamines. Moreover, it should be noted that the overall yield (taking into account the synthesis of the 2,2′-diphenylbenzidine precursor) is extremely low.

2,2′-Bis(3-sulfobenzoyl)benzidine (2,2′-BSBB) [97], another wholly aromatic side chain diamine without any ether or aliphatic groups, is synthesized in good overall yields (63%). As represented below (Fig. 19), this monomer is prepared through a nitration reaction, a Friedel–Crafts acylation, a reduction, and finally a post-sulfonation. Notwithstanding the presence of benzoyl groups which are more stable than alkoxy linkages, SPI membranes based on these monomers are quite sensitive toward hydrolysis. One probable reason is related to the electron-withdrawing effect of the benzoyl group which reduces the diamine basicity.

3,5-Diamino-3′-sulfo-4′-(4-sulfophenoxy)benzophenone (DASSPB) and 3,5-diamino-3′-sulfo-4′-(2,4-disulfophenoxy)benzophenone (DASDSPB) [98] were synthesized in a three-step process (respectively 24 and 43% yields). After a Friedel–Crafts reaction on diphenyl ether, two or three sulfonic acid groups were grafted, depending on the sulfonation reaction conditions (6 h at 40 °C or 8 h at 60 °C). A reduction by stannous chloride dehydrate gave

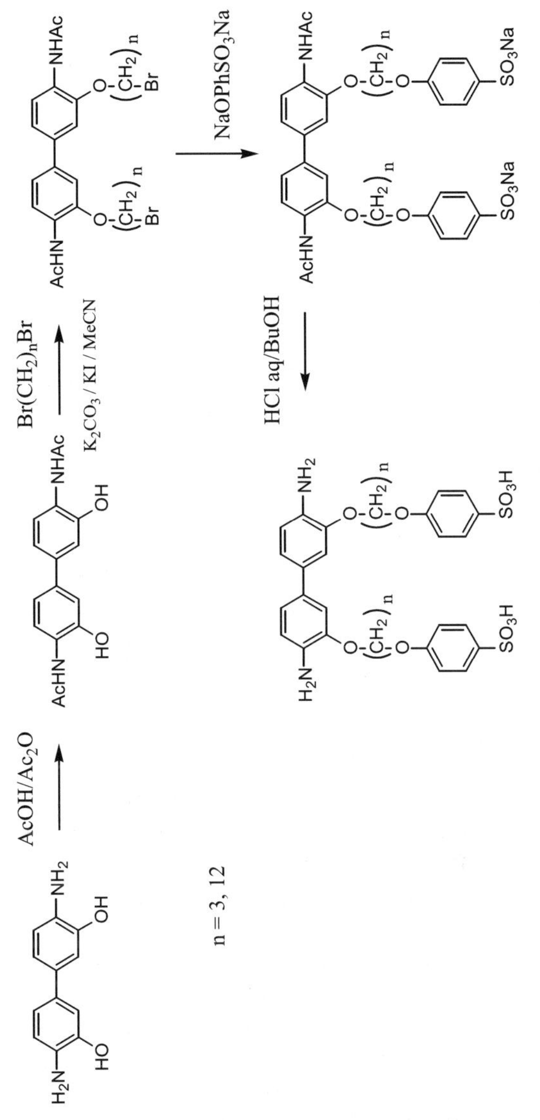

Fig. 16 Synthesis of 3,3′-bis[3-(4-sulfophenoxy)propoxy]benzidine and 3,3′-bis[12-(4-sulfophenoxy)dodecyloxy]benzidine

the corresponding diamines (Fig. 20). The incorporation of several sulfonic acid groups on the same monomer side chain was intended to favor the formation of hydrophilic nanodomains. The resulting SPI membranes effectively present high water uptake at low relative humidity, and consequently high proton conductivities even in the low RH range. More striking is the isotropic

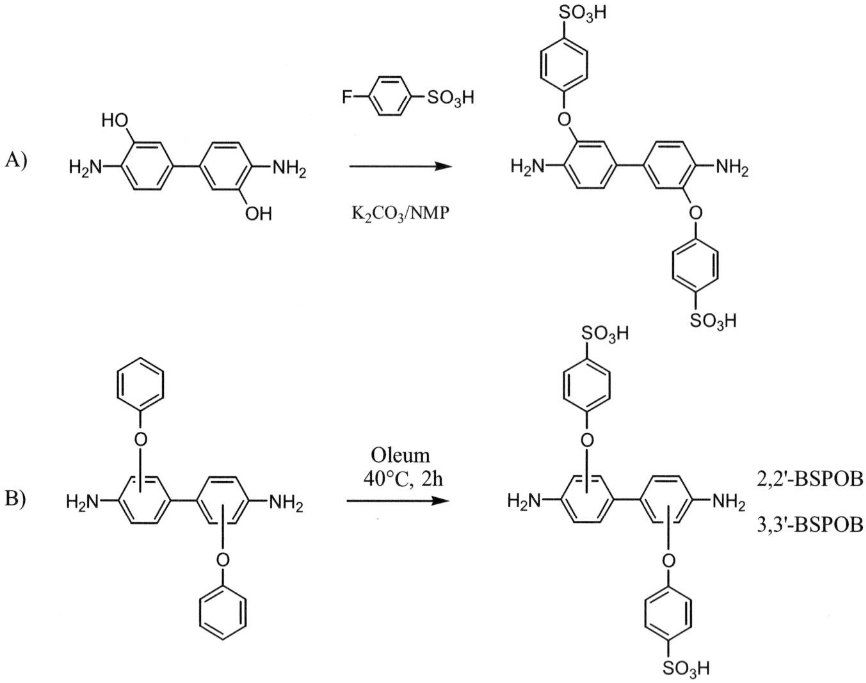

Fig. 17 Synthesis of BSPOB

swelling in water of these membranes, a phenomenon which might be related to the specific nature of the monomers involved.

According to a common procedure (aromatic nucleophilic substitution involving a binaphthyl phenol, reduction of the nitro groups, and sulfonation), a binaphthyl-containing diamine (2,2′-bis(*p*-aminophenoxy)-1,1′-binaphthyl-6,6′-disulfonic acid (BNDADS)) has been synthesized by Li et al. [99] (Fig. 21). Binaphthyl moieties induce a kinked chain structure which is supposed to increase the polymer solubility, inhibit interchain interactions and chain packing, and therefore increase the free volume accessible to water, thus helping in the formation of the observed microphase-separated structures.

A multistep reaction involving a lithiation, carbonation, Friedel–Crafts reaction, sulfonation, and aromatic nucleophilic substitution (overall yield 30%) (Fig. 22) was designed to synthesize a sulfonated diamine similar to *p*-BAPPS-2DS, but whose sulfonic acid groups are removed from the main monomer backbone [100]. The basicity of this diamine is expected to be high as no electron-withdrawing groups are present on the phenyl ring bearing the amino groups. The flexibility of this monomer is provided by the two phenoxy bridges.

Based on the different above-mentioned synthetic pathways, it is possible to design a great variety of sulfonated diamines. In this respect, the synthe-

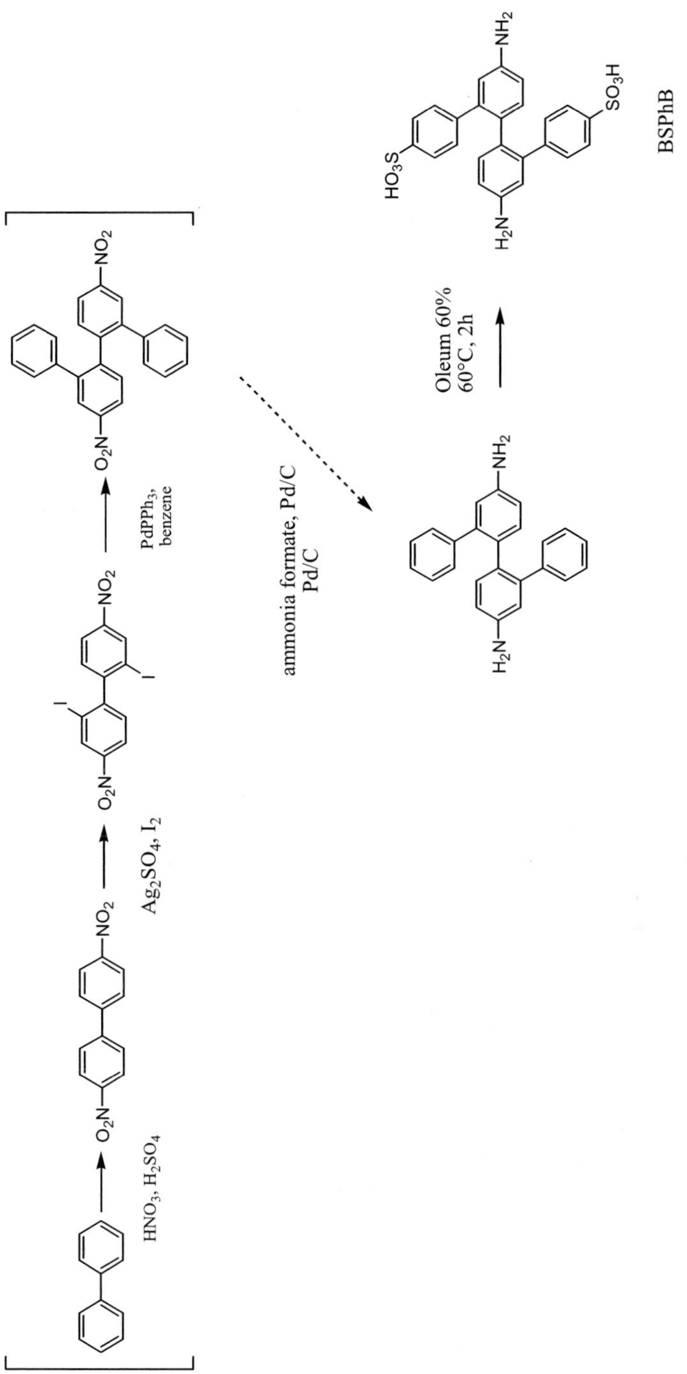

Fig. 18 Synthesis of BSPhB

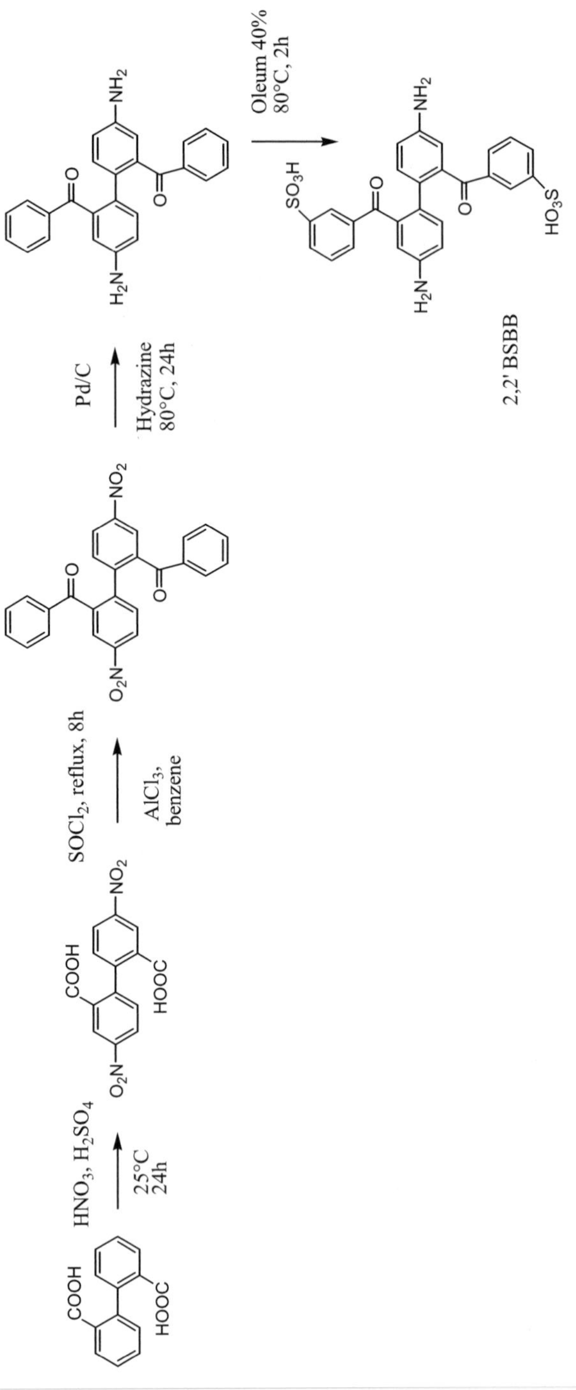

Fig. 19 Synthesis of 2,2′-BSBB

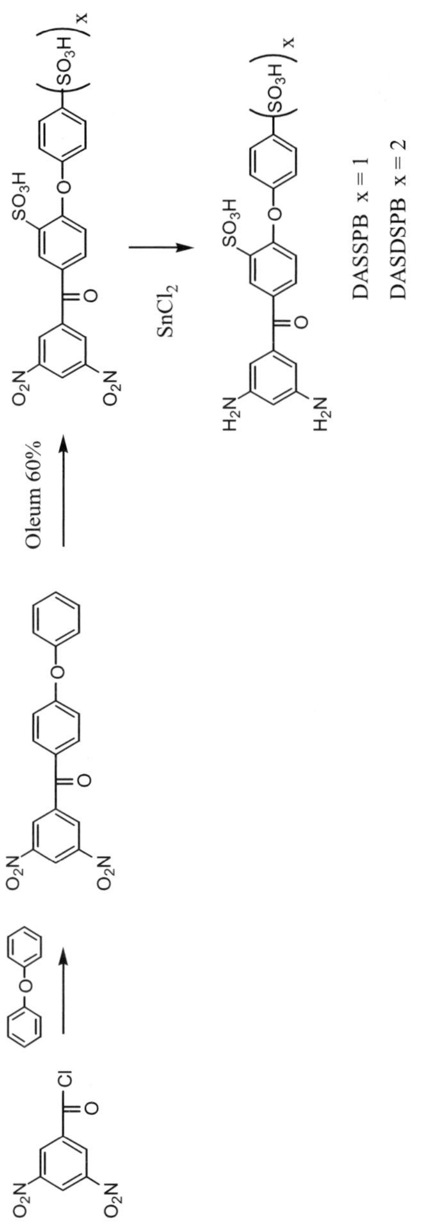

Fig. 20 Synthesis of DASSPB and DASDSPB

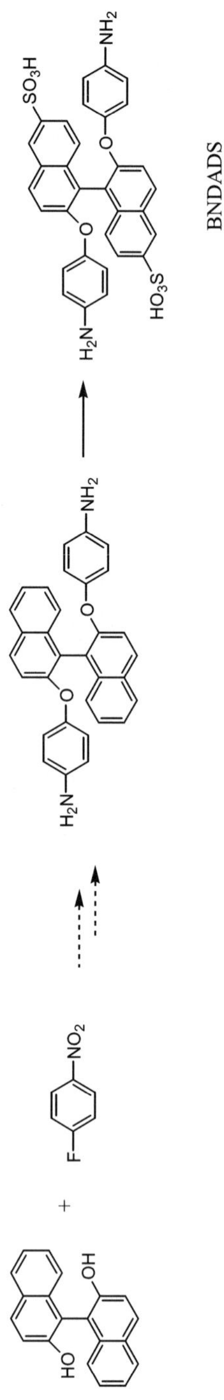

Fig. 21 Synthesis of BNDADS

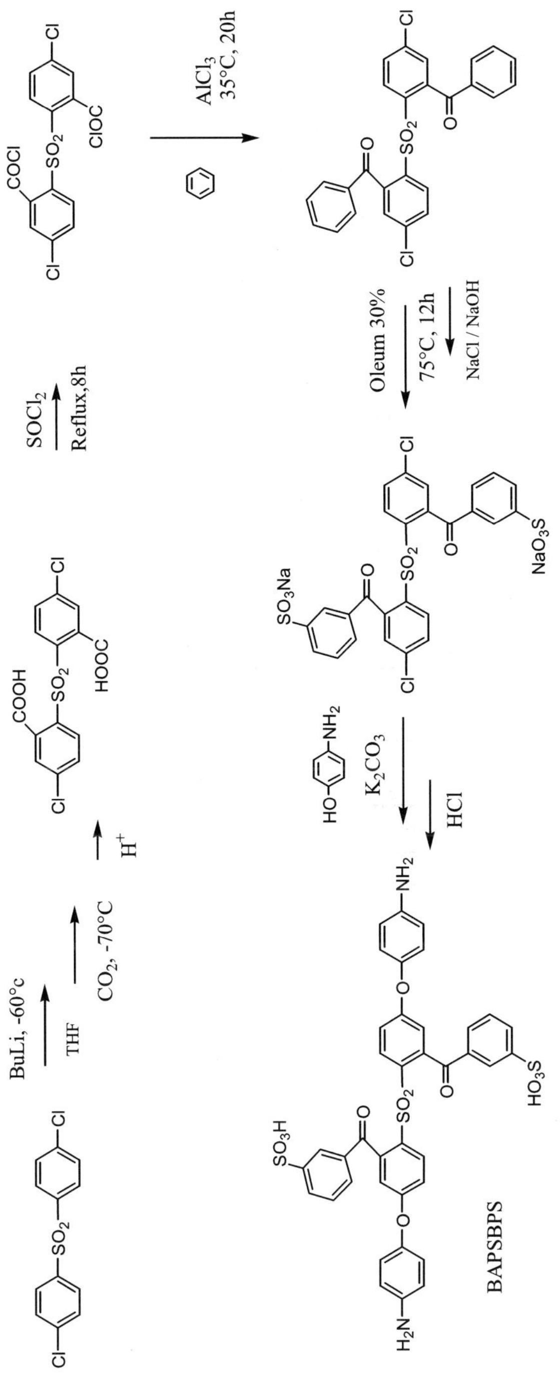

Cl–SO_2–Cl

BuLi, -60°c

THF

CO_2, -70°C

H^+

Cl, COOH, SO_2, HOOC, Cl

$SOCl_2$

Reflux,8h

Cl, COCl, SO_2, ClOC, Cl

$AlCl_3$

35°C, 20h

Cl, O, SO_2, O, Cl

Oleum 30%

75°C, 12h

NaCl / NaOH

SO_3Na, Cl, O, SO_2, O, Cl, NaO_3S

HO–NH_2

K_2CO_3

HCl

SO_3H, H_2N, O, O, SO_2, O, O, NH_2, HO_3S

BAPSBPS

Fig. 22 Synthesis of BAPSBPS

sis of various isomers (such as *m*-, *p*-, *o*-, or *i*-BAPBDS) [80] is particularly interesting as the resulting SPIs have the same chemical composition and IEC. However, depending on the respective positions of the sulfonic acid groups or on the specific structure of the monomer (*meta* or *para* isomers), these polymers do have different macroscopic properties (in terms of solubility, hydrolytic stability, proton conductivity, membrane morphology, or membrane swelling behavior). Comparing the properties of SPI membranes bearing SO_3H in the main chain or spaced from the backbone by alkyl or aryl groups is also very meaningful. Therefore, in-depth structure–property studies result in a better understanding of the phenomena involved during a fuel cell test and suggest how the polyelectrolyte chemical structure should be modified in order to improve the membrane performance.

Amongst the different sulfonated diamines recently reported, 2,2′- and 3,3′-BSPB are the most extensively studied. Interestingly, 2,2′- and 3,3′-BSPB based membranes have shown a much longer water stability at 100 °C compared to BAPBDS membranes [101]. This was attributed to the higher basicity of the side-chain sulfonated monomer and to the well microphase-separated structure morphology induced by this monomer [101, 102] (nanosized hydrophilic sulfonic acid domains and hydrophobic polyimide domains). However, at higher temperatures (130 °C), sulfopropoxy groups are cleaved, resulting in a proton conductivity decrease (around 20%) [103]. Nevertheless, these membranes seem very promising for lower temperature applications as witnessed by long-term fuel cell tests under H_2/O_2 [33], H_2/air [104], or in DMFCs [105–107]. It is worth mentioning that 5000 h at 80 °C is up to now the best long-term test reported for a SPI (and for any sulfonated polyaromatic membranes as well).

2.2 Cross-Linked Sulfonated Polyimides

One route to improve the ionic conductivity of these polymers is to increase the IEC. The presence of a large number of sulfonic groups on the polymer chain has, however, a detrimental effect on the swelling properties of the membrane as an excessive water uptake leads to a lower conductivity and weaker mechanical properties. To overcome such a problem, the idea of cross-linking the membranes seems particularly attractive, as far as they do not become brittle.

Different strategies have been described in the literature for obtaining cross-linked aromatic proton conductive membranes based on polyphosphazene [108], polysulfone [109], and polymer blends [110]. However, as mentioned by Kerres in a review article [111], ionically cross-linked membranes are not suitable for high-temperature applications because of excessive water membrane swelling. Accordingly, particular interest has been devoted to obtaining covalent cross-linked SPI membranes. Different strategies have been described.

Miyatake et al. [112] have reported the cross-linking of polyimide membranes by electron-beam irradiation (Fig. 23). These membranes are obtained by classical solvent-casting methods from *m*-cresol solutions. After acidification (HNO_3/EtOH), the 50-μm-thick membranes are irradiated with an electron beam at room temperature and under air. The experimental conditions are designed to induce a cross-linking reaction only in a small depth (10 μm) on each face, or to affect the whole membrane thickness. However, the presented results do not show any effect on the molecular weight and the water uptake, which highly suggests the absence of cross-linking by electron-beam irradiation.

Watanabe et al. used different branching agents in order to synthesize branched/cross-linked SPIs. They introduced either melamine [110, 112] in a fully aromatic SPI (Fig. 24) or tris(aminoethyl)amine in an aromatic/aliphatic SPI backbone [33, 113]. In both cases, the proportion of trifunctional branching agent (leading to a stoichiometric proportion of amine and anhydride groups) was limited to 2 mol %. The authors studied the effect of this new polymer architecture on the membrane water uptake and maximum tensile stress at break.

Another triamine monomer (1,3,5-tris(4-aminophenoxy)benzene, TAPB) has been used as cross-linking agent by Yin et al. [114, 115] (Fig. 25). Typically, anhydride-terminated sulfonated oligomers are prepared from BAPBDS and NTDA in *m*-cresol at 180 °C for 20 h. After adding some triamine monomer, in a second step, the reaction medium is kept at moderate temperature (50 °C), resulting in a polyamic acid intermediate. By a thermal treatment at high temperature or in the presence of an acetic anhydride/pyridine mixture, complete imidization is performed during film formation.

Another way to obtain cross-linked SPI membranes consists of cross-linking linear high molecular weight SPIs. From this perspective, a specific comonomer is introduced into the SPI. The addition of bifunctional

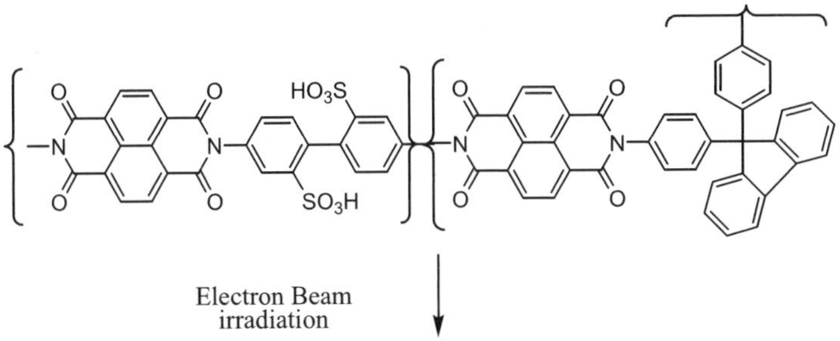

Electron Beam irradiation

Fig. 23 Electron beam cross-linking

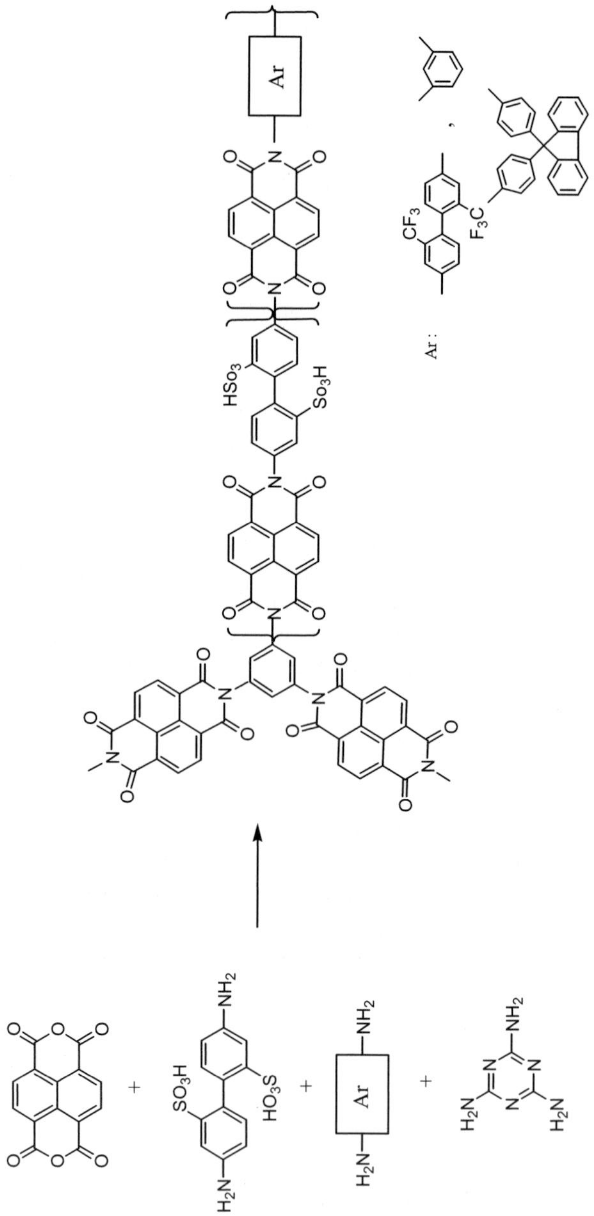

Fig. 24 Cross-linked SPI in the presence of melamine

cross-linkers enables the formation of covalent bridges between the SPI chains. A first example is given by Sundar and coworkers [116]. In this case, the cross-linking is based on the quaternization reaction between acridine groups of the SPI chain and a dibromoalkane reagent. Series of cross-linked

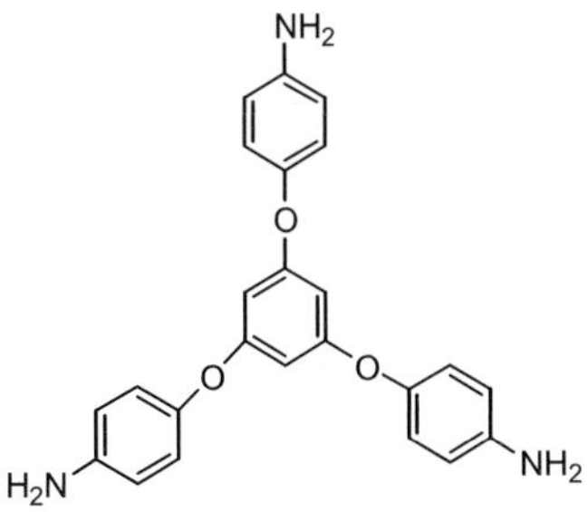

Fig. 25 Triamine as cross-linking agent (TAPB)

membranes were prepared by modifying the length of the dibromo compounds. For this purpose, the authors firstly synthesized high molecular weight SPIs based on NTDA/BDSA, 3,6-diaminoacridine (DAA), and 2-bis[4-(4-aminophenoxy)phenyl]hexafluoropropane (HFBAPP). Unusually, the reaction is stirred at room temperature in *m*-cresol until it becomes homogeneous before being chemically imidized. The polymer is then further dissolved in cresol before the introduction of dibromoalkanes (Fig. 26). After casting the solution, the quaternization reaction is realized by curing the films at 80 °C for 1 h and 120 °C for 12 h. Although no details concerning the reticulation extent are given, one can suppose that a high cross-linking rate could be obtained with this procedure, according to the acridine monomer proportions involved (20 wt. %).

According to a similar approach, Park et al. [117] described the synthesis of cross-linked SPIs based on carboxylic acid containing structures and the formation of ester bridges using various alkane diols. The low hydrolytic stability of ester linkages in aqueous acidic media precludes any long-term use of such materials for fuel cell applications. However, using different diols (HO–$(CH_2)_n$–OH, $n = 2$–10) the authors studied the effect of the cross-linker chain length on the final membrane properties (water uptake, proton conductivity, methanol permeability). An optimum cross-linker chain length ($n = 5$–6) was determined (Fig. 27). The same authors [118] also report the use of *N*,*N*-bis(2-hydroxyethyl 2-aminoethanesulfonic acid) (BES) as a sulfonic acid cross-linker in order to further improve the proton conductivity of the cross-linked membranes.

Yang et al. [119] have proposed a cross-linking method based on the thermal polymerization of SPI oligomers end-capped with a thermal reaction function. These authors synthesized sulfonated oligoimides end-capped with maleimide functions. These telechelic oligomers are either self-cross-linked or cross-linked with poly(ethylene glycol) diacrylates (PEGDAs). Self-cross-linked SPI films are obtained from a triethylammonium salt SPI solution in NMP containing 3 wt. % of 4,4′-azobis(4-cyanovaleric acid) (ACVA). Thin

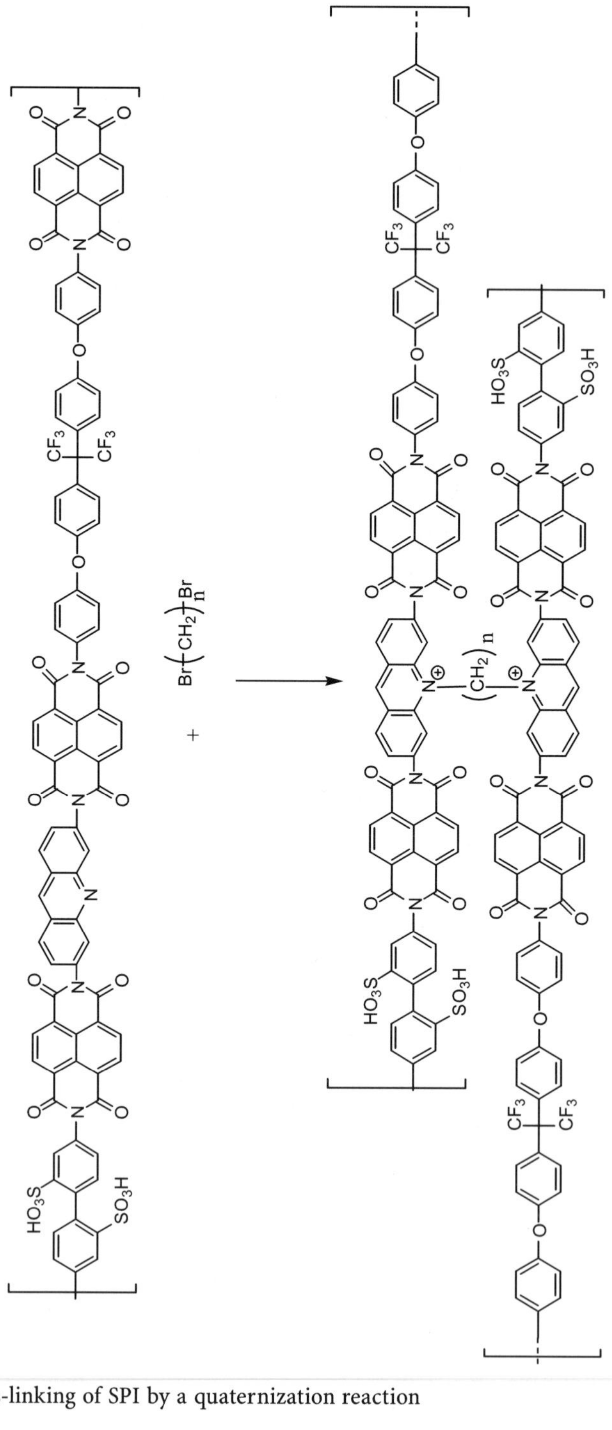

Fig. 26 Cross-linking of SPI by a quaternization reaction

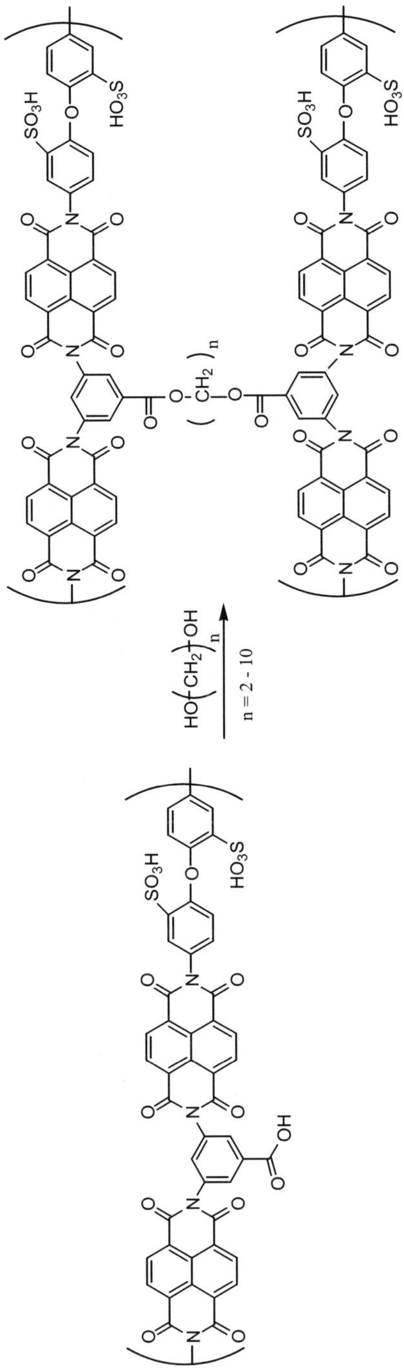

Fig. 27 Cross-linking of SPI by the formation of ester linkages

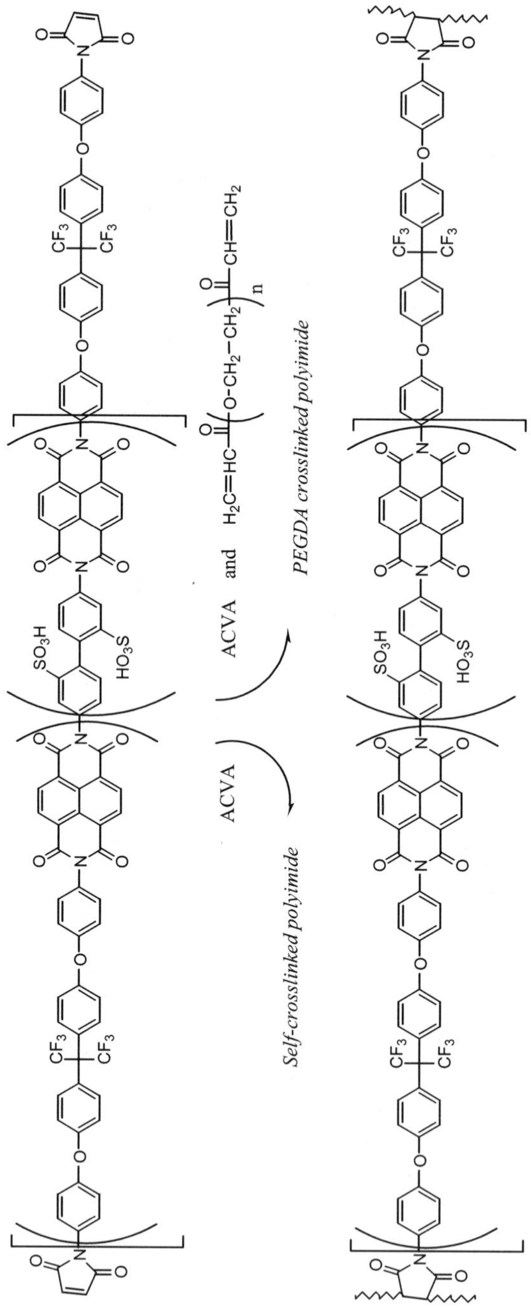

Fig. 28 Cross-linkable telechelic oligomers

films are obtained by spin coating and then cured at 80 °C for 4 h, 120 °C for 6 h, and 180 °C for 3 h.

The same procedure was used for obtaining the cross-linked SPI containing poly(ethylene glycol) diacrylates (Fig. 28). This approach is interesting because of the hydrophilic and flexible nature of PEG sequences. The mechanical properties as well as the water uptake of the cross-linked systems are kept, thus improving the conductivities at high temperatures. However, the stability of the PEG unit in fuel cell conditions is somewhat questionable.

High IEC and high molecular weight linear SPIs can be easily cross-linked in the presence of phosphorus pentoxide/methanesulfonic acid (PPMA) or phosphorus pentoxide. Two different ways were recently described by Okamoto and his group:

1. Dry SPI membranes (in their acidic form) are immersed into PPMA at 80 °C [120, 121].
2. SPIs (in their acidic form) in DMSO solution containing 5 wt. % of phosphorus pentoxide are cast into films which are thermally cured at high temperature and under vacuum [121].

In both cases, the cross-linking reaction is based on the formation of sulfonyl linkages, as represented in Fig. 29. As the sulfonic groups are involved in the cross-linking process, such a reaction leads to a reduction of the SPI initial IEC. Surprisingly, the resulting cross-linked SPIs show an improved water stability and a rather high proton conductivity, in spite of a 10 to 20% IEC decrease.

Another strategy adopted to reduce the degree of swelling of the membranes without significantly losing their proton conductivity consists in the elaboration of semi-interpenetrating networks (IPNs) based on SPI. Lee and coworkers [122] polymerized PEGDA in the presence of SPI to synthesize IPN membranes with various ionic contents (Fig. 30). Thanks to the incorporation of hydrophilic PEG groups, the proton conductivity was shown to be improved in spite of rather low IEC values. Moreover, the better structural stability of such materials resulted in higher water stability compared to that of the pure SPI membrane.

Organic–inorganic (SPI–SiO_2) interpenetrating networks appear to be very promising materials as solid electrolytes for fuel cell applications. Lee et al. [123] showed that the presence of silica reduces the membrane water uptake and methanol permeability, while it increases the membrane selectivity (Fig. 31). Besides, the formation of an organic IPN improves the material hydrolytic stability. Hence, the mechanical properties (tensile strength and elongation at break) are increased by a factor of 100.

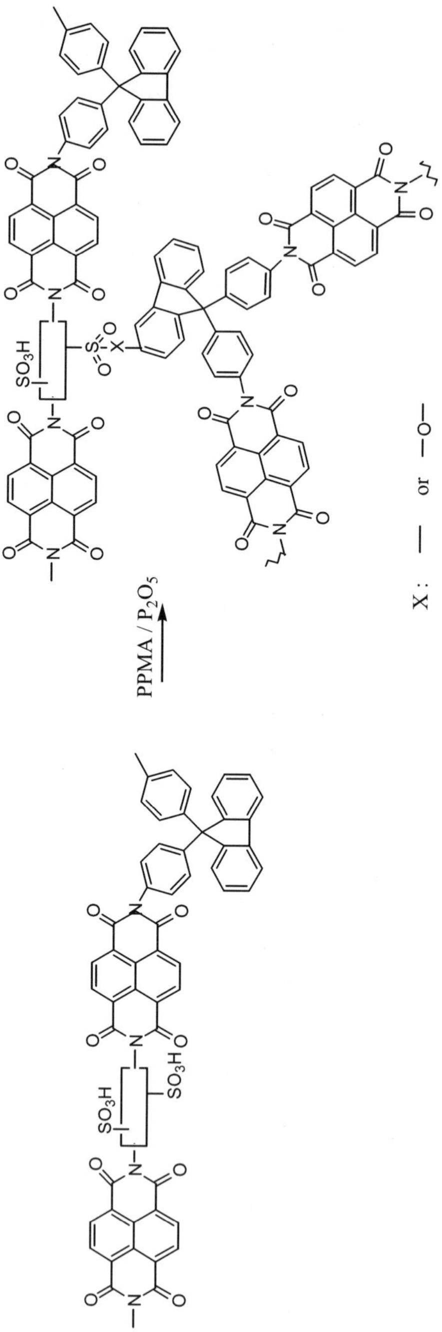

Fig. 29 SPIs by the formation of sulfonyl linkages in $PPMA/P_2O_5$

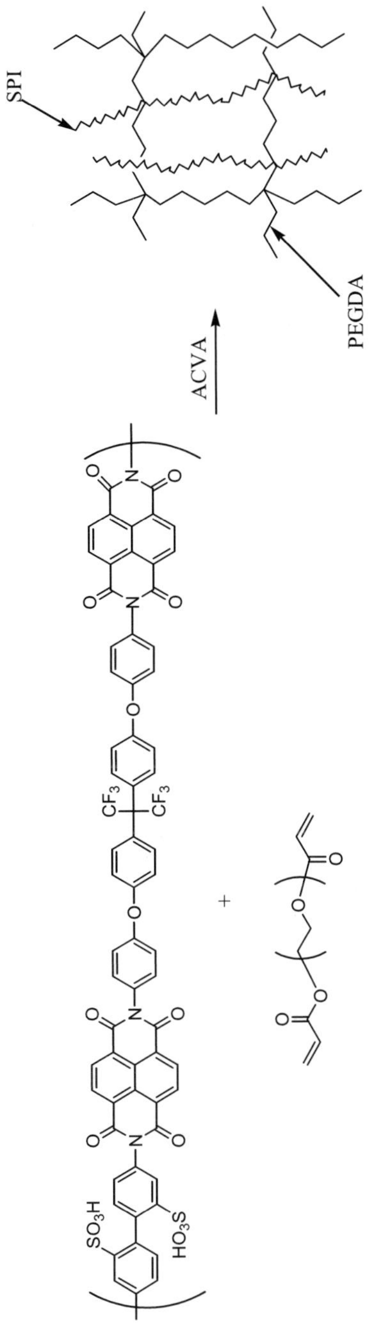

Fig. 30 Synthesis of SPI and poly(ethylene glycol) diacrylate based semi-IPN membranes

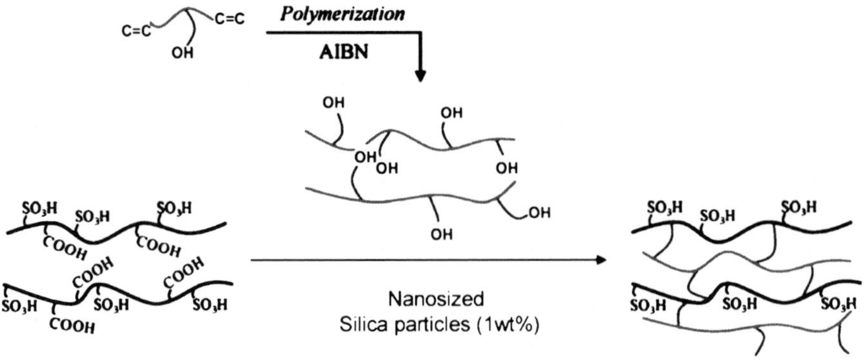

Fig. 31 SPI–silica nanocomposite containing interpenetrating polymer network

3 Sulfonated Polyimide Membrane Properties and Structure

3.1 Ion Content

The first membrane characterization is the determination of its ion content. It is usually characterized by the IEC, which corresponds to the number of protons per gram of dry polymer. Typical values for ion-exchange membranes range from 5×10^{-4} to 5×10^{-3} mol/g and so are expressed in millimoles per gram (or milliequivalents per gram). In the specific case of PEMs, a too low IEC does not permit a sufficient fuel cell performance to be obtained, while a too large value leads to an excessive water uptake and insufficient mechanical properties. The useful range is thus typically restricted to 1.2–2.5 meq/g for most of the sulfonated aromatic polymers depending on their swelling properties. The ion content of ionomer membranes is also often defined as the equivalent weight (EW), which corresponds to the inverse of the IEC value (EW = 1000/IEC in eq/g). These values are somewhat misleading when compared to Nafion because the dry polymer density can be notably different from one polymer to another (typical polymer densities are 1 for aliphatic, 1.4 for aromatic, and 2.1 for perfluorinated polymers). Therefore, a more relevant value should be the equivalent volume expressed in cm^3/eq. For example, the EW for Nafion is 1100 g/eq (IEC = 0.91 meq/g) and one extensively studied SPI exihibits an IEC of 1.3 meq/g (EW = 770 g/eq). These values seem significantly different while they are very close when expressed as equivalent volume (524 and 550 cm^3/eq, respectively). However, the polymer matrix density is not easy to determine experimentally compared to the dry weight and the equivalent volume is never used.

A theoretical IEC is set during the polymer synthesis through the ratio of disulfonated and neutral diamines. It can be experimentally confirmed by acid–base titration. The acid membrane is neutralized with a large excess of sodium chloride in order to release the protons in the solution for titration. While this measurement is easy and reproducible for Nafion, it becomes more difficult for SPI. The titration curve is no longer characterized by a single wave and some of the protons are not released when the membrane is neutralized in the concentrated sodium chloride solution. This behavior can be due to both a difficulty in accessing some protons and to a lower acidity of the sulfonate group linked to a phenyl group and embedded in a highly charged environment.

For sulfonated aromatic polymers, the degree of sulfonation can be determined by the integration of some specific ^{1}H NMR peaks, which gives the ratio of substituted aromatic rings to unsubstituted ones [124]. However, this technique requires a good solubility in common deuterated solvents and it can hardly be applied to SPIs [76] since broad lines are often obtained due to polymer chain aggregation [31, 63]. Specific infrared bands, such as the SO_3 vibration bands [50, 63, 125], could be used to evaluate the number of sulfonic groups. Since the absorption coefficients are not known, a totally sulfonated homopolymer can be used as reference material or any specific bands as internal reference [68, 126]. However, these bands are pretty weak and overlap with other bands. Therefore, this technique does not allow a precise determination of the SO_3 content. Transmission infrared spectroscopy can be performed on very thin films (around 5 μm) [68, 126] or by dispersing a SPI powder in KBr pellets [63]. The attenuated total reflection (ATR) technique can also be used directly on membranes. In this case, the analysis is restricted to the first few microns (the penetration depth) and it is necessary to assume a chemical homogeneity between the membrane surface and bulk.

The last method is the determination of the sulfur content either by elemental analysis or by scanning electron microscopy with energy-dispersive X-ray analysis (SEM-EDX) [126]. The intensity related to the degree of sulfonation is linear as a function of the desired IEC over a wide range of ion content (Fig. 32). This experiment demonstrates that the synthesis using functionalized monomers is a very efficient method to control the ion content. This latter method also permits the sulfur concentration profile across the membrane to be accessed. Flat profiles were obtained indicating a homogeneous distribution of the ionic groups, at least on a micrometer scale. Sulfur determination by elemental analysis also confirmed the efficiency of the polycondensation reaction, since the slope of the actual sulfonation level versus the designed one was found to be very close to 1 [52]. It has to be pointed out that, except for the titration method, the IEC determinations give the overall number of sulfonate groups but they cannot differentiate between SO_3 groups isolated in the hydrophobic matrix and those which participate effectively in the ion conduction.

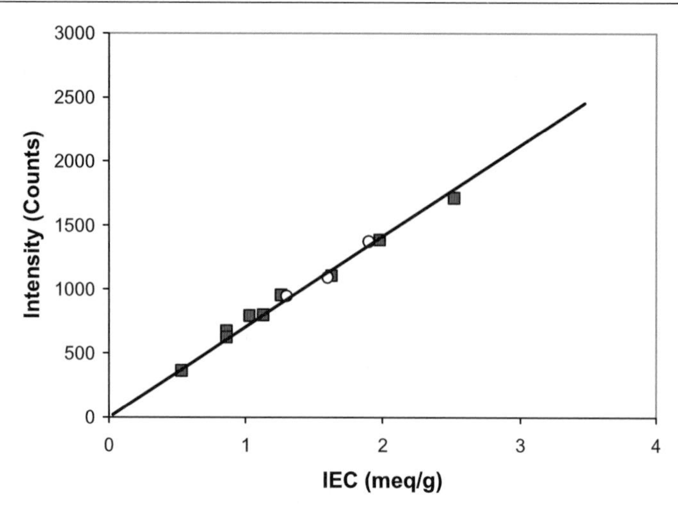

Fig. 32 SPI SEM-EDX sulfur content data (■) obtained for various degrees of sulfonation. Data obtained in the same conditions for sulfonated polyether ether ketones prepared by post-sulfonation (○) are given for comparison

3.2 Membrane Water Uptake and Swelling Properties

Most of the membrane properties are directly related to the membrane water content which thus appears as one of the most important characteristics. It can be easily measured gravimetrically by either immersing the membrane in water at room temperature or equilibrating the membrane in saturated vapor in a closed vessel. The water uptake is then expressed in weight % with respect to the dry polymer weight. Polyimides are subject to water sorption due to the presence of hydrophilic carboxyl groups in the imide functions. However, this water uptake is limited at 1.5 to 5% w/w depending on the imidization rate and the chemical structure [127, 128]. The introduction of sulfonate groups greatly enhances the water uptake up to 80% w/w in saturated vapor [129] and to a possible dissolution in liquid water for highly sulfonated materials [130]. Since the water uptake is strongly related to the sulfonate content and to the polymer density, a normalized value is now commonly used: the number of water molecules per sulfonate group, λ. The λ values were shown to be constant over a wide range of ion content and for different structures, as presented in Fig. 33 [33–35, 65, 100, 131–133]. This behavior is specific to SPIs since sulfonated hydrocarbons or perfluorinated materials usually present a divergence of the water uptake for a critical value of the ion content [134, 135]. Surprisingly, Myatake et al. observed a nonlinear behavior of the water uptake as a function of the ion content with a maximum for fluorenyl-containing SPIs [112]. This discrepancy is probably attributable to a problem encountered during the polymer synthesis since this effect was

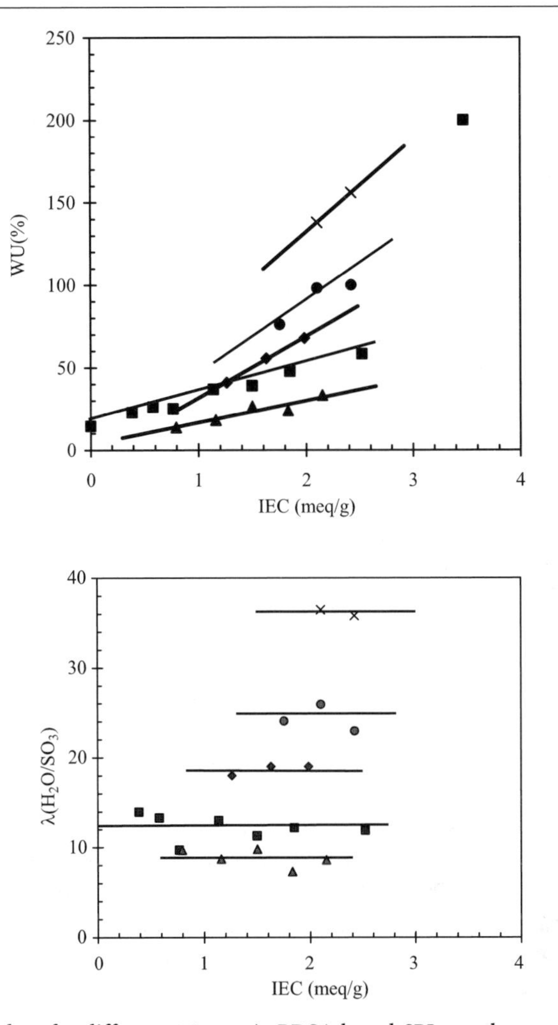

Fig. 33 Swelling data for different 1.3 meq/g BDSA-based SPI membranes: phthalic ODA (▲) and naphthalenic ODA/fluorenyl 1 : 1 (■); ODA (♦); diaminooctane (DAO) with five sulfonated monomers in the ionic sequence (●; and DAO with nine sulfonated monomers (×). Swelling data are expressed as water uptake in w/w% (*up*) and as the number of water molecules per sulfonate group, λ

not reproduced with similar systems (Fig. 15) [136]. Other discrepancies can be observed for very high IEC values, probably due to a loss of the mechanical properties or partial dissolution [36, 97]

The λ value variation in dependence on the SPI chemical structure is more difficult to analyze. Lee et al. [54] claimed a significant effect of the chemical structure on the water uptake, while their values appear similar when normalized by the IEC. Polymers with similar ion content and ion distribution along the polymer chain and using different aromatic monomers in the hydrophobic sequence were compared and the data of Geniès et al. [31] suggest similar swelling behaviors. However, a large variation can be observed when significantly different diamines are used (Fig. 15). For example, aliphatic monomers such as diaminooctane (DAO) lead to large water uptakes compared to aromatic monomers. However, the effect of aliphatic sequences can be counterbalanced by the use of ionic pendant side chains [33]. It is more difficult to reach a conclusion when taking into account the numerous data published from the Yamaguchi University [35, 59, 70, 71, 75, 87, 137, 138], probably because the chemical modifications are operated in both the ionic and neutral part of the polymer. Geniès et al. [31] observed a nonnegligible effect of the ion distribution along the polymer chain, since random copolymers exhibit systematically smaller water uptakes compared to block copolymers. For very long sequences, larger water uptakes are often observed (Fig. 33). In the case of ODA SPIs, this difference can be attributed to a macrophase separation in ion-rich and ion-poor phases (see Sect. 3.3). In addition, the poor solubility and the high glass transition temperature of these polymers probably induce a strong effect of the casting conditions on the structure, the existence of porosity, and the swelling properties.

A surprising result also arises from the comparison of the water uptake values obtained after membrane equilibration in pure water and in a saturated vapor atmosphere. The obtaining of different values in these conditions where the water activity is supposed to be identical is called Schroeder's paradox. This effect was mainly observed at elevated temperature for Nafion membranes [139] and was shown to be related to the membrane pretreatment [3, 140]. In the case of SPIs, a wide difference close to a factor of 2 ($\lambda = 19$ in liquid water and 11 at 100% RH) was observed at room temperature [141]. Moreover, this effect was shown to be reversible. A membrane equilibrated in liquid water and placed in a saturated atmosphere will slowly lose some weight and reach the equilibrium value determined at 100% RH in around 15 days.

Nowadays, PEMFCs operate at 80 °C and one of the main objectives is to increase this temperature as much as possible. Thus, the water uptake has to be measured as a function of temperature or at least at elevated temperature. For most of the ionomer membranes, the water uptake increases almost exponentially with temperature up to partial or total dissolution [135, 142]. In the early work of Faure et al. [29], the SPI water uptake was shown to be independent of the temperature in the range of room temperature to 80 °C for naphthalenic SPIs, while a slight increase was observed for phthalic SPIs as confirmed recently [50]. However, phthalic SPIs are very sensitive to hydrolysis, which induces a fast polymer degradation [29, 50]. This sensitivity increases with temperature and the swelling data recorded at elevated temperature should be considered with care [68].

The swelling in liquid water could be a nonrelevant value to describe the membrane behavior in fuel cells. As an example, the sulfonated polyether ketones suffer from an excessive water uptake when immersed in liquid water at elevated temperatures [135], while a high-temperature fuel cell test over 1400 h has been performed with fully hydrated gases [15]. Therefore, the swelling in liquid water at elevated temperature is of limited value. Most of the authors prefer nowadays to determine the water uptake in defined temperature and RH conditions. Water sorption isotherms have been measured and analyzed [59, 129, 143]. These results confirmed the previous data since the sorption isotherms are superimposable when expressed in λ values for different ion content and chemical structures [59, 71, 75, 87]. The sorption isotherms were classically analyzed as the superposition of different phenomena (Henry, clustering, dual-mode, Flory–Huggins, BET II processes) [143]. The water sorption isotherms exhibit a distinct sorption–desorption hysteresis which was attributed to the very low desorption process [59]. The authors considered that the equilibrium was not attained within the experimental timescale (1 h of equilibration for each experimental point). However, they did not present longer experiments to validate this conclusion. The same experiments were carried out varying the equilibration time for each data point from 60 to 720 min (Fig. 34). While the observed hysteresis is slightly reduced on increasing the equilibration time, the extrapolation of the kinetic data to infinite equilibration time clearly indicates that the sorption–desorption hysteresis is a true phenomenon. This unexplained behavior is an extension of Schroeder's paradox when the water activity is different from 1, since different water uptakes are obtained for the same water activity.

Water management is an important issue in fuel cell operation and consequently fuel cell modeling. Water transfer through the membrane is usually treated using Fick's law [144] and a diffusion coefficient which depends on both the water content and the temperature [145]. Since the resolution of Fick's law at short times implies a $t^{1/2}$ evolution of the water uptake, the diffusion coefficients are often extracted from the initial slope of the swelling kinetics [59, 143]. These values, which are determined in the presence of a concentration gradient and which take into account the surface adsorption processes, can be considered as more representative than the self-diffusion coefficients determined using pulse field gradient nuclear magnetic resonance (PFG-NMR), diffusion of radiotracers, or conductivity data through the Nernst–Einstein equation [146]. Despite it being shown that ionomer membranes exhibit highly non-Fickian behavior [147], this data treatment was applied to SPI systems and the results suggest that the diffusion coefficients present a maximum value around 50% RH [59]. However, this model is not able to reproduce the entire swelling kinetics using the nonapproximated equation. These discrepancies can be attributed to a series of polymer matrix relaxations upon swelling which are not encountered in this simplified model.

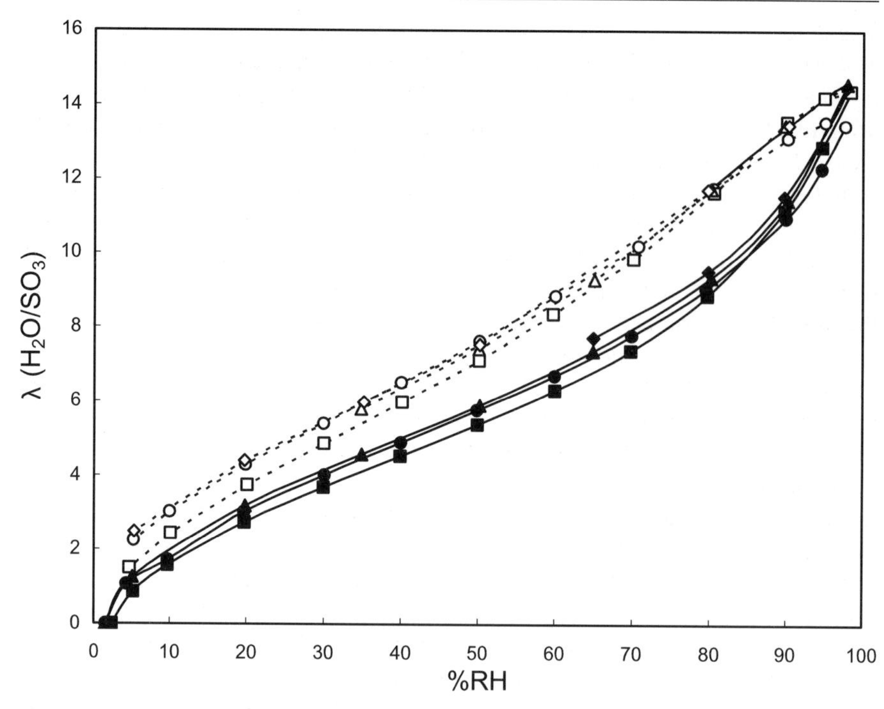

Fig. 34 Water sorption isotherm of a naphthalenic SPI for various equilibration times for each data point: (■) 60 min; (●) 120 min; (▲) 240 min, and (♦) 720 min. Full and empty symbols correspond to absorption and desorption, respectively

Maréchal et al. [148–150] have studied the hydration mechanisms by infrared spectrometry on both acid and neutralized forms and on homopolymers and block copolymers as a function of the relative humidity at room temperature. The analysis of the hydration spectra reveals different processes associated with the ionization of the sulfonated groups, the counterion solvation, the hydration of the carboxylated groups in the imide functions, and the condensation of H_2O on other water molecules [150]. Since the analysis is quantitative for each identified process, water sorption isotherms were built (Figs. 35 and 36) and led to an understanding of the hydration better than with the Langmuir, clustering, and Henry concepts. The energy transfers associated with water sorption have been measured [151]. The first step is interpreted as the water sorption in unrelaxed holes and interaction on specific sites. The second step in the range of activity larger than 0.6 is attributed to a swelling process. It is characterized by a progressive decrease of the interactions between the water molecules and the specific sites for the benefit of water–water interactions.

The membrane swelling can also be studied through the macroscopic dimensional changes. For most ionomer membranes, the linear expansion is found to be identical within the three directions (isotropic swelling), except when an orientation is generated by the membrane preparation process (typical case of extruded membranes). On the contrary, SPI membranes exhibit a strong swelling anisotropy with dimensional changes observed mainly along the membrane thickness. Cornet et al. [141] have shown the dimensional change along the thickness can vary from 30 to 60% depending on the ion content and the block character, while the in-plane swelling is limited to 5–8%. This swelling anisotropy, which was confirmed by the study of new polymers [34, 79, 97], was attributed to a foliated structure packed along the membrane thickness. It was also shown that the swelling process is strongly related to the membrane preparation procedure and it is isotropic for membranes equilibrated in saturated vapor. In such a case, the dimensional changes are very small despite the absorption of ten water molecules per sulfonic group. It reveals that the first step of the hydration process corresponds to the filling of nanopores (no porosity is observable by SEM) created during the casting procedure [141]. The system becomes glassy during solvent evaporation, trapping some solvent molecules in a rigid structure. After solution casting and solvent evaporation at high temperature, the membranes are washed to remove the residual solvent and acidified. Some empty space is then created by the residual solvent extraction and by the replacement of the large alkylammonium by small protons. This porosity cannot be relaxed due to the high glass transition temperature and polymer rigidity.

As for most polyaromatic polymers, SPI membranes exhibit a low methanol crossover and these membranes were considered as promising ma-

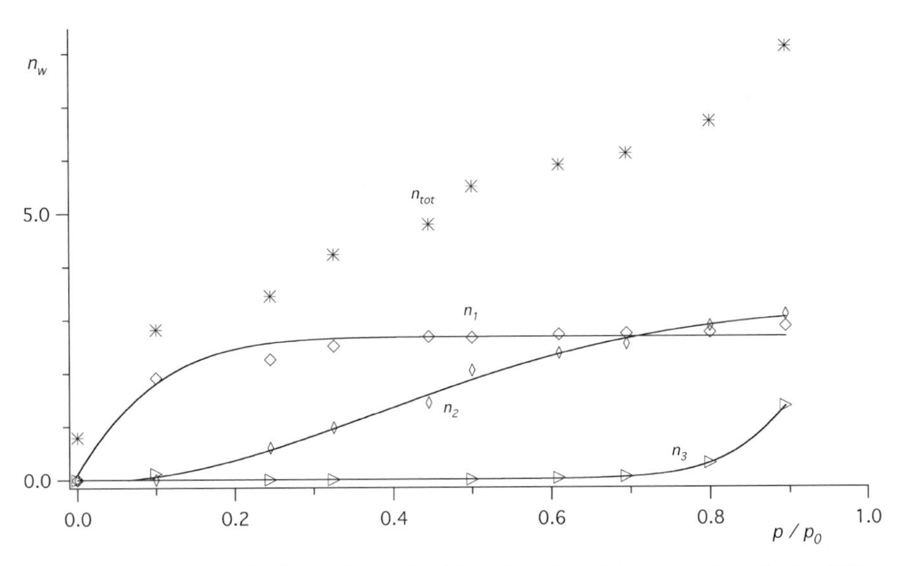

Fig. 35 Water sorption isotherm determined by infrared and decomposition in the different hydration mechanisms (reproduced from [150])

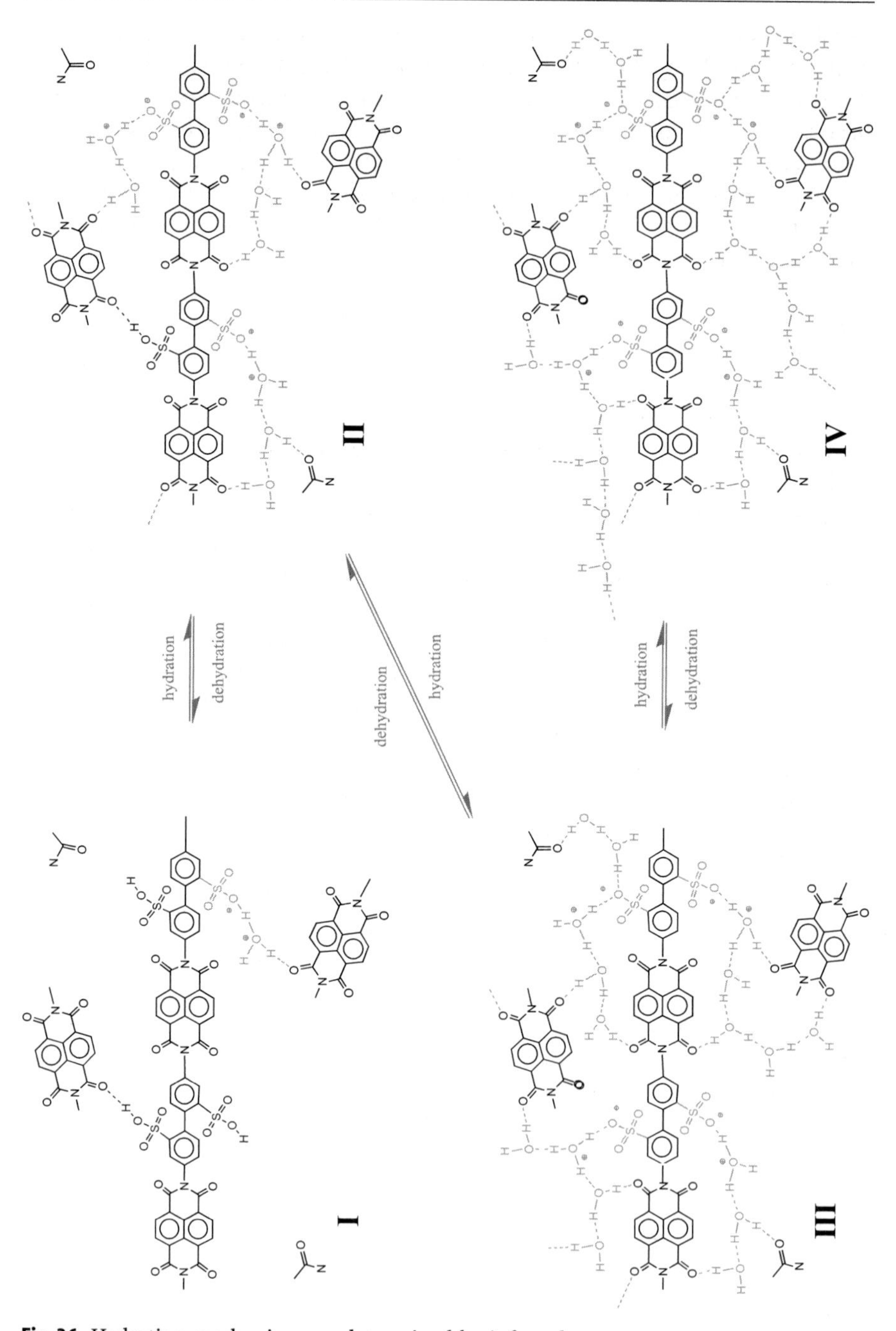

Fig. 36 Hydration mechanisms as determined by infrared spectroscopy

terials for direct methanol fuel cells (DMFCs) [52, 76, 87, 112]. However, it is difficult to find any data in the literature about the solvent uptake either in methanol or in water/methanol mixtures. On the contrary to Nafion membranes, which exhibit a high affinity to methanol [142], pure methanol is obviously a poor solvent for SPI since it can be used to precipitate the polymer from cresol solutions. The methanol uptake is roughly twice smaller than the water uptake in the same conditions [152], while it is five times larger in the case of Nafion [142].

3.3 Structure and Morphology

It is commonly accepted that the membrane microstructure has a large impact on the transport properties and should be controlled. The complex multiscale structure of ionomer membranes has required different series of complementary experiments and the progress in the experimental setups over more than 20 years to be partly elucidated [153]. Moreover, only one type of Nafion membrane was commercially available. It was thus only possible to vary the external parameters (swelling agent, temperature, counterion, mechanical constraints ...) and not the structural ones. The development of new membranes should have been the opportunity to progress in this field, but surprisingly only scarce structural studies have been performed on these new materials and they were mainly conducted on SPI [153]. The ionomer membranes are usually characterized by a nanophase separation of the highly hydrophilic ionic groups embedded in the hydrophobic polymer matrix. These domains are often considered as spherical to minimize the interfacial energy [4], and they have to be connected to each other to explain the high values of the ionic conductivity [9]. Nevertheless, this commonly accepted statement is not straightforward since each ionic group is linked to the polymer chains and the energy of the chains should also be minimized to avoid too large packing constraints. Recently, a new model was proposed which seems able to deal with most of the experimental data. This model is based on elongated particles composed of the fluorinated matrix with the ionic groups at the interface [7, 8]. However, it is far from being straightforward that such a model can be directly transposed to non-fluorinated materials. The small-angle X-ray and neutron scattering (SAXS and SANS) techniques are the most suitable tools to explore the membrane microstructure [153]. Nevertheless, the image of the structure is obtained in the reciprocal (Fourier) space, and an adequate model including the shape and the spatial distribution of the ionic domains is necessary to interpret the data. A direct image of the structure can be obtained by electron or atomic force microscopy techniques, but in the absence of marked structural features the analysis is often subject to interpretation in addition to many possible artifacts that can arise from the sample preparation.

The first SANS analysis of SPI membranes was performed on both naphthalenic and phthalic ODA/BDSA block copolymers with five repeat units in the ionic sequence [29]. The SANS spectra revealed a broad maximum located at very low angles compared to Nafion (Fig. 18). This low-angle scattering maximum was also observed by SAXS and called the ionomer peak since such a peak is commonly observed with ionomer materials [52, 154]. Its origin is still subject to controversy [153] but its position is usually related to the characteristic size of the scattering objects (assumed to be ionic clusters). The low-angle position of the ionomer peak suggests large interdomain distances and consequently a large domain size (typically 5 to 10 times larger than in Nafion). This large size was attributed to the block character of the polymer (five repeat units in the ionic sequence in this case). In the case of random highly charged SPIs, Myatake et al. did not observe any scattering peak [110] and concluded the absence of ionic aggregation due to a less pronounced hydrophilic/hydrophobic separation and to the weaker flexibility of polyaromatic chains. Despite the data being recorded in a very restricted angular range, they confirm the results previously observed by Piroux et al. who found that the ionomer peak vanishes for highly charged SPIs.

In the SAXS data obtained for phthalic block copolymers where the ionomer peak is well defined, the peak position was found to be proportional to the number of repeat units in the ionic sequence [154]. In addition, the intensity of the ionomer peak was shown to be directly related to the electron density of the counterions, confirming that this low-angle maximum has to be associated with the ionic domains [154]. The SAXS patterns of fully hydrated phthalic SPIs were not found to be sensitive to the ion content [52], and the authors concluded that the number of water clusters increases as ion content increases without modification of their size. Essafi et al. found a slight variation with similar systems, which could also be explained by the increase of a strong upturn in intensity at low angles related to large-scale heterogeneities [154]. Piroux et al. [151, 155] studied the effect of the ion content on the structure of fluorenyl-based naphthalenic SPIs by SANS from a totally sulfonated polymer to a non-sulfonated one. Two regimes were observed: at low ion content the peak slowly shifts toward large angles as the ion content increases, while this shift is significantly more pronounced at high degrees of sulfonation. This transition was correlated to the water permeability, which is very low in the former case and drastically increases at high ion content. The proposed interpretation is that the ionic domains are dispersed in the polymer matrix at low ion content, while they form a continuous and preponderant phase for a degree of sulfonation larger than 50% [151].

In the case of rigid naphthalenic polymers, the structural features appear less marked due to the extreme polymer chain rigidity. The ionomer peak is restricted to a shoulder on a huge small-angle scattering at very low angles (Fig. 37). The introduction of flexible segments and bulky groups was shown to favor a better organization [31, 155]. The surprising result is probably that

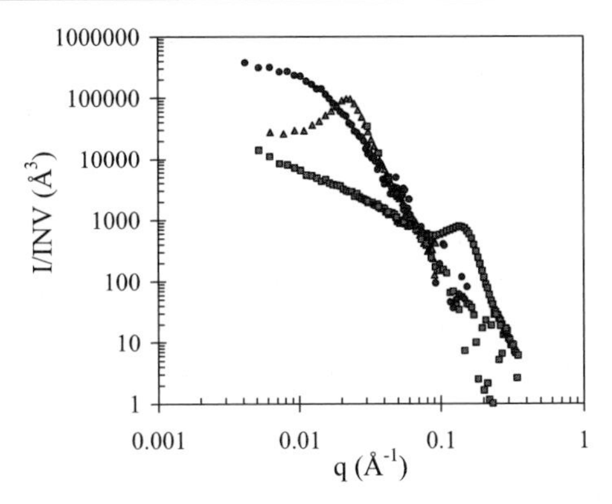

Fig. 37 SANS spectra of flexible phthalic (▲) and rigid naphthalenic (●) SPI membranes soaked in deuterated water. The SANS spectrum of a water-swollen Nafion membrane is given as reference (■). The data were normalized by the scattering invariant for comparison

the small-angle scattering spectra only change in intensity when the water content is varied. The ionomer peak position does not shift toward small angles, as is usually observed for most of the proton conducting membranes such as Nafion [154]. It confirms that the swelling mainly corresponds to the filling of a preexisting porosity created by the membrane preparation process as explained previously (see Sect. 3.2).

The membrane swelling observed mainly along the membrane thickness suggested an anisotropic microstructure. Micro-SAXS experiments were conducted using synchrotron radiation with a special optical device to focus the beam on the sample (1 μm half width at half maximum) [156]. The structure was then studied with the membrane plane parallel and perpendicular to the X-ray beam. The data revealed a highly anisotropic structure in agreement with a foliated structure packed along the membrane thickness (Fig. 38). Recently, quadrupolar splitting NMR experiments were performed to quantify the degree of anisotropy, and it was found that SPI membranes present an outstandingly high level of orientation on a molecular scale (similar to inorganic clays) [157]. Scanning electron microscopy pictures recorded after cryofracture of swollen membranes and freeze drying also confirmed a foliated structure but on a micrometric scale [158]. Transmission electron microscopy pictures of ultrathin membranes neutralized with cesium ions to stain the ionic domains also revealed a lamellar structure in the intermediate range (50 to 100 nm in Fig. 39) [158]. This result seems to have been confirmed recently by Myatake et al. [113]. However, they also published STEM images on the same system which suggest spherical ionic domains [33, 105].

The micro-SAXS data also revealed a new scattering peak at 0.45 $Å^{-1}$ (average distance 15 Å) which corresponds to the average distance between ionic groups in the ionic sequence. Its intensity is anisotropically distributed suggesting an orientation on the polymer chains and that the anisotropy extends from the molecular level to the macroscopic size.

For block copolymers presenting long ionic and hydrophobic sequences, a macroscopic phase separation occurs with a lenticular ion-rich phase dispersed in an ion-poor matrix, as demonstrated by the electron contrast when stained by cesium ions (Fig. 39). This phase separation was attributed to a nonnegligible quantity of triblock copolymers formed of either two ionic sequences spaced by one hydrophobic sequence or the opposite structure. These polymers are not miscible and then phase separate.

The micro-SAXS experiments were conducted on membranes at different levels of hydration. On the contrary to the expected behavior, the "ionomer peak" does not shift as the water content increases when observed along the membrane thickness [158]. Two main effects are observed in the spectra upon swelling. First, the level of intensity increases at very large angles due to the filling of the nanoporosity without any change in the local structure. This

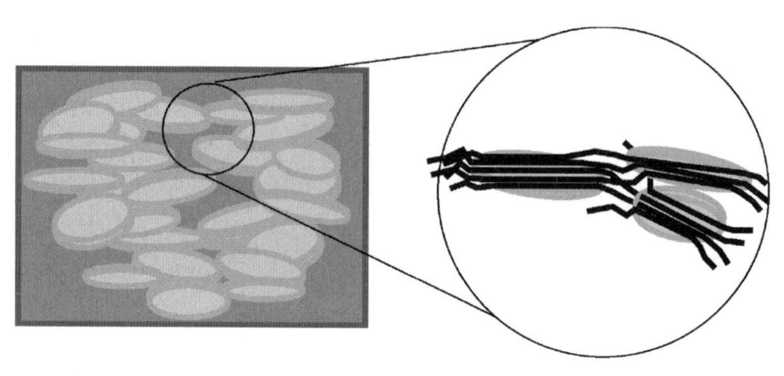

Fig. 38 Foliated structure of rigid SPI

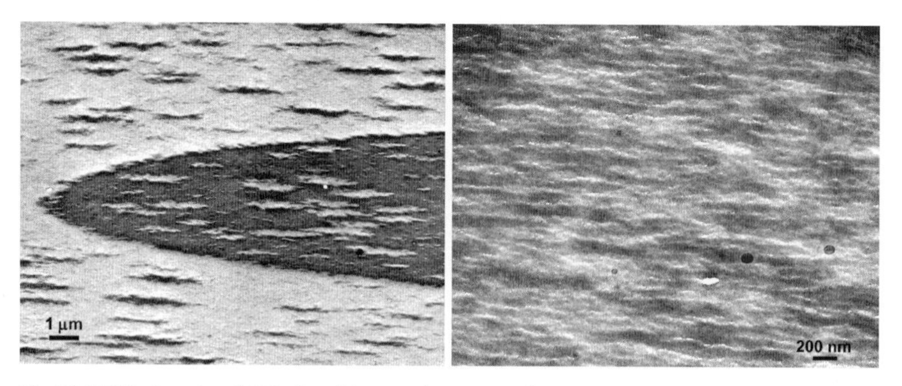

Fig. 39 TEM pictures of SPI ultrathin membranes. *Left*: SPI with very long ionic sequences and low ion content inducing a macroscopic phase separation in ion-rich (*black*) and ion-poor (*white*) domains. *Right*: homogeneous SPI (short ionic sequences) revealing the foliated structure on a nanometric scale

nanoporosity thus corresponds to empty space between the sulfonated sequences inside the large ionic domains. This result reconciles SAXS data, which suggest the presence of very large ionic domains, with the electron spin resonance (ESR) analysis of the counterion dynamics, which suggests that these ions evolve in significantly smaller domains than in Nafion (see Sect. 3.4) [152]. The second effect of swelling on SAXS data is the appearance of an intense small-angle scattering at very low angles suggesting the formation of large water domains as observed by SEM. These micrometric domains are responsible for the thickness increase when immersed in water [158].

3.4 Transport Properties

The multiscale structure of ionomer membranes should induce multiscale transport properties [6, 159]. In other words, the mobility of molecules inside the membrane will depend on the timescale of observation. At very short timescales (typically in the range of a few picoseconds), the local dynamics within the ionic domains was experimentally studied by quasi-elastic neutron scattering (QENS) [160, 161] or numerically by molecular dynamics [162] in the case of Nafion. The local diffusion coefficient of the water molecules determined by QENS is just slightly lower than in pure water, but the motion is restricted inside small domains with some jumps from one domain to another [160, 161]. However, these techniques have not yet been applied to SPIs and according to our knowledge, the only insight into the local structure of the ionic aggregates was obtained through an ESR study of the dynamics of VO_2^+ counterions [152]. The ESR spectra and rotational correlation time deduced from the simulation are sensitive to the size of the solvent cluster because of the dynamical effects. The size of the water cluster in Nafion was found to be 30–40 Å, in agreement with the scattering data [9, 163]. The same analysis performed with SPI block copolymers leads to a size smaller than 30 Å, which is roughly one order of magnitude smaller than the correlation distance deduced from the scattering data for the same materials. This result suggests that the swollen ionic domains in SPI cannot be seen as water droplets with the ionic groups at the water/polymer interface, but present a nonhomogeneous composition with a more or less segregated distribution of water molecules and of the sulfonated parts of the polymer chains.

At the opposite side of the timescale (in the range of seconds to minutes), the macroscopic diffusion coefficient of water in swollen Nafion membranes, as determined by the diffusion of tritiated water through the membrane, is lower by a factor of 10 compared to the local diffusion coefficient or the self-diffusion in bulk water. This high value integrates all the restricted motions, which shows that the Nafion morphology is favorable to obtain a high ionic conductivity [160]. One important issue is the identification of the typical time and length scales of the mobility restriction and thus the correlation with the structure features. Two intermediate ranges were recently explored by NMR experiments. In the range of milliseconds, the dynamical behavior can be studied using PFG-NMR with a water molecule mean free path of around several microns [164]. This technique was then used to extract the diffusion coefficients of water within the membrane [165], which is an important parameter for mass transfer models in operating fuel cells [144]. The water molecules explore many ionic domains and an average contribution to the dynamics is obtained corresponding to the macroscopic diffusion. However, when less mobile solvent molecules or counterions are used, the mean free path correlated with the same NMR sequence is smaller and the diffusion coefficients determined by PFG-NMR in Nafion are no longer constant over the timescale of observation [165, 166]. This result was attributed to some heterogeneities in the range from a few tenths of microns to microns [166]. PFG-NMR was applied to SPI and the same behavior was observed on combining the data obtained with tetramethylammonium and lithium counterions as NMR probes of the motion (Fig. 40) [146, 158]. In addition, the diffusion experiments conducted in the membrane plane and along the membrane thickness revealed some transport anisotropy as expected with a multiscale foliated structure. Block copolymers with different equivalent weights and ionic group distributions along the polymer chain were studied using PFG-NMR, radiotracer diffusion, and conductivity measurements [146]. The in-plane diffusion is not affected by the length of the ionic sequences, while a strong effect of the ion content is observed in the transverse direction.

More recently, another intermediate timescale was explored by NMR relaxometry [167, 168]. The spin–lattice relaxation times of water molecules were determined in the range of 20 ns to 20 μs by varying the magnetic field fre-

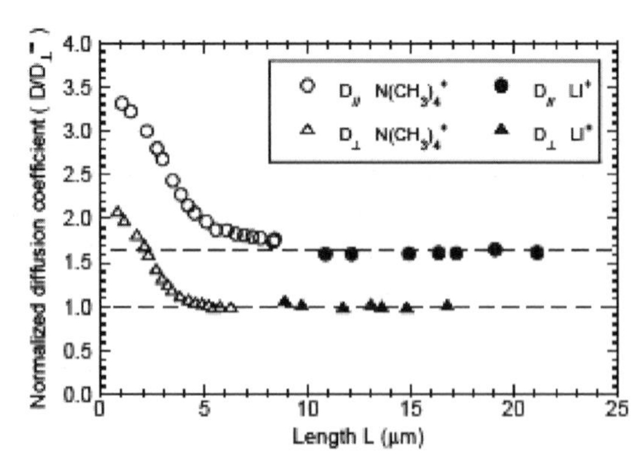

Fig. 40 Diffusion coefficients of lithium and tetramethylammonium ions with SPI membranes determined by pulse field gradient NMR (reproduced from [157])

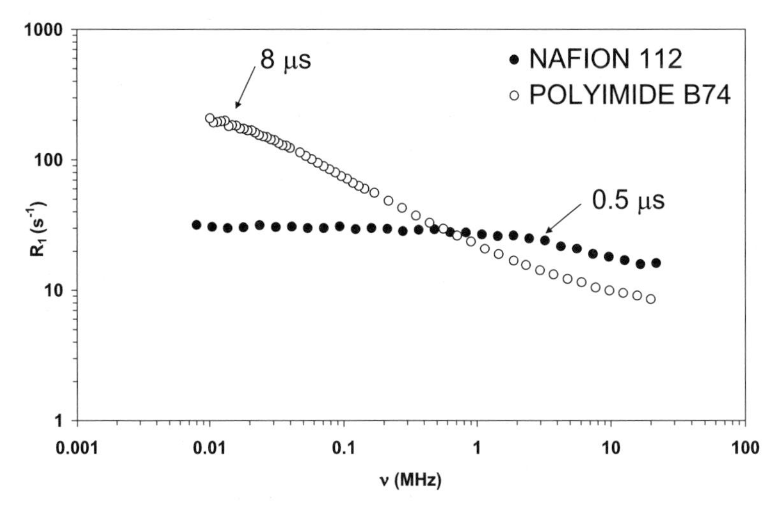

Fig. 41 Dispersion curves of fully hydrated Nafion and SPI membranes measured by NMR relaxometry [167]

quency from 10 kHz to 20 MHz and depending on the water content. This technique is well suited for the study of ionomer membranes because of its extreme sensitivity to water–polymer interactions, but it requires a structural and dynamical model to extract characteristic features. The authors observed significantly different dispersion curves for Nafion and SPI membranes with a 40 times slower dynamic in the SPI (Fig. 41). The effect of confinement is predominant in polyimides even at high water content (algebraic law with a slope of - 0.5 characteristic of porous materials), whereas the diffusion quickly reaches a bulk behavior in Nafion (a plateau is observed at low magnetic fields).

3.5 Ionic Conductivity

A high ionic conductivity is required for use in fuel cells. Among the specifications for PEMs, the desired conductivity is 0.1 S/cm over a wide range of humidity and temperature. However, its determination is far from being straightforward. As an example, the values published for various SPIs vary from a few 10^{-3} to 1.67 S/cm [31, 62]. It is difficult to differentiate between the effects originating from the chemical structure, the ion content, the membrane morphology, and the use of different measurement protocols. It is worth noting that normalization using Nafion as reference is not really convincing, since these data are even more scattered due to a strong effect of the membrane pretreatment on the conductivity values. Two main types of conductivity cells are used in the literature: the mercury electrode cell, in which the membrane is sandwiched between two mercury electrodes [31, 141, 146, 158, 159], and the two (or four-probe) platinum electrode cell [52, 61, 71, 76, 112]. In the former cell, the conductivity corresponds to the transversal value (along the membrane thickness, while the platinum cell gives the conductivity along the membrane plane (longitudinal conductivity). Due to the structural anisotropy originating from the foliated structure [72, 141, 158], the diffusion of ionic species is expected to be significantly larger in the membrane plane compared to the transversal value, as experimentally demonstrated by PFG-NMR [146]. With respect to fuel cell application, the transversal value is obviously more relevant. However, the longitudinal cell seems to be easier to use for a study depending on water content and temperature, because the membrane is free to equilibrate in the desired medium and the mercury cannot be heated for security reasons.

The effect of ion content, block length, chemical structure, and temperature on the ionic conductivity was studied. The data are often compared to Nafion despite SPI membranes usually presenting a larger ion content. For similar ion contents (even taking into account the polymer density), the conductivity is smaller for SPI compared to Nafion [141]. This is commonly observed for most of the aromatic polymers and, as a consequence, it cannot be only attributed to a difference in the membrane morphology. It is more likely to be essentially due to a lower dissociation of the ionic groups. The pK_a of perfluorinated ionomers is estimated to be highly negative (-6) while the value for an aryl sulfonic acid is only -1 [169]. Nevertheless, very high conductivity values (larger than 0.1 S/cm) can be found for SPIs but they are restricted to highly sulfonated materials (IEC larger than 2 meq/g) [32, 35, 62, 63, 71, 75, 112]. Myatake et al. [62] found 1.67 S/cm at 120 °C for a fluorenyl-containing SPI. This value is probably somewhat significantly overestimated since it is the largest value never observed for an ionomer membrane and it is very close to the maximum theoretical value, as calculated from the equivalent conductance of free protons taking into account the concentration and temperature effect. Another group recently published very high conductivity values (0.95 S/cm at 80 °C) [58]. This value obtained for a very high ion content (100% sulfonated) suggests that the loss of the mechanical properties due to large water uptake leads to nonsignificant values. In other words, these values suggest a complete absence of tortuosity restriction to proton motion, which is surprising in such materials. A random distribution of the ionic groups along the polymer chains is not favorable for the formation of well-defined ion conducting pathways, and a too large charge segregation increases the distance between ionic domains and the resistance created by the interdomain phase. As a consequence, an optimum value of four repeat units in the ionic sequences was experimentally observed for various structures (Fig. 42). However, some authors claim they have obtained large conductivity values using blends of sulfonated homopolyimides and sulfonated copolyimides [170].

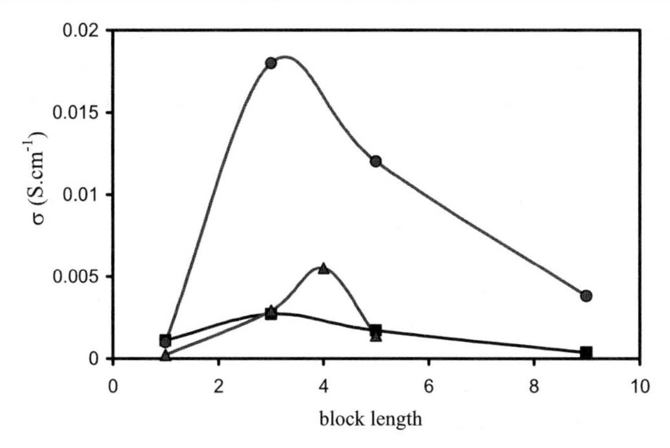

Fig. 42 Effect of the ionic sequence length on the ionic conductivity of phthalic (▲) and naphthalenic ODA (●) and BAHF (■)

The effect of temperature at 100% RH on the ionic conductivity depends on the ion content (Fig. 43). For highly conductive SPI systems, a linear behavior in an Arrhenius plot was observed similar to the data obtained for Nafion, which indicates a similar activation energy for proton conduction (21 kJ mol^{-1}) [62, 63, 112]. It was deduced that both systems share similar conduction mechanisms involving hydronium ions [62]. Increasing the ion content not only increases the ionic conductivity but also, after a critical value, the conductivity no longer increases exponentially with the inverse of temperature and a maximum value is observed (Fig. 43) [62, 112]. Most of the less conducting systems also exhibit a linear behavior in Arrhenius plots but with significantly larger slopes (higher activation energies) [54, 76, 87]. The introduction of flexible and bulky groups into the hydrophobic sequence of

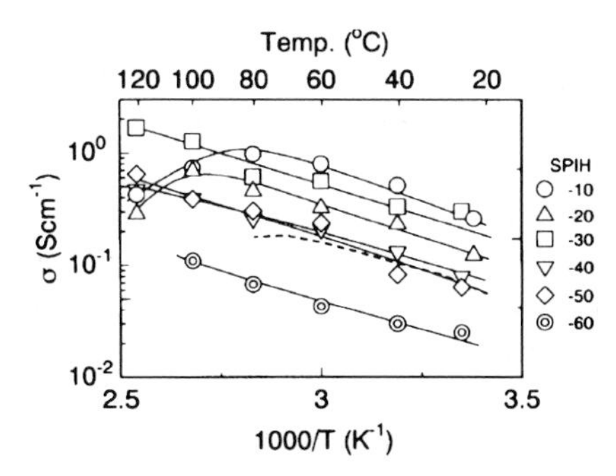

Fig. 43 Ionic conductivity at 100% RH of fluorenyl-based SPI membranes as a function of ion content and temperature. The data obtained for Nafion are also reported (*dashed line*)

block copolymers induces a drop of the ionic conductivity by a factor close to 10, probably because of a less favorable membrane morphology [31, 141]. On the contrary, the introduction of flexibility into the ionic part of the polymer seems to favor the conducting properties [71]. The highly conductive SPI membranes also exhibit a similar behavior to Nafion as a function of relative humidity [63, 112], while the other systems lose more rapidly their conducting properties at low relative humidity [65, 71, 75]. Some of these systems were then proposed for use as humidity sensors [171].

Most of the conductivity measurements were performed on the SPI acidic form since it is the relevant value for the fuel cell application. Nevertheless, a few data were obtained on neutralized forms [141, 146, 159]. Rollet et al. used sodium and tetramethylammonium ions to study the transport processes within the membranes and a transport anisotropy was clearly observed from longitudinal and transversal measurements [146].

Cornet et al. [141] used ammonium ions with different sizes in order to check the occurrence of a critical size for transport restriction. The ionic conductivity first decreases as the ammonium size increases (from 5 to 30 $Å^3$) due to a lower mobility compared to protons. For larger counterion sizes, the conductivity is roughly constant until 1000 $Å^3$ where a cutoff is observed. This result suggests a radius of 10 Å for the conductive pathways, at least between two ionic domains. Despite SPI membranes being designed for fuel cell applications, these membranes can be used efficiently as the separator in electrodialysis experiments. For example, SPI membranes appear to be promising materials for separating copper or chromium ions from acidic solutions [172].

3.6 Mechanical and Thermal Properties

For most of the alternative membranes to Nafion, the thermal stability is studied by thermogravimetric analysis (TGA). The different weight losses depending on temperature have been determined and identified in some cases by gas chromatographic analysis [71, 138]. The first weight loss is related to the membrane dehydration between room temperature and 150 °C. The second one between 250 and 350 °C was identified as a desulfonation process (SO_2 and SO losses) followed by polymer matrix degradation (CO and CO_2 losses). From these experiments SPI membranes were considered as thermally stable up to at least 200 °C. However, it is somewhat questionable how these experiments are representative of the fuel cell environment. Indeed, it is difficult to consider these dynamic experiments performed with a temperature increase rate of 10 to 20 °C/min as indicative of the long-term stability in fuel cells, especially when TGA experiments are conducted in the presence of an inert gas. These experiments can be used to extract information on the sulfonate content from the integration of the weight loss related to the desulfonation process, since this process does not overlap with other weight losses [31]. In order to increase the measurement precision, the membranes can be neutralized with heavy ammonium counterions which will also degrade in the same temperature range.

Aromatic heterocyclic polymers, and especially polyimides, are known to exhibit high glass transition temperatures (T_g) [173]. The T_g is shifted toward high temperatures by the introduction of ionic groups along the polymer chain. As a consequence, the thermal analysis of SPIs does not reveal any T_g or melting temperature below the decomposition temperature ($\approx 250\,°C$) [112]. A glassy polymer matrix should induce a lower sensitivity to physical ageing which is obviously an advantage for the fuel cell application. Indeed, the membrane should be less subject to creeping under mechanical constraints and should exhibit minimal dimensional changes upon swelling and during temperature cycles. Nevertheless, it is difficult to prepare reproducible SPI membranes because the structure is quenched in a nonequilibrium state. Therefore, the morphology and properties should be very sensitive to slight modifications of the casting process.

The obtaining of good mechanical properties is obviously an important issue because the end of a fuel cell test is directly related to a general or local membrane breaking. To our knowledge, the study of SPI mechanical properties is limited in the literature to the measurement of stress–strain curves in the dry and water-swollen states [39, 110, 112, 126]. These curves present an elastic behavior limited to few percent of deformation followed by a plastic linear behavior (Fig. 44). Despite the polymer being in a glassy state, the water sorption induces plasticization. The maximum tensile stress at break is of the order of 70 MPa for SPI compared to less than 20 MPa for Nafion and the elon-

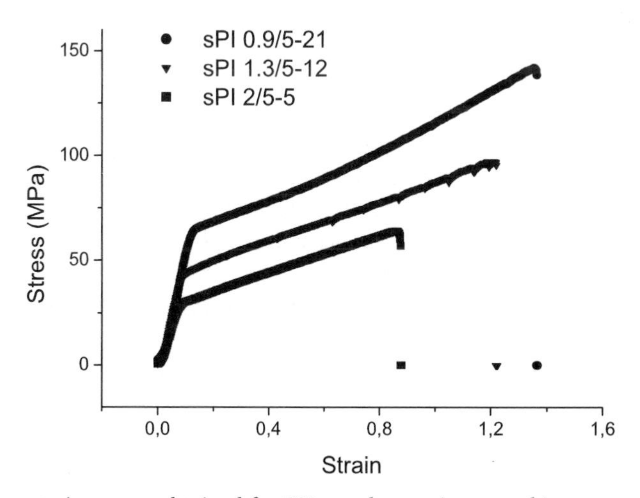

Fig. 44 Stress–strain curves obtained for SPI membranes immersed in water as a function of ion content [126]

gation at break is significantly smaller (less than 20% for SPI and close to 300% for Nafion) [39, 110]. Despite the fact that mechanical properties are now often determined for new ionomers [97, 132], a complete study of the mechanical properties is still necessary including the combined effect of water content and temperature.

3.7 Gas and Methanol Permeation

As previously reported for Nafion, the membrane thickness has been reduced in order to decrease the ohmic drops within the membrane and to enhance the fuel cell performance. However, it requires a low gas permeation [141]. The gas barrier properties of Nafion are very good in the dry state [174], but they decrease when hydrated due to a higher gas solubility in the water phase than in the perfluorinated one. The gas permeation properties of SPIs have been determined as a function of the ion and water content [33, 151, 155, 175, 176]. While large differences in diffusivity and selectivity are observed in the dry state depending on the ion content [155], the gas permeation is significantly reduced when hydrated, which suggests the existence of a closed nanoporosity [151]. This porosity located in the ionic domains is then filled by water molecules, thus reducing the gas permeation. While the gas barrier properties of SPI can be considered as favorable for its use as a fuel cell membrane, they become a serious drawback for the use of this ionomer to prepare fuel cell electrodes (see Sect. 4) [177].

Nafion membranes should not be adapted for DMFCs because of an enhanced solvent affinity in the presence of alcohols which solvates the perfluoroether side chain [142]. This solvent uptake induces a large methanol crossover through the membrane which depolarizes the electrodes. As a consequence, the power density obtained with DMFCs is significantly lower than that of hydrogen feed fuel cells (roughly by a factor of 10). Most of the alternative membranes including SPIs present a low affinity to methanol and consequently a low methanol permeation. Methanol permeabilities were determined for different chemical structures and ion contents [52, 76, 87]. The values were found to be systematically smaller than those for Nafion (typically between 1 and 8×10^{-7} cm^2/s compared to 1 to 2.5×10^{-6} cm^2/s for Nafion). However, the performance in DMFCs is not significantly improved because of a significant decrease of the ionic conductivity in the presence of a large methanol concentration [76, 105, 107].

4 Fuel Cell Tests and Membrane Stability

While the objective is an application in fuel cells, most of the studies performed on new materials do not include a fuel cell evaluation. Moreover, when some fuel cell tests are presented, they are restricted to a polarization curve to estimate the fuel cell performance in comparison to Nafion and only scarce works concern the long-term stability. The specifications for automotive application are 5000 h of operation at 80 °C over 5 to 10 years and more than 10 000 start–stop cycles (typically three cycles per day over 10 years). This latter constraint is probably the most difficult to achieve since Nafion membranes are able to operate for more than 10 000 h under stationary load and temperature conditions but the lifetime is reduced to a few hundred hours when operating under cycling conditions [11]. The membrane lifetime is defined as the duration of fuel cell operation until a total or partial rupture induces gas mixing. It is well known that the membrane stability can be significantly enhanced by increasing the membrane thickness or decreasing the ion content, but this stability would be obtained at the expense of the fuel cell performance which is not acceptable.

4.1 Fuel Cell Performance

The fuel cell performance of a new membrane material is directly related to both its ionic conductivity and its gas permeation properties. It is usually evaluated through the recording of polarization curves, which correspond to the evolution of the cell voltage against the current density (see an example in Fig. 45). These curves can be arbitrarily divided into three main parts: (1) at low current densities the cell voltage depends mainly on the behavior of the electrodes (activation zone); (2) in the intermediate zone, an almost linear behavior is observed related to the ohmic drops in the cell; and finally (3) at large current densities, the cell voltage often drops due to either the cell flooding, the membrane drying, or gas starvation. Cell flooding occurs when the water produced at the cathode by the electrochemical reaction and transported by electroosmosis cannot be completely removed and accumulates. Gas starvation occurs when the gas consumption is larger than the reactant diffusivity inside the active layer of the electrodes and cannot feed in a sufficient amount of the catalyst particles. In addition, the experimental conditions, such as the gas humidification, pressure, temperature, and the use of oxygen or air, are important parameters which influence differently the fuel cell performance depending on the type of membrane. As an example, Wainright et al. [178] showed the dramatic effect of air humidification (dry or RH = 38%) on the SPI-based fuel cell behavior and they obtained only 60 mA/cm^2 at 0.5 V while Faure et al. got 900 mA/cm^2 at 0.5 V with similar

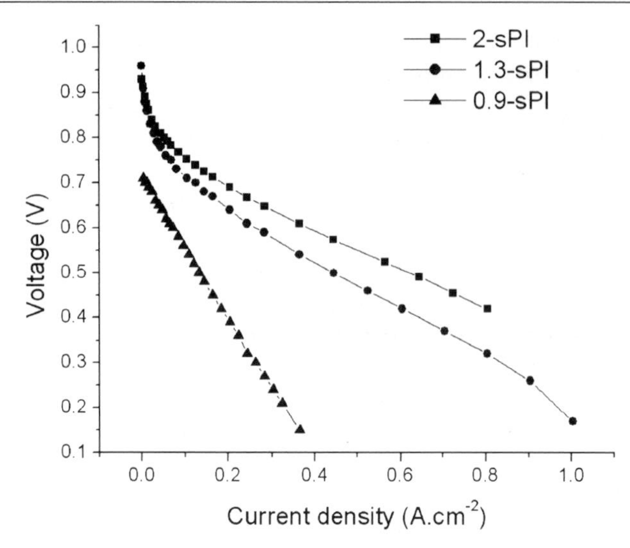

Fig. 45 Polarization curves obtained for SPI membranes depending on the ion content; H_2/O_2 fuel cell test at 80 °C with non-hot-pressed Nafion containing electrodes [68]

SPI membranes using fully humidified gases [29]. Therefore, it is always difficult to evaluate the fuel cell performance obtained for various membranes, since the optimal conditions can be significantly different from one membrane to another.

The former paper on the development of SPIs for fuel cell application displayed results on both fuel cell performance and long-term stability [29]. It was shown that significantly different performance and lifetime can be obtained for phthalic and naphthalenic structures with the same ion content. For naphthalenic structures a fuel cell test was run over 3000 h at 60 °C with a fuel cell performance close to that obtained with a Nafion membrane presenting a similar thickness. On the contrary, the phthalic SPI lifetime was limited to 70 h with a relatively poor performance, indicating that there are no direct correlations between lifetime and performance in these systems. The fuel cell performance is also highly influenced by the quality of the electrodes. Commercial electrodes are the result of a long-term optimization of Nafion and this work cannot be reasonably repeated for each membrane under study. As a consequence, new membrane materials are evaluated using Nafion-based electrodes. Besse et al. have presented the difficulties in building electrodes for alternative membranes based on SPI and their effect on the fuel cell performance [177]. The electrodes were composed of an active layer made from platinum on carbon powder and a polytetrafluoroethylene dispersion on which a Nafion layer was deposited by a spraying method. The optimized quantity of Nafion is around 0.7–0.9 mg/cm^2. The same procedure was used for SPI membranes using *m*-cresol solutions of

SPI instead of Nafion. Best performances were obtained for 1.3 mg/cm^2. The results were very similar to those for electrodes impregnated with Nafion. To our knowledge, this is the first Nafion-free MEA that has been tested. It can be considered as an advantage since it is useless to replace the Nafion from the membrane while it is still used in the electrodes. However, the lower gas permeation of SPI compared to Nafion does not favor the gas supply on the catalyst particles. A third generation of electrode was prepared by double impregnation: a Nafion solution directly on the active layer and a SPI solution to ensure the quality of the membrane/electrode interface. In such a case, the fuel cell performance was significantly improved. This work also demonstrated that the use of organic solvents such as *m*-cresol to dissolve the aromatic polymer has no poisoning effect on the Pt catalyst properties. A significant improvement of the performance as the IEC increases from 1.3 to 2 meq/g was also evidenced. In two recent papers [68, 72], the fuel cell behavior of SPIs based on *p*-BAPBDS/NTDA/*p*-BAPB and BDSA/NTDA/ODA was investigated. The totally sulfonated homopolymer BAPBDS/NTDA (IEC = 2.63 meq/g) and a 2 : 1 copolymer (IEC = 1.89 meq/g) were shown to present slightly better fuel cell performance than Nafion. However, the SPI membrane presented lower thickness (25–40 μm) compared to Nafion 112 (55 μm). The conductivity of SPI membranes increases exponentially up to 1 meq/g and then reaches a ceiling for larger IEC values [158]. The fuel cell performances of BDSA/NTDA/ODA 0.9, 1.3, and 2 meq/g were shown to be strongly related to the ionic conductivity and 0.9 meq/g appears to be a practical lower limit for the IEC value to extract reasonable current densities (Fig. 45) [68]. This result was attributed to the anisotropic structure which favors a percolation threshold at low ion content, as confirmed by the similar fuel cell performance obtained for the half and totally sulfonated BAPBDS/NTDA/BAPB SPIs [72]. The fuel cell performance of *o*-BAPBDS/NTDA/ODA 1.3 meq/g was found to be larger than both 1.3 and 2 meq/g BDSA/NTDA/ODA as a consequence of an increase of the ionic conductivity and water uptake by 40% [68]. In a recent work, a fuel cell performance similar to Nafion 112 was published which can be considered as a very promising result [33, 104].

SPI membranes have also been recently evaluated in DMFCs [76, 106]. The DMFC performance in identical conditions and with similar thicknesses is slightly improved with SPI compared to Nafion 112 [106]. The methanol crossover is twice lower than the other explored range of current densities, corresponding to a lower permeability as evidenced by the higher open circuit voltage. However, the current densities are still limited (0.15 A/cm^2 at 0.5 V) and the membrane also exhibited unsatisfactory stability in DMFC conditions [76].

4.2 In Situ Stability

The fuel cell stability of alternative membranes is a key issue for their industrial use in fuel cells. However, the polymer degradation in fuel cells has been actually studied for only two different systems: the sulfonated polystyrenes (SPSs) and the SPIs [68, 179–181]. SPS is hydrolytically stable even at high temperature, and the degradation in fuel cell conditions was shown to be mainly due to radical attacks on the α-carbon which generate a series of polymer chain scissions. The molecular weight decreases inducing a loss of the mechanical properties and the elution of sulfonated oligomers out from the membrane. A continuous loss of the conducting properties is then observed. The SO_3 loss was observed by infrared spectroscopy and through the analysis of the sulfur concentration profile in the membrane cross section. The authors concluded that the degradation mainly takes place at the cathode side of the cell because the oxygen reaction proceeds through some peroxide intermediates that have strong oxidative ability. On the contrary, Büchi et al. suggested that the degradation mechanism is due to the formation of $HO_2^{\cdot}$ radicals at the anode due to oxygen diffusion through the membrane and an attack of the tertiary hydrogen at the α-carbon of SPS [179]. This contradictory interpretation reveals how it is difficult to analyze the degradation even in an extensively studied system.

The former SPIs used in fuel cells were based on phthalic SPIs and their lifetime was limited to a few tens of hours [40]. Naphthalenic SPIs were then synthesized and exhibited a more pronounced stability since 3000 h of operation at 60 °C were attained. However, this structure was only poorly soluble in very expensive organic solvents such as 3-chlorophenol. In order to favor the solubility, the 4,4′-ODA was replaced by a 1 : 1 mixture of 4,4′- and 3,4′-ODA inducing a decrease of the membrane stability. The SPI fuel cell degradation has been recently studied in more detail [68]. While the fuel cell performance continuously decreases upon ageing with SPS membranes [181], the fuel cell voltage under stationary electric load is remarkably constant over hundreds of hours when SPI membranes are used [29, 68]. All the polarization curves recorded every 50 h were similar until the end of the test, which was very sudden and caused by the membrane rupture and gas mixing. The lifetime was determined for different ion contents and temperatures and it was shown that the degradation is a thermoactivated process (Fig. 46). The post mortem membrane analysis revealed that the sulfur concentration profile in the membrane cross section is not flat any more, as for the pristine membrane, but is decreased on the cathode side. This finding was analyzed to be a result of the elution of sulfonated oligomers by the water produced by the cell. The obtaining of a constant performance upon ageing appears somewhat contradictory with a continuous loss of sulfonated oligomers, but a possible explanation can be that the loss of ionic groups is compensated by higher proton mobility due to a decrease of the obstruction in the ionic pathways. An infrared study was

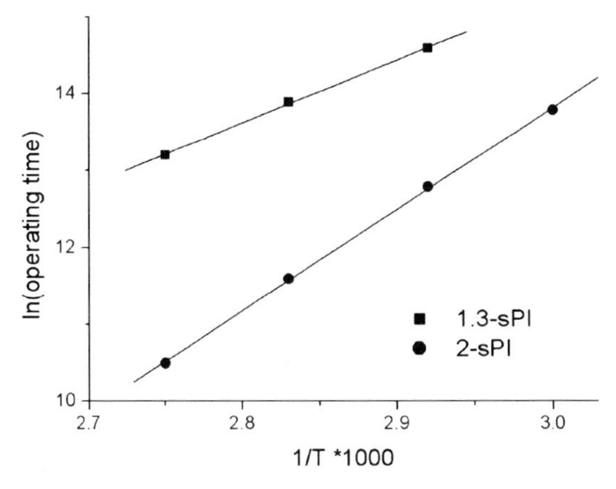

Fig. 46 Arrhenius plot of the membrane lifetime in fuel cells obtained for the 1.3 and 2 meq/g SPI membranes with five repeat units in the ionic sequences and ODA as non-charged monomer (the operating times are expressed in seconds) [68]

conducted on both sides of the membrane after a fuel cell test of 600 h. Since the number of chemical modifications is pretty low, the spectra before and after the fuel cell tests are apparently similar and the IR analysis was performed on difference spectra. At the cathode side, the difference spectrum presents only negative bands characteristic of sulfonated oligomers, indicating that the main process is chain scission by hydrolysis and elution. On the contrary, the difference spectrum recorded at the anode side is more complex with a superposition of positive and negative bands. This spectrum suggests the occurrence of an additional process, which is likely to be an oxidation process due radical attack as previously suggested by Büchi et al. [179]. More recently, a fuel cell test was performed [33, 104] and fairly stable performances over 5000 h were obtained at 80 °C and 0.2 A/cm^2 which up to now is the best result obtained with a sulfonated polyaromatic membrane. In addition, the open circuit voltage (OCV) does not vary significantly over these 5000 h, indicating the absence of gas permeation through the membrane [104].

4.3 Ex Situ Studies of the Stability

Since a lot of new SPI structures have been prepared, it was necessary to develop a fast and easy to reproduce stability test to avoid a time- and material-consuming fuel cell test. Polyimides are known to be sensitive to hydrolysis [182]. The introduction of sulfonate groups along the polymer chains increases the overall hydrophilicity and thus the water diffusion within the structure, which favors the hydrolytic process. Two different tests are now commonly used in the literature: (1) the membrane is immersed in liquid water at 80 °C to test the stability against hydrolysis [35, 61]; and (2) the membrane is soaked in Fenton's reagent (3% H_2O_2 solution containing 2 ppm of $FeSO_4$ at 80 °C [110] or 30% H_2O_2 and 30 ppm of $FeSO_4$ at 30 °C [71]) to evaluate the stability in oxidative media [35, 62]. The stability is then controlled through a simple mechanical test (the membranes were broken when lightly bent) or when they start to dissolve [35]. In some recent papers, more quantitative data have been produced that measure either the weight loss [34] or the decrease of the mechanical properties [58, 79, 183]. However, these data are limited to a few points are not an actual kinetic analysis. The stability can vary from 5 up to 200 h depending on the chemical structure and the ion content. Selected chemical modifications can increase the hydrolytic stability up to 1000 h [58, 132]. The Yamaguchi University group defined two classes of sulfonated diamine monomers (see Sect. 2.1.2 and Fig. 3): the sulfonated diamines in which the amine and sulfonated groups are located on the same phenyl ring and those in which these groups are separated by at least one phenyl ring [35, 75]. A higher basicity of the amine group is invoked in the second case to explain a higher hydrolytic stability. It is well known that aromatic diamines with higher basicity are more reactive with dianhydrides. Therefore, since hydrolysis is the reverse reaction to polycondensation, these monomers should lead to more stable polymers as experimentally observed [35, 75].

In order to quantify the loss of the mechanical properties, stress–strain curves were recorded as a function of the ageing time in hot water [79, 126, 183]. As expected, the membranes become more brittle and the plastic deformation decreases and completely disappears in the stress–strain curves (Fig. 47). This effect is faster as temperature and ion content increase. The same analysis was performed on a NTDA/*o*-BAPBDS SPI, which confirmed that the mechanical stability was improved compared to NTDA/ODA SPI. A significantly longer degradation time (roughly by a factor of 3) was necessary to produce a 50% loss of the plastic deformation [68]. However, a shorter lifetime was obtained when tested in fuel cell conditions. This example reveals the difficulty in developing ex situ degradation tests which are really representative of the lifetime in fuel cells. The discrepancy can be attributed to the fact that the polymer probably presents a lower molecular weight since the polycondensation reaction has not been optimized for this material. The membrane rupture in fuel cell conditions will appear at a critical value of the molecular weight. This value can be rapidly reached because of a high degradation rate (high number of polymer chain scissions per hour of operation) or because of a low value of the initial molecular weight (low number of possible scissions until reaching the critical value). Improved stability could then be obtained by the choice of the monomers and increasing the total molecular weight.

The analysis of both the membrane material after ageing and the degradation products extracted from the membrane revealed that the number of sulfonated species within the membrane decreases [126]. These sulfonated

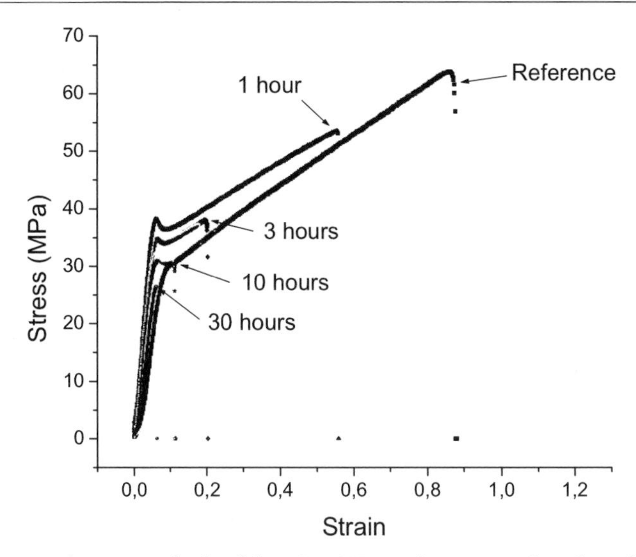

Fig. 47 Stress–strain curves obtained for 2/55 SPI membrane as a function of ageing time in water at 70 °C [126]

species were mainly identified as sulfonated oligomers with a very low proportion of the sulfonated monomer. To interpret this result it is necessary to take into account the fact that water solubility significantly increases with the length of the oligomers because of the increasing number of free sulfonic acid compared to terminal amine groups. Sulfur content analysis by scanning electron microscopy reveals that the concentration profiles across the membrane are homogeneous whatever the degradation time, as confirmed by infrared analysis [126]. This result indicates that the elution of soluble sulfonated oligomers is very fast. The degradation kinetics suggests that the block copolymers are more stable than random ones. However, above a critical length of the blocks a macroscopic phase separation occurs (see Fig. 39). In this case, the very fast degradation of the ion-rich phase will induce a fragility of the overall material which will break selectively along this phase. The effect of degradation on the membrane microstructure was analyzed by SAXS and SANS. A strong effect is observed on the SAXS data [126], since the ionomer peak of Cs^+ neutralized membranes continuously vanishes while the SANS spectra are only slightly modified for membranes soaked in deuterated water (Fig. 48). The difference is attributable to the origin of the contrast. With X-rays, the contrast mainly arises from the difference in the electron density between a phase containing the ionic group and the polymer matrix. With neutrons, the contrast is due to the higher hydrogen density in the hydrocarbon matrix compared to the ionic domains filled by the deuterated water, and this contrast is decreased by the presence of ionic hydrocarbon polymer sequences within the ionic domains. The elution of

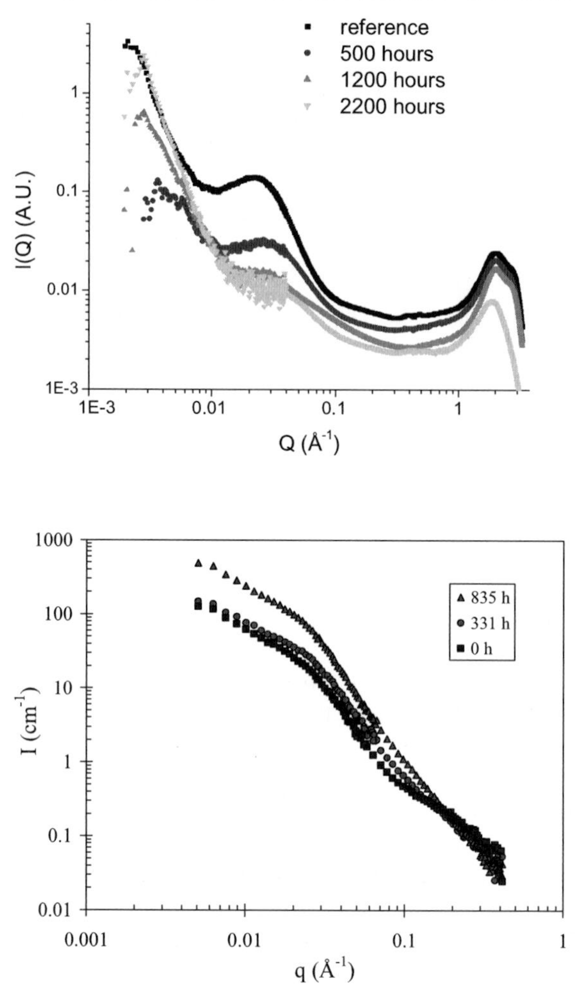

Fig. 48 SAXS (*top*) and SANS (*bottom*) spectra of ODA 1.3 meq/g SPI membranes as a function of the immersion time in water at 80 °C

sulfonated oligomers decreases the density of ionic groups within the ionic domains and thus the signal in X-ray experiments. On the contrary, these sulfonated oligomers are replaced by some D_2O molecules because of the absence of reorganization (glassy structure) and the contrast increases in neutron experiments. Another effect is a continuous change of the behavior at large angles, which can be interpreted as a more marked interface in the aged samples once again related to the elution of the sulfonated component and the creation of more defined water domains.

The addition of 0.5% H_2O_2 to the aggressive medium induces an accelerated degradation, as revealed by the fast kinetics (Fig. 49). In a first approx-

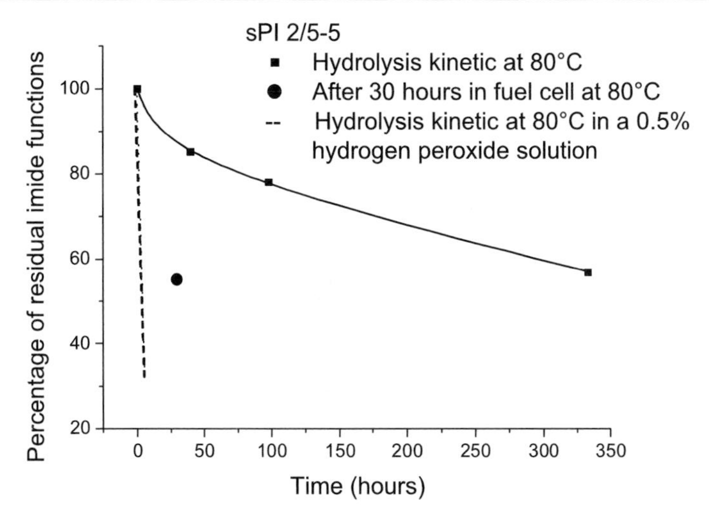

Fig. 49 Kinetics of the loss of imide function determined by infrared in pure water and 0.5% H_2O_2 solution [126]

imation, the infrared study does not indicate any new degradation process. The ageing is still limited to the hydrolysis of the imide rings [126]. The concentration was then adjusted to reproduce the degradation observed in fuel cell conditions at the same temperature and the corresponding concentration was found to be 0.05%. However, further experiments will be necessary to evaluate the relevant concentration depending on the fuel cell conditions (temperature, current density, gas humidification ...).

Geniès et al. have studied the degradation mechanisms using model compounds [48]. Such a study permits the identification and quantification of the degradation products as the reaction proceeds. Indeed, the spectra obtained with polymers are more complex, the effect of the chemical degradation is restricted to the modification of only a few chemical bonds, and the degradation products are more numerous and difficult to isolate and identify. Once modifications are identified on model compounds, it is then easier to look for them in the data obtained with aged polymers. Different degradation products of phthalic models were identified by NMR as issued from a partial and complete hydrolysis of the imide rings, namely amide–acid and diacid compounds, respectively. The kinetics reveals that the amide–acid appears at the beginning of degradation and then its concentration presents a maximum after 2 h of ageing, while a continuously increasing content of diacid material is observed. On the contrary to phthalic structures, the degradation of the naphthalenic compounds does not lead to the formation of an amide–acid structure (absence of characteristic peaks in ^{13}C NMR data). The analysis of the NMR spectra suggests that the degradation of the naphthalene imide after 120 h at 80 °C leads to the complete hydrolysis of only one imide function. It is worth noting that, after 1200 h of ageing at 80 °C, the NMR spectrum is identical to the one obtained after only 120 h. This result was interpreted through the existence of several equilibriums allowing the regeneration of the imide ring. Despite it being surprising to observe the formation of imide and anhydride functions in water solution, this process explains the fact that some plateaus are observed in the degradation kinetics of membranes followed by infrared, because two scissions on the same chain are necessary to produce a sulfonated oligomer that can be eluted out from the membrane [126]. The regeneration of the imide rings will also be favored by the rigid structure in the membrane which allows the two segments to stay in close vicinity. It would be interesting to synthesize and analyze model compounds corresponding to the two amine types defined by the Yamaguchi group in order to compare the hydrolysis kinetics and the rate of recombination. The multiplication and the combination of new monomers offer a large number of possibilities that cannot be completely evaluated. Moreover, the estimation of the stability strongly depends on the initial molecular weight and therefore on the synthesis conditions. The determination of the degradation kinetics through the study of model compounds is the only way to actually quantify the polymer stability. In addition, these model compounds can also be used to study and optimize the polymer synthesis in order to increase the molecular weights [55].

5 Conclusion

Sulfonated polyimides are considered as promising materials to be used as proton exchange membranes for fuel cells. Different series of polymers were synthesized by varying the chemical structure, the ion content, and the ion distribution along the polymer chain. The main route for the functionalization is direct polymerization using sulfonated monomers. The two main developments in this field are the synthesis of new monomers to increase the membrane stability and the use of trifunctionalized monomers to initiate some cross-links. The properties of SPI materials have been widely studied. The swelling properties are unusual since (1) the swelling is observed mainly along the membrane thickness, (2) the water uptake appears constant when expressed as the number of water molecules per sulfonic group over a wide range of ion content, (3) the effect of temperature is very limited, and (4) a strong difference in the water uptake is observed on the sorption isotherms between sorption and desorption experiments and for the membrane equilibrated in liquid water and saturated vapor. The ionic conductivity of highly charged SPIs is similar to that of Nafion. The structure was studied by small-angle scattering and electron microscopy techniques and a multiscale foliated structure packed along the membrane thickness was evidenced.

The transport properties are strongly influenced by this structural anisotropy and can be analyzed as the diffusion inside a nanoporous material. Finally, fuel cell tests in PEMFCs and DMFCs indicate a pretty good performance but a low stability. In situ and ex situ membrane degradation was studied and the main degradation process is the hydrolysis of the imide rings, which is accelerated when used in fuel cells due to the presence of radicals. As a consequence, these materials will be a reasonable alternative to Nafion when the stability is improved by at least a factor of 10, which appears to be a realistic goal according to the improvements of the chemical stability using specially designed monomers.

References

1. Coulter MO (ed) (1980) Modern chlor-alkali technology. Ellis Horwood, Chichester
2. Eisenberg A, Yeager HL (eds) (1982) Perfluorinated ionomer membranes. ACS Symposium Series 180. American Chemical Society, Washington DC
3. Broka K, Ekdunge P (1997) J Appl Electrochem 27:117
4. Eisenberg A (1970) Macromolecules 3:147
5. Mauritz KA, Moore RB (2004) Chem Rev 104:4535
6. Kreuer KD, Paddison SJ, Spohr E, Schuster M (2004) Chem Rev 104:4637
7. Rubatat L, Rollet A-L, Gebel G, Diat O (2002) Macromolecules 35:4050
8. Rubatat L, Gebel G, Diat O (2004) Macromolecules 37:7772
9. Gierke TD, Munn GE, Wilson FC (1981) J Polym Sci B Polym Phys 19:1687
10. Kerres JA (2001) J Membr Sci 185:3
11. Escribano S, Morin A, Solan S, Sommacal B, Capron P, Rougeaux I, Gebel G (2005) Proceedings of the 3rd European PEFC forum, Lucerne, 2005, p PA25
12. Xie J, Wood DL, Wayne DM, Zawodzinski TA, Atanassov P, Borup RL (2005) J Electrochem Soc 152:A104
13. Borup R, Meyers J, Pivovar B, Kim YS, Mukundan R, Garland N, Myers D, Wilson M, Garzon F, Wood D, Zelenay P, More K, Stroh K, Zawodzinski T, Boncella J, McGrath JE, Inaba M, Miyatake K, Hori M, Ota K, Ogumi Z, Miyata S, Nishikata A, Siroma Z, Uchimoto Y, Yasuda K, Kimijima KI, Iwashita N (2007) Chem Rev 107:3904
14. Rikukawa M, Sanui K (2000) Prog Polym Sci 25:1463
15. Rozière J, Jones D (2003) Annu Rev Mater Res 33:503
16. Hickner MA, Ghassemi H, Kim YS, Einsla BR, McGrath JE (2004) Chem Rev 104:4587
17. Jagur-Grodzinski J (2007) Polym Adv Technol 18:785
18. Jannasch P (2003) Curr Opin Colloid Interface Sci 8:96
19. Fontanella JJ, Wintersgill MC, Wainright JS, Savinell RF, Litt M (1998) Electrochim Acta 43:1289
20. He RH, Li QF, Xiao G, Bjerrum NJ (2003) J Membr Sci 226:169
21. Rikukawa M, Inagaki D, Kaneko K, Takeoka Y, Ito I, Kanzaki Y, Sanui K (2005) J Mol Struct 739:153
22. Kobayashi T, Rikukawa M, Sanui K, Ogata N (1998) Solid State Ionics 106:219
23. Glipa X, El Haddad M, Jones DJ, Rozière J (1997) Solid State Ionics 97:323
24. Jouanneau J, Mercier R, Gonon L, Gebel G (2007) Macromolecules 40:983

25. Einsla B, Tchatchoua CN, Kim YJ, McGrath JE (2003) Abstr Pap Am Chem Soc 226:U391
26. Xiao GY, Sun GM, Yan DY, Zhu PF, Tao P (2002) Polymer 43:5335
27. Gao Y, Robertson GP, Guiver MD, Jian XG (2003) J Polym Sci A Polym Chem 41:497
28. Sua Y-H, Liu Y-L, Suna Y-M, Lai J-Y, Wang D-M, Gaoe Y, Liu B, Guiver MD (2007) J Membr Sci 296:21–28
29. Faure S, Cornet N, Gebel G, Mercier R, Pineri M, Sillion B (1997) Proceedings of the second international symposium on new materials for fuel cells and modern battery systems, Montréal, 1997, p 818
30. Zhang Y, Litt M, Savinell RF, Wainrigth JS (1999) Polym Prepr 40:480
31. Geniès C, Mercier R, Sillion B, Cornet N, Gebel G, Pineri M (2001) Polymer 42:359
32. Watari T, Fang J, Tanaka K, Kita H, Okamoto K-I (2001) Polym Mater Sci Eng 85:334
33. Asano N, Aoki M, Suzuki S, Miyatake K, Uchida H, Watanabe M (2006) J Am Chem Soc 128:1762
34. Hu Z, Yin Y, Kita H, Okamoto K-I, Suto Y, Wang H, Kawasato H (2007) Polymer 48:1962
35. Guo XX, Fang JH, Watari T, Tanaka K, Kita H, Okamoto KI (2002) Macromolecules 35:6707
36. Rodgers M, Yang Y, Holdcroft S (2006) Eur Polym J 42:1075
37. Genova-Dimitrova P, Baradie B, Foscallo D, Poinsignon C, Sanchez JY (2001) J Membr Sci 185:59
38. Sauviat M, Salle R, Sillion B (1969) In: Polyimides sulfonés utilisables comme échangeur de cations, leurs procédés de préparation, French Patent 6 923 249
39. Solomin VA, Lyakh EN, Zhubanov BA (1992) Polym Sci USSR 34:274
40. Faure S (1996) Synthèse et caractérisation de nouvelles membranes polyimides sulfoniques pour pile à combustible H_2/O_2. Université Joseph Fourier, Grenoble
41. Lakshmi VV, Choudhary V, Varma IK (2004) Macromol Symp 210:21
42. Shibuya N, Porter RS (1992) Macromolecules 25:6495
43. Bailly C, Williams DJ, Karasz FE, MacKnight WJ (1987) Polymer 28:1009
44. Xing PX, Robertson GP, Guiver MD, Mikhailenko SD, Wang KP, Kaliaguine S (2004) J Membr Sci 229:95
45. Jeon JY, Shim BS (2002) J Appl Polym Sci 85:1881
46. Kim IC, Lee KH, Tak TM (2003) J Appl Polym Sci 89:2483
47. Noshay A, Robeson LM (1976) J Appl Polym Sci 20:1885
48. Geniès C, Mercier R, Sillion B, Petiaud R, Cornet N, Gebel G, Pineri M (2001) Polymer 42:5097
49. Gunduz N, McGrath JE (2000) Abstr Pap Am Chem Soc 219:U373
50. Jang W, Lee C, Sundar S, Shul Y, Han H (2005) Polym Degrad Stab 90:431
51. Kim YK, Park HB, Lee YM (2002) Desalination 145:389
52. Woo Y, Oh SY, Kang YS, Jung B (2003) J Membr Sci 220:31
53. Yin Y, Yamada O, Tanaka K, Okamoto K-I (2006) Polym J (Tokyo, Jpn) 38:197
54. Lee C, Sundar S, Kwon J, Han H (2004) J Polym Sci A Polym Chem 42:3621
55. Kim HJ, Litt MH, Nam SY, Shin EM (2003) Macromol Res 11:458
56. Chen S, Yin Y, Tanaka K, Kita H, Okamoto K-I (2006) High Perform Polym 18:637
57. Li N, Cui Z, Zhang S, Li S, Zhang F (2007) J Power Sources 172:511
58. Yan J, Liu C, Wang Z, Xing W, Ding M (2007) Polymer 48:6210
59. Watari T, Wang HY, Kuwahara K, Tanaka K, Kita H, Okamoto K (2003) J Membr Sci 219:137
60. Gunduz N, McGrath JE (2000) Polym Prepr 41:182
61. Lee C, Sundar S, Kwon J, Han H (2004) J Polym Sci A Polym Chem 42:3612

62. Miyatake K, Zhou H, Uchida H, Watanabe M (2003) Chem Commun 368
63. Miyatake K, Asano N, Watanabe M (2003) J Polym Sci A Polym Chem 41:3901
64. Ye X, Bai H, Ho WSW (2006) J Membr Sci 279:570
65. Asano N, Miyatake K, Watanabe M (2004) Chem Mater 16:2841
66. Jang W, Kim D, Choi S, Shul YG, Han H (2006) Polym Int 55:1236
67. Cornet N (1999) Relation entre la structure et les propriétés de membranes en polyimide sulfoné pour pile à combustible H_2/O_2. PhD thesis, Université Joseph Fourier, Grenoble
68. Meyer G, Gebel G, Gonon L, Capron P, Marsacq D, Mercier R (2006) J Power Sources 157:293
69. Li Y, Jin R, Cui Z, Wang Z, Xing W, Qiu X, Ji X, Gao L (2007) Polymer 48:2280
70. Fang JH, Guo XX, Harada S, Watari T, Tanaka K, Kita H, Okamoto K (2002) Macromolecules 35:9022
71. Watari T, Fang J, Tanaka K, Kita H, Okamoto K-I, Hirano T (2004) J Membr Sci 230:111
72. Yamada O, Yin Y, Tanaka K, Kita H, Okamoto K-I (2005) Electrochim Acta 50:2655
73. Islam MN, Zhou WL, Honda T, Tanaka K, Kita H, Okamoto K (2005) J Membr Sci 261:17
74. Zhou WL, Watari T, Kita H, Okamoto K (2002) Chem Lett 534
75. Guo XX, Fang JH, Tanaka K, Kita H, Okamoto KI (2004) J Polym Sci A Polym Chem 42:1432
76. Einsla BR, Kim YS, Hickner M, Hong Y-T, Hill ML, Pivovar BS, McGrath JE (2005) J Membr Sci 255:141
77. Zhai F, Guo X, Fang J, Xu H (2007) J Membr Sci 296:102
78. Shobha HK, Sankarapandian M, Glass TE, McGrath JE (2000) Polym Prepr 41:1298
79. Fang J, Guo X, Xu H, Okamoto K-I (2006) J Power Sources 159:4
80. Yin Y, Chen S, Guo X, Fang J, Tanaka K, Kita H, Okamoto K-I (2006) High Perform Polym 18:617
81. Shang Y, Xie X, Jin H, Guo J, Wang Y, Feng S, Wang S, Xu H (2006) Eur Polym J 42:2987
82. Karlsson LE, Jannasch P (2004) J Membr Sci 230:61
83. Jannasch P (2005) Fuel Cells 5:248
84. Hen S, Yin Y, Kita H, Okamoto K-I (2007) J Polym Sci A Polym Chem 45:2797
85. Bae JM, Honma I, Murata M, Yamamoto T, Rikukawa M, Ogata N (2002) Solid State Ionics 147:189
86. Yasuda T, Li Y, Miyatake K, Hirai M, Nanasawa M, Watanabe M (2006) J Polym Sci A Polym Chem 44:3995
87. Yin Y, Fang J, Cui Y, Tanaka K, Kita H, Okamoto K-I (2003) Polymer 44:4509
88. Yin Y, Fang JH, Kita H, Okamoto K (2003) Chem Lett 32:328
89. Yin Y, Fang JH, Watari T, Tanaka K, Kita H, Okamoto K (2004) J Mater Chem 14:1062
90. Yin Y, Yamada O, Suto Y, Mishima T, Tanaka K, Kita H, Okamoto K (2005) J Polym Sci A Polym Chem 43:1545
91. Jönsson NA, Merenyi F, Lars-Erik W (1971) US Patent 3,859,341
92. Jönsson NA, Merenyi F, Svahn CM, Gulander J (1978) Acta Chem Scand B 32:317
93. Yasuda T, Miyatake K, Hirai M, Nanasawa M, Watanabe M (2005) J Polym Sci A Polym Chem 43:4439
94. Miyatake K, Yasuda T, Hirai M, Nanasawa M, Watanabe M (2006) J Polym Sci A Polym Chem 45:157
95. Fang J, Guo X, Litt M (2004) Trans Mater Res Soc Japan 29:2541
96. Suto Y, Yin Y, Kita H, Okamoto K-I (2006) J Photopolym Sci Technol 19:273

97. Li N, Cui Z, Zhang S, Xing W (2007) J Membr Sci 295:148
98. Hu Z, Yin Y, Chen S, Yamada O, Tanaka K, Kita H, Okamoto K-I (2006) J Polym Sci A Polym Chem 44:2862
99. Li Y, Jin R, Wang Z, Cui Z, Xing W, Gao L (2007) J Polym Sci A Polym Chem 45:222
100. Chen S, Yin Y, Tanaka K, Kita H, Okamoto K-I (2006) Polymer 47:2660
101. Yin Y, Yamada O, Okamoto K-I (2005) Trans Mater Res Soc Jpn 30:387
102. Okamoto K, Yin Y, Yamada O, Islam MN, Honda T, Mishima T, Suto Y, Tanaka K, Kita H (2005) J Membr Sci 258:115
103. Yin Y, Suto Y, Sakabe T, Chen S, Hayashi S, Mishima T, Yamada O, Tanaka K, Kita H, Okamoto K-I (2006) Macromolecules 39:1189
104. Aoki M, Asano N, Miyatake K, Uchida H, Watanabe M (2006) J Electrochem Soc 153:A1154
105. Higuchi E, Asano N, Miyatake K, Uchida H, Watanabe M (2007) Electrochim Acta 52:5272
106. Song JM, Asano N, Miyatake K, Uchida H, Watanabe M (2005) Chem Lett 34:996
107. Song JM, Miyatake K, Uchida H, Watanabe M (2006) Electrochim Acta 51:4497
108. Guo QH, Pintauro PN, Tang H, O'Connor S (1999) J Membr Sci 154:175
109. Kerres J, Ullrich A, Meier F, Haring T (1999) Solid State Ionics 125:243
110. Miyatake K, Zhou H, Matsuo T, Uchida H, Watanabe M (2004) Macromolecules 37:4961
111. Kerres JA (2005) Fuel Cells 5:230
112. Miyatake K, Zhou H, Watanabe M (2004) Macromolecules 37:4956
113. Miyatake K, Watanabe M (2006) J Mater Chem 16:4465
114. Yin Y, Hayashi S, Yamada O, Kita H, Okamoto K (2005) Macromol Rapid Commun 26:696
115. Yin Y, Yamada O, Hayashi S, Tanaka K, Kita H, Okamoto K-I (2006) J Polym Sci A Polym Chem 44:3751
116. Sundar S, Jang WB, Lee C, Shul Y, Han H (2005) J Polym Sci B Polym Phys 43:2370
117. Park HB, Lee CH, Sohn JY, Lee YM, Freeman BD, Kim HJ (2006) J Membr Sci 285:432
118. Lee CH, Park HB, Chung YS, Lee YM, Freeman BD (2006) Macromolecules 39:755
119. Yang SJ, Jang WB, Lee C, Shul YG, Han H (2005) J Polym Sci B Polym Phys 43:1455
120. Kido M, Hu Z, Ogo T, Suto Y, Okamoto K-I, Fang J (2007) Chem Lett 36:272
121. Fang J, Zhai F, Guo X, Xu H, Okamoto K-I (2007) J Mater Chem 17:1102
122. Lee S, Jang W, Choi S, Tharanikkarasu K, Shul Y, Han H (2007) J Appl Polym Sci 104:2965
123. Lee CH, Hwang SY, Sohn JY, Park HB, Kim JY, Lee YM (2006) J Power Sources 163:339
124. Gao Y, Robertson GP, Guiver MD, Ean XG, Mikhailenko SD, Wang KP, Kaliaguine S (2003) J Membr Sci 227:39
125. Maréchal Y, Jamroz D (2004) J Mol Struct 693:35
126. Meyer G, Perrot C, Gebel G, Gonon L, Morlat S, Gardette J-L (2006) Polymer 47:5003
127. Han H, Gryte CC, Ree M (1995) Polymer 36:1663
128. Seo J, Cho KY, Han H (2001) Polym Degrad Stab 74:133
129. Piroux F, Espuche E, Mercier R, Pinéri M (2003) J Membr Sci 223:127
130. Guo X, Tanaka K, Kita H, Okamoto K-I (2004) J Polym Sci A Polym Chem 42:1432
131. Galatanu AN, Rollet AL, Porion P, Diat O, Gebel G (2005) J Phys Chem B 109:11 332
132. Li N, Cui Z, Zhang S, Xing W (2007) Polymer 48:7255
133. Peckham TJ, Schmeisser J, Rodgers M, Holdcroft S (2007) J Mater Chem 17:3255
134. Ghielmi A, Vaccarono P, Troglia C, Arcella V (2005) J Power Sources 145:108

135. Xing PX, Robertson GP, Guiver MD, Mikhailenko SD, Kaliaguine S (2004) J Polym Sci A Polym Chem 42:2866
136. Cornet N, Geniès C, Gebel G, Jousse F, Mercier R, Pinéri M (2000) Proc 11th annual Conference of the North American Membrane Society, NAMS 2000, Boulder Colorado
137. Guo X, Fang J, Watari T, Tanaka K, Kita H, Okamoto K-I (2002) Macromolecules 35:6707
138. Yin Y, Fang J, Watari T, Tanaka K, Kita H, Okamoto K-I (2004) J Mater Chem 14:1062
139. Choi PH, Datta R (2003) J Electrochem Soc 150:E601
140. Cornet N, Gebel G, de Geyer A (1998) J Phys IV France 8:Pr5 63
141. Cornet N, Diat O, Gebel G, Jousse F, Marsacq D, Mercier R, Pineri M (2000) J New Mater Electrochem Syst 3:33
142. Gebel G, Aldebert P, Pineri M (1993) Polymer 34:333
143. Detallante V, Langevin D, Chappey C, Métayer M, Mercier R, Pinéri M (2001) J Membr Sci 190:227
144. Weber AZ, Newman J (2004) Chem Rev 104:4679
145. Zawodzinski TA, Neeman M, Sillerud LO, Gottesfeld S (1991) J Phys Chem 95:6040
146. Rollet A-L, Diat O, Gebel G (2004) J Phys Chem 108:1130
147. Rivin D, Kendrick CE, Gibson PW, Schneider NS (2001) Polymer 42:623
148. Maréchal Y (2004) J Mol Struct 700:217
149. Jamroz D, Maréchal Y (2004) J Mol Struct 693:35
150. Jamroz D, Maréchal Y (2005) J Phys Chem B 109:19664
151. Piroux F, Espuche E, Mercier R, Pineri M (2003) J Membr Sci 223:127
152. Motyakin MV, Cornet N, Schlick S, Gebel G (2000) Bull Polish Acad Sci Chem 48:273
153. Gebel G, Diat O (2005) Fuel Cells 5:261
154. Essafi W, Gebel G, Mercier R (2004) Macromolecules 37:1431
155. Piroux F, Espuche E, Mercier R, Pineri M, Gebel G (2002) J Membr Sci 209:241
156. http://www.esrf.fr/exp_facilities/ID13/index.html
157. Rollet A-L, Porion P, Delville A, Diat O, Gebel G (2005) Magn Reson Imaging 23:367
158. Blachot JF, Diat O, Putaux J-L, Rollet A-L, Rubatat L, Vallois C, Müller M, Gebel G (2003) J Membr Sci 214:31
159. Rollet A-L, Blachot J-F, Delville A, Diat O, Guillermo A, Porion P, Rubatat L, Gebel G (2003) Eur Phys J E Soft Matter 12:131
160. Volino F, Pinéri M, Dianoux AJ, de Geyer A (1982) J Polym Sci B Polym Phys 20:481
161. Pivovar AA, Pivovar BS (2005) J Phys Chem B 109:785
162. Paddison SJ, Pratt LR, Zawodzinski T, Reagor DW (1998) Fluid Phase Equilib 151:235
163. Gebel G, Lambard J (1997) Macromolecules 30:7914
164. Johnson CS (1999) Prog Nucl Magn Reson Spectrosc 34:203
165. Gong X, Bandis A, Tao A, Meresi G, Wang Y, Inglefield PT, Jones AA, Wen WY (2001) Polymer 42:6485
166. Rollet AL, Simonin JP, Turq P, Gebel G, Kahn R, Vandais A, Noel JP, Malveau C, Canet D (2001) J Phys Chem B 105:4503
167. Perrin J-C, Lyonnard S, Levitz P, Guillermo A, Diat O, Gebel G (2005) Magn Reson Imaging 23:409
168. Perrin J-C, Lyonnard S, Guillermo A, Levitz P (2006) J Phys Chem B 110:5439
169. Kreuer KD (2001) J Membr Sci 185:29
170. Okazaki Y, Nagaoka S, Kawakami H (2007) J Polym Sci B Polym Phys 45:1325
171. Ueda M, Nakamura K, Tanaka K, Kita H, Okamoto KI (2007) Sens Actuators B Chem 127:463
172. Vallejo E, Pourcelly G, Gavach C, Mercier R, Pineri M (1999) J Membr Sci 160:127

173. Qian ZG, Ge ZY, Li ZX, He MH, Liu JG, Pang ZZ, Fan L, Yang SY (2002) Polymer 43:6057
174. Grossi N, Espuche E, Escoubes M (2001) Sep Purif Technol 22:255
175. Piroux F, Espuche E, Mercier R (2004) J Membr Sci 232:115
176. Piroux F, Espuche E, Mercier R, Pineri M (2002) Desalination 145:371
177. Besse S, Capron P, Diat O, Gebel G, Jousse F, Marsacq D, Pinéri M, Marestin C, Mercier R (2002) J New Mater Electrochem Syst 5:109
178. Wainright JS, Litt MH, Zhang Y, Liu CC, Savinell RF (2000) Electrochem Soc Proc 14
179. Büchi FN, Gupta B, Haas O, Scherrer GG (1995) Electrochim Acta 40:345
180. Gode P, Ihonen J, Strandroth A, Ericson H, Lindbergh G, Paronen M, Sundholm G, Walsby N (2003) Fuel Cells 3:21
181. Yu J, Yi B, Xing D, Liu F, Shao Z, Fu Y, Zhang H (2003) Phys Chem Chem Phys 5:611
182. DeIasi R, Russell J (1971) J Appl Polym Sci 15:2965
183. Jang W, Choi S, Lee S, Shul Y, Han H (2007) Polym Degrad Stab 96:1289

Subject Index

Printing: Krips bv, Meppel, The Netherlands
Binding: Stürtz, Würzburg, Germany